T0260166

GALEN

HYGIENE

II

THRASYBULUS

ON EXERCISE WITH
A SMALL BALL

LCL 536

GALEN

HYGIENE

BOOKS 5–6

THRASYBULUS

ON EXERCISE WITH
A SMALL BALL

EDITED AND TRANSLATED BY

IAN JOHNSTON

HARVARD UNIVERSITY PRESS

CAMBRIDGE, MASSACHUSETTS

LONDON, ENGLAND

2018

Copyright © 2018 by the President and Fellows
of Harvard College
All rights reserved

Library of Congress Control Number 2017940308
CIP data available from the Library of Congress

ISBN 978-0-674-99713-4

Composed in ZephGreek and ZephText by
Technologies 'N Typography, Merrimac, Massachusetts.
Printed on acid-free paper and bound by
Maple Press, York, Pennsylvania

CONTENTS

ΓΑΛΗΝΟΥ ΥΓΙΕΙΝΩΝ ΛΟΓΟΣ

HYGIENE

E

1. Πέμπτον ἐνιστάμενος λόγον ὑπὲρ τῆς ὑγιεινῆς
πραγματείας παραμυθήσασθαι βούλομαι πρότερον,
εἴ τινες ἄρα δυσχεραίνουσι τῷ μήκει τῆς διδασκαλίας.
οὐχ ἡμέτερον ἔγκλημα τοῦτ᾽ ἐστίν, ἀλλὰ τῆς φύσεως
αὐτοῦ τοῦ προκειμένου πράγματος ἴδιον. εἰ μὲν γὰρ
οἷόν τέ ἐστιν μακρὰν θεωρίαν ἐν βραχέσι διελθόντα
μηδὲν τῶν χρησίμων παραλιπεῖν, ἡμεῖς ἁμαρτάνο-
μεν, οὐκ ἀναγκαίως μηκύνοντες· εἰ δ᾽ οὐκ ἐγχωρεῖ
σαφέστερόν τε ἅμα καὶ θᾶττον ὑπὲρ τῶν αὐτῶν εἰπεῖν, οὐχ ἡμῖν μέμφεσθαι προσῆκεν, ἀλλὰ τοῖς παρα-
λείπουσιν οὐκ ὀλίγα τῶν ἀναγκαίων.

ἔστι δ᾽ ἐν τοῖς ἀναγκαιοτάτοις τε ἅμα καὶ οἷον
στοιχείοις ἁπάσης τῆς ὑγιεινῆς πραγματείας, ὡς
πάμπολλαι φύσεις τῶν ἀνθρώπων εἰσίν, ὅπερ ὡμολό-
γηται μὲν ἅπασιν ἰατροῖς τε καὶ γυμνασταῖς. οὕτω δὲ
γράφουσιν οἱ πλείους αὐτῶν ὑπὲρ τῆς ὑγιεινῆς ἀγωγῆς, ὡς ὑπὲρ ἑνὸς ἀνθρώπου διαλεγόμενοι, καὶ ταῦτα
μηδὲ ἐπιδειξάμενοί ποτε κἂν ἕνα τινὰ πρὸς αὐτῶν
ὠφεληθέντα· καίτοι γ᾽ οὐδ᾽ ἕνα δίκαιον ἦν, ἀλλὰ
παμπόλλους ἐπιδείξαντας ἔμπροσθεν μὲν νοσοῦντας
συνεχῶς, ὕστερον δ᾽ ἀνόσους ἔτεσι πολλοῖς ὑπ᾽ αὐ-

BOOK V

1. In beginning the fifth book on the subject of hygiene, I 305K wish first to reassure those who are disquieted by the length of the teaching. This is not my fault; it is a feature specific to the nature of the actual matter before us. If it is possible to cover a lengthy subject briefly without leaving anything useful out, then I am at fault and not perforce delayed. On the other hand, if it is not possible to speak about these matters relatively clearly and at the same time quite quickly, it is inappropriate for me to be blamed; instead the blame should fall on those who leave out many 306K of the essentials.

Among the most essential components—elements as it were—of the whole subject of hygiene is that there are very many natures of people, as is agreed by all doctors and gymnastic trainers. And yet the majority of them write about the training in hygiene as if discoursing on one person, without showing at any time whether even one person was benefitted by them. Indeed, it was not really one but many whom they showed to be continually diseased before but later free of disease for many years, being pre-

τῶν διαφυλαχθέντας, οὕτως ἐπιχειρεῖν γράφειν. ἀλλ᾽ ἡμεῖς γε τοῖς ἔργοις αὐτοῖς ἐπιδειξάμενοι τοῦτο καὶ πολλοὺς τῶν ἔμπροσθεν συνεχῶς νοσούντων ἀνόσους φυλάξαντες ἐκ τοῦ πεισθῆναι τοῖς ὑγιεινοῖς παραγγέλμασιν, οὕτως ἐπεχειρήσαμεν ὑπὲρ αὐτῶν γράφειν. ἐνίους μὲν γὰρ ὅλως ἐκωλύσαμεν γυμνάζεσθαι, ταῦτα δὴ τὰ ἐκ τῶν ἐπιτηδευμάτων γυμνάσια, μόναις ταῖς κατὰ τὸν βίον ἐνεργείαις ἀρκεῖσθαι συμβουλεύσαντες· ἐνίους δὲ τοῦ πλήθους ἀφελεῖν τῶν γυμνασίων ἐκελεύσαμεν, ὡς εἰς ἐλάχιστον συστεῖλαι τὸ πᾶν· ἐνίους δὲ τὰς ποιότητας ὑπαλλάξαι μόνον ἢ τὴν τάξιν ἢ τὸν καιρόν· ἐνίους δὲ τὴν σύμπασαν αὐτῶν ἰδέαν μεταβαλεῖν.

ὡσαύτως δὲ καὶ περὶ τῶν λουτρῶν ὑποθήκας δόντες, ὡς ἤτοι μὴ λούεσθαι παντάπασιν[1] ἢ πλεῖον ἢ ἔλαττον καὶ πρὸ τροφῆς μόνον ἢ καὶ μετὰ τροφὴν ἢ θερμοτέροις τῶν πρόσθεν ἢ χλιαρωτέροις ὕδασιν ἢ ψυχροῖς παντάπασιν, ἀνόσους ἐφυλάξαμεν ἔτεσι παμπόλλοις. ἀλλ᾽ οὐκ ἂν τούτων οὐδὲν ἐποιήσαμεν, εἰ μήτε τὰς φυσικὰς τῶν σωμάτων διαφορὰς ἠπιστάμεθα μήτε τὴν ἑκάστῳ προσήκουσαν ὑγιεινὴν δίαιταν. ἔνιοι δὲ τῶν ὑγιεινὰ συγγράμματα γραφόντων ἢ καὶ χωρὶς γραμμάτων ὑποθήκας διδόντων οὐδὲ σφᾶς αὐτοὺς ἀνόσους ἠδυνήθησαν φυλάξαι, κἄπειθ᾽ ὅταν ἐπισκώπτωνται πρός τινων ἄλλα τέ τινα λεγόντων πρὸς αὐτοὺς καὶ προφερομένων ἐκεῖνο τὸ ἔπος "ἄλλων ἰατρὸς αὐτὸς ἕλκεσι βρύων." οἱ μὲν εἰς ἀσχολίας δή τινας ἀναφέρουσι τὴν αἰτίαν, οἱ δὲ καὶ δι᾽ ἀκολασίαν

served by them—this is how they attempt to write. But I, in fact, have shown this by these same actions: many who were previously continuously diseased were kept free of disease, being persuaded by the precepts of hygiene—and this is how I attempted to write about them. Thus, there were some whom I prevented from exercising at all, even with those exercises that are suitable, wishing them to be satisfied with the activities of life only. Some, however, I ordered to set aside the majority of the exercises, so as to   307K reduce the totality of exercise to a minimum. Some I directed to change the qualities only, or the order, or the time. Some, however, I directed to change their whole form [of exercise].

In the same way too, by giving suggestions about bathing—either not to bathe at all, or more or less, or before food only, or after food, or with waters that are hotter than before, or more lukewarm or colder altogether—I preserved them free of disease for many years. But I would have done none of these things, if I had known neither the physical differences of bodies, nor the hygiene regimen appropriate to each. However, some of those who write treatises on hygiene, or give advice apart from writing, have not been able to keep themselves free of disease, and then, when they are ridiculed by others, they say some other thing to them or they offer that statement: "the doctor to others is himself full of ulcers."[1] There are some who attribute the cause to overwork and some who accept they

---

[1] On this apparent quote from Euripides, see Koch 136.

---

[1] post παντάπασιν: ἢ πλεῖον Ko; ἢ πλείω τῶν ἔμπροσθεν Ku

5

308K ὁμολογοῦσι νοσεῖν. ἀλλὰ τούτοις μὲν ἡ ἀπολογία
πολὺ χείρων ἐστὶ τῆς κατηγορίας, ἐμοὶ γοῦν κριτῇ.

τοῖς δ᾽ εἰς τὰ κατὰ τὸν βίον ἑαυτοῖς συμπίπτοντα
τὴν αἰτίαν ἀναφέρουσιν, εἰ μὲν ἐφήμερόν τινα πυρετὸν πυρέξειαν ἐπ᾽ ἐγκαύσει καὶ ψύξει καὶ κόπῳ καί
τισιν ἑτέροις τοιούτοις αἰτίοις, ἀφίστασθαι χρὴ τῶν
ἐγκλημάτων, εἰ δέ τινα τῶν ἄλλων, οὐκ ἀφίστασθαι.
οὐδὲ γὰρ ἡμεῖς τὸ πάμπαν ἀπύρετοι διεμείναμεν,
ἀλλὰ διὰ κόπους τινὰς ἐπυρέξαμεν ἁπάντων τῶν ἄλλων νοσημάτων ἀπαθεῖς διατελοῦντες ἐτῶν ἤδη παμπόλλων, καὶ μέντοι καὶ πληγέντες τινὰ μέρη τοῦ
σώματος, ἐφ᾽ οἷς ἕτεροι φλεγμοναῖς τε ἅμα καὶ βου-
βῶσιν ἁλόντες ἐπύρεξαν, οὔτε βουβῶνα ἔσχομεν οὔτ᾽
ἐπυρέξαμεν, οὐκ ἄλλοθέν ποθεν ἢ ἐκ τῆς ὑγιεινῆς θεωρίας τῶν τηλικούτων ἐπιτυχόντες, καὶ ταῦτα μήτε
κατασκευῆς σώματος ὑγιεινῆς εὐτυχήσαντες ἐξ ἀρχῆς μήτε βίον ἀκριβῶς ἐλευθέριον ἔχοντες, ἀλλὰ καὶ
ταῖς τῆς τέχνης ὑπηρεσίαις δουλεύοντες καὶ φίλοις
καὶ συγγενέσι καὶ πολίταις ὑπηρετοῦντες εἰς πολλὰ
καὶ τῶν νυκτῶν τὸ πλεῖστον ἀγρυπνοῦντες, ἐνίοτε μὲν
309K ἀρρώστων ἔνεκα, διαπαντὸς δὲ τῶν ἐν παιδείᾳ καλῶν.

ἀλλ᾽ ὅμως οὐκ ἐνοσήσαμεν οὐδὲν νόσημα τῶν ἐκ
τοῦ σώματος ὁρμωμένων ἐτῶν ἤδη παμπόλλων, ὅτι
μή, καθάπερ ἔφην, ἐφήμερόν ποτε πυρετὸν ἐν τῷ
σπανιωτάτῳ διὰ κόπον γενόμενον. καίτοι κατά γε τὴν
τῶν παίδων ἡλικίαν καὶ προσέτι τῶν ἐφήβων τε καὶ
μειρακίων οὐκ ὀλίγαις οὐδὲ μικραῖς ἑάλωμεν νόσοις.
ἀλλὰ μετά γε τὸ εἰκοστὸν ὄγδοον ἔτος ἀπὸ γενετῆς

6

are sick through intemperance. But in them, the defence 308K
is much worse than the accusation—at least in my judgment.

We must absolve from these charges those who refer
to the things befalling them in life as the cause of their
being febrile with some ephemeral fever due to heat-
stroke, cold, fatigue or some other such causes, even if we
must not excuse any of the others. For I have not remained
entirely free of fever, but have been febrile due to certain
fatigues, while continuing to be unaffected by all other
diseases for many years now. And yet, although I have
been struck in certain parts of the body in which others,
being attacked by inflammations and buboes, have been
febrile, I have neither had buboes nor been febrile; and I
have attained such an age from no other source than the
theory of hygiene. And I have attained this without either
having a healthy constitution of the body from the start,
or a life entirely free, but being enslaved by the duties of
the art, assisting friends, relatives and citizens to a great
degree, remaining sleepless for many nights, sometimes
for the sake of the sick, but throughout for the delights of 309K
study.

Nonetheless, I have not suffered any of those diseases
arising from the body for many years now, apart, as I said,
from an occasional ephemeral fever arising very rarely due
to fatigue. And yet, during childhood, and besides this,
during puberty and adolescence, I was attacked by diseases neither few in number nor minor in magnitude. But
after I reached the age of twenty-eight, having persuaded

ἐμαυτὸν πείσας, ὡς ἔστι τις ὑγιεινὴ τέχνη, τοῖς προσ-
τάγμασιν αὐτῆς ἠκολούθησα παρ' ὅλον τὸν ἑξῆς
βίον, ὡς μηκέτι νοσῆσαι νόσημα μηδέν, ὅτι μὴ σπά-
νιόν που πυρετὸν ἐφήμερον. ἔστι δὲ δήπου καὶ τοῦτον
αὐτὸν φυλάξασθαι τελέως, ἐλεύθερον ἑλόμενον βίον,
ὡς ἔν τε τοῖς ἔμπροσθεν ἤδη γέγονε δῆλον ἔτι τε μᾶλλον ἔσται σαφὲς ἐν τοῖς ἐφεξῆς εἰρησομένοις, εἰ θέλοι
τις προσέχειν τὸν νοῦν. ἐγὼ γάρ φημι μηδὲ βουβῶνα
δύνασθαι γενέσθαι τοῖς ἀκριβῶς παρεσκευασμένοις
εἰς ὑγείαν, εἴ γ' ἀπέριττον αὐτοῖς ἐστι τὸ σῶμα τοῦ
γένους τῶν περιττωμάτων ἑκατέρου, τοῦ τε κατὰ τὸ
ποσὸν καὶ τοῦ κατὰ τὸ ποιόν.

310K πρὸς μὲν δὴ τοὺς νῦν ἐπαγγελλομένους ἢ λέγειν ἢ
γράφειν ὑγιεινὰ παραγγέλματα καὶ ταῦθ' ἱκανά· πρὸς
δὲ τοὺς ἔμπροσθεν, ὑπὲρ ὧν οὔτε ἐκ τῆς τῶν ἄλλων
ἱστορίας ἠκούσαμεν, ὡς ἤτοι σφᾶς αὐτοὺς ἀνόσους
διεφύλαξαν ἢ τοὺς πειθομένους αὐτοῖς, οὔτ' ἐτόλμη-
σαν ἐν τοῖς συγγράμμασιν ἀλαζονεύεσθαί τι τοιοῦτον,
οὐδ' ἀντιλέγειν ἀναγκαῖον, ἀλλὰ θαυμάζειν μόνον, εἴ
τινες αὐτῶν ἐν ἑνὶ βιβλίῳ τὴν ὑγιεινὴν ἅπασαν θεωρίαν ἐπηγγείλαντο περιλαβεῖν, οὐ μὰ Δία οὕτω γρά-
φοντες ὑπὲρ αὐτῆς, ὡς Ἱπποκράτης, ἐν τοῖς πρώτοις
τε καὶ γενικωτάτοις κεφαλαίοις ἐνδεικνύμενος τὴν
μέθοδον, ἀλλ' ἐξεργαζόμενοι πᾶν ἀκριβῶς, ὥσπερ
ἡμεῖς. ἡ μὲν δὴ τοῦ μήκους τῶν λόγων αἰτία καὶ δὴ
λέλεκται. χρὴ δ', ὅστις ἐστὶ φιλόκαλός τε ἅμα καὶ
φιλόπονος ἐπὶ τοῖς ἀρίστοις, οὐ πρὸς τὸ μῆκος ἀπ-
αγορεύειν τῆς ὑγιεινῆς τέχνης, ἀλλὰ τοῦ μεγέθους

myself that there is an art of hygiene, I followed its precepts for the whole of my subsequent life, and was never sick with any disease apart from the occasional ephemeral fever in some degree. It is, of course, possible for someone who chooses a free life to preserve himself completely, as is clear from what has already been said, and it will be clearer still in those things that will be said in what follows, if he wishes to direct his attention to them. Thus, I say, no bubo can come into existence in those perfectly prepared for health, if in fact their body is free of superfluity of each class of the superfluities, both in quantity and to quality.

With regard to those who now proclaim to either say 310K or write the precepts of hygiene, these things are enough. Regarding our predecessors, however, we have not heard from the accounts of others how they kept either themselves or those prevailed upon by them free of disease. Since they did not venture to brag about such a thing in their writings, it is not necessary to refute them, but only to wonder if some of them claimed to encompass the whole theory of hygiene in one book, and not, by the gods, writing about it in the way Hippocrates did when he showed the method under the primary and most generic headings, but going through the whole subject in its entirety, as I am doing. And so the reason for the length of the discussions has now been stated. It is necessary, however, for someone who is a lover of truth and at the same time diligent among the best not to flag in the face of the length of the art of hygiene, but to marvel at the magni-

ἄγασθαι τῶν ἐπαγγελμάτων αὐτῆς. πῶς γὰρ οὐ μεγάλα καὶ θαυμαστὰ τῆς τέχνης ταύτης ἐστὶν ἔργα,
311K γηράσαντα μέχρι πλείστου, ταῖς αἰσθήσεσιν ἀπαθῶς
ὑγιαίνοντα, διὰ παντὸς φυλάττεσθαι ἄνοσον ἀνώδυ-
νον ὁλόκληρον, εἴ γε δὴ μὴ παντάπασιν ἐξ ἀρχῆς
νοσώδους εἴη τετυχηκὼς σώματος; ἀλλ᾽ ἐγὼ καὶ πάνυ
μοι δοκῶ τινας ὑγιεινοὺς φύσει θεάσασθαι παμπόλλαις νόσοις ἁλισκομένους καὶ τελευτῶντας δὴ κατά
γε τὸ γῆρας ἀνιάτοις περιπεσεῖν πάθεσιν, οὓς ἐνῆν,
ὅσον ἐπὶ τῇ φυσικῇ τοῦ σώματος ἕξει, ταῖς τ᾽ αἰσθήσεσιν ἁπάσαις ἀπηρώτους διατελέσαι μέχρι γήρως
ἐσχάτου καὶ τοῖς ἄλλοις ἅπασι μορίοις τοῦ σώματος
ὑγιαίνοντας.

πῶς οὖν οὐκ αἰσχρόν ἐστιν ἀρίστης φύσεως τυ-
χόντα βαστάζεσθαι μὲν ὑπ᾽ ἄλλων διὰ ποδάγραν,
κατατείνεσθαι² δὲ ταῖς ὀδύναις λιθιῶντα καὶ κόλον
ἀλγοῦντα καὶ κατὰ κύστιν ἕλκος ἐκ κακοχυμίας
ἔχοντα; πῶς δ᾽ οὐκ αἰσχρόν ἐστι διὰ τὴν θαυμαστὴν
ἀρθρῖτιν ἀδυνατοῦντα χρῆσθαι ταῖς ἑαυτοῦ χερσὶν
ἑτέρου δεῖσθαι τοῦ προσφέροντος τὴν τροφὴν τῷ
στόματι καὶ τοῦ τὴν ἕδραν ἀπονίζοντος ἐν τῷ ἀπο-
πάτῳ; ἄμεινον γάρ, ὅστις μὴ παντάπασιν εἴη μαλακός, ἑλέσθαι δὴ μυριάκις τεθνάναι, πρὶν τοιοῦτον
312K ὑπομεῖναι βίον. εἰ δὲ δὴ καὶ τοῦ κατ᾽ αὐτὸν αἴσχους
τις ὑπερορᾷ δι᾽ ἀναισχυντίαν τε καὶ μαλακίαν, ἀλλὰ
τῶν γε πόνων οὐκ ἐχρῆν ὑπερορᾶν, οὓς νύκτωρ τε καὶ
μεθ᾽ ἡμέραν ἔχουσιν, ὥσπερ ὑπὸ δημίων στρεβλού-
μενοι τῶν παθῶν. καίτοι τούτων ἁπάντων ἢ ἀκολα-

10

tude of its precepts. For how are the actions of this art not
great and wondrous—actions that allow a person to grow
to a great age, maintaining himself continuously healthy,   311K
unaffected in the senses, free of disease, free of pain and
sound in all parts, unless he happened to be altogether
diseased in the body from the beginning? But it seems to
me I saw some who were very healthy in nature seized by
very many diseases, and finally in old age encountering
incurable affections, for whom it was possible, to the extent that this will be due to the physical state of the body,
to continue unimpaired in all the senses to the extreme of
age, being healthy in all the other parts of the body.

How then is it not shameful for someone who has the
best nature to be carried around by others due to gout, or
to be brought undone by the pains of stone, or pains in the
colon, or to have an ulcer in the bladder from *kakochymia*?
How is it not shameful for someone to be unable to use
his own hands due to severe arthritis and to need someone
else to bring his food to his mouth or wash his fundament
after defecation? If he were not altogether a coward, it
would be a thousand times better for him to choose to die
before enduring such a life. If, however, someone over-   312K
looks his own shame due to shamelessness and faintheart-
edness, he must not overlook the sufferings he has day
and night, just like those who are suffering public torture.
And indeed, intemperance or ignorance or both are in-

---

2 κατατείνεσθαι Ko; κατατήκεσθαι Ku

11

σίαν ἢ ἄγνοιαν ἢ ἀμφοτέρας ἀναγκαῖον αἰτιάσασθαι.
ἀλλὰ τὴν μὲν ἀκολασίαν οὐκ ἦν καιρὸς ἐπανορθοῦσθαι, τὴν δ' ἄγνοιαν ὧν χρὴ ποεῖσθαι ἐλπίζω διὰ
τῆσδε τῆς πραγματείας ἰάσασθαι, καθ' ἑκάστην φύσιν σώματος ἰδίαν ἀγωγὴν ὑγιεινὴν θέμενος.

2. Ἄρξασθαι δὲ δήπου³ δίκαιον ἂν εἴη ἀπὸ τῆς
ἀρίστης, ἧς ὁ σκοπός ἐστιν οὐκ ἐπανόρθωσις, ἀλλὰ
φυλακὴ τῶν ὑπαρχόντων, ὅπερ ἡμεῖς ἐποιήσαμεν εἰπόντες ὑπὲρ αὐτῆς ἄχρι δεῦρο καθ' ἑκάστην τῶν ἡλικιῶν πλὴν τῆς παρακμαστικῆς καλουμένης, ἧς τὸ
τελευταῖον μέρος ἰδίως ὀνομάζεται γῆρας, ἔχοντός
τινα καὶ τούτου τομὴν ἰδίαν, ὡς ὕστερον εἰρήσεται
κατ' ἐκεῖνο τοῦ λόγου τὸ μέρος, ἔνθα τὸ καλούμενον
τῆς τέχνης γηροκομικὸν μέρος εἰς διδασκαλίαν ἄξο-
313K μεν. ἀλλὰ νῦν γε τὰ κυριώτατα τῶν προειρημένων
ἀναλαβόντες ἐπὶ τὰς μοχθηρὰς φύσεις τῶν σωμάτων
μεταβησόμεθα δεικνύντες, ὅπως ἑκάστην αὐτῶν ἐν
ὑγείᾳ μάλιστα φυλακτέον ἐστίν.

τὸ τοίνυν ἄριστον σῶμα σκοποὺς ἔχει κατὰ μὲν
τὰς ποσότητας καὶ ποιότητας καὶ δυνάμεις ἐν μὲν
τοῖς γυμνασίοις τὰ μέτριά τε καὶ σύμμετρα προσαγο-
ρευόμενα, μετὰ τοῦ πᾶσιν ὁμοτίμως τοῖς μορίοις τοῦ
σώματος προσάγεσθαι, φυλαττομένων ἡμῶν ἅπασαν
ὑπερβολήν, εἰ δέ ποθ' ἁμαρτηθείη καθ' ὁτιοῦν τῶν
εἰρημένων, ἐπανορθούντων τὸ σφάλμα. κατὰ δὲ τὴν
τῶν ἐσθιομένων τε καὶ πινομένων φύσιν ἐν ποσότητι
καὶ ποιότητι καὶ δυνάμει σκοπὸς πάλιν ἐστὶ κἀνταῦθα
τὸ σύμμετρον, ὡς μήτε πλείω μήτ' ἐλάττω λαμβάνειν,

evitable causes of all these. But now is not the time to correct intemperance, although I do hope to cure the ignorance of those things that must be done, establishing through this book a method of hygiene for each specific bodily nature.

2. Perhaps it may be right to begin from the best [constitution], the objective of which is not correction but preservation of what exists, which is what I did, speaking about this up until now in relation to each of the stages of life, apart from that which is called "past the prime" ("decay"). The final part of this is termed specifically "old age." Since this also has a specific division, I shall speak about it later in that part of the discussion where I shall introduce the so-called geriatric part of the art to the instruction. But now, as we are taking up the most important 313K aspects of what has been previously said, we shall pass on to abnormal natures of bodies, showing how we must maintain each of them most in health.

Accordingly, the best body has objectives in respect of quantities, qualities and capacities in the exercises that are called moderate and balanced applied uniformly to all parts of the body, when we guard against every excess. If, however, a mistake is made at any time in any one whatsoever of the things mentioned, the fault is corrected. The objective in respect of the nature of the foods and drinks in terms of quantity, quality and capacity is here again moderation, so as to not to take too much or too little, but

---

3 δήπου Ko; μοί που Ku

ἀλλ' ἢ ὅσα πεφθέντα καὶ ἀναδοθέντα καὶ θρέψαντα
τὸ σῶμα καλῶς (εἰ δέοι καὶ τοῖς ἔτ' αὐξανομένοις τι
προστεθῆναι σύμμετρον) οὐδὲν ἐάσει περιττὸν οὐδ'
ἐνδεές. οὕτω δὲ καὶ κατὰ τοὺς ὕπνους τε καὶ τὰς ἐγρη-
γόρσεις καὶ τὰ λουτρὰ καὶ τὰς τῆς ψυχῆς ἐνεργείας
ὅσα τ' ἄλλα τοιαῦτα τὴν συμμετρίαν δηλονότι φυ-
314K λάττειν προσήκει, εἰ δέ ποθ' ἁμάρτοι τις καθ' ὁτιοῦν
τῶν εἰρημένων, ἐπανορθοῦσθαι τὸ σφάλμα.

κοινὸς δ' ἔστω σοι σκοπὸς ἁπάσης ἐπανορθώσεως
ἡ τῆς ἐναντίας ἀμετρίας χρῆσις, εἰ μὲν ἐπὶ πλέον
ἐπόνησε τὸ σῶμα τῇ προτεραίᾳ, καθαιροῦντι τὸ πλῆθος τῶν γυμνασίων, εἰ δ' ἐνδεέστερον, αὐξάνοντι,
οὕτω δὲ καὶ ὀξυτέραις εἰ ἐχρήσατο ταῖς κινήσεσιν,
ἀνιέντι μετρίως, εἰ δ' ἐκλελυμέναις, ἐπιτείνοντι. κατὰ
δὲ τὸν αὐτὸν τρόπον ἐπὶ μὲν τοῖς εὐτονωτέροις τὰ μα-
λακώτερα γυμνάσια παραλαμβάνων, ἐπὶ δὲ τοῖς ἀτο-
νωτέροις τὰ ἰσχυρότερα, καὶ τὰ σφοδρὰ δὲ τοῖς ἀμυ-
δροῖς ἀντεισάγων καὶ τοῖς σφοδροῖς τἀναντία, καὶ
συλλήβδην εἰπεῖν ἅπασαν ἀμετρίαν ἐπανορθούμενος
διὰ τῆς ἐναντίας ἀμετρίας ὑγιαίνοντα διαφυλάξεις
τὸν ἄνθρωπον. εἰς δὲ τὸ μηδὲν ταῖς ἐπανορθώσεσι
σφάλλεσθαι πρῶτον μὲν χρὴ διαγινώσκειν ἀκριβῶς
τὰς διαθέσεις τοῦ σώματος, εἶτα μεμνῆσθαι τῶν ἐν τῇ
προτεραίᾳ γενομένων ἁπάντων. αἱ μὲν γὰρ διαθέσεις
ἐνδείξονται τὸ πλημμεληθέν, ἡ μνήμη δὲ τῶν προγε-
γενημένων, εἰς ὅσον χρὴ μετακινῆσαί τι τῶν συνήθων,
ὑπαγορεύσει. κατὰ μὲν δὴ τὰς διαθέσεις αἱ τοιαίδε
315K εἰσὶν ἀμετρίαι, ὡς ἰσχνότερον ἢ εὐογκότερον ἢ σκλη-

14

as much as is concocted, distributed and nourishes the body well (if necessary something may also be added to moderation for those who are still growing), allowing neither excess nor deficiency. In this way too, in sleep, wakefulness, baths and the functions of the soul, and other such things, it is obviously appropriate to maintain a balance, while if at any time someone errs in any one whatsoever   314K
of these things mentioned, the fault is corrected.

Let your general objective in every correction be the use of the opposite excess. If the body has labored too much on the previous day, do this by reducing the amount of exercises; if it labored too little, do this by increasing the amount of exercise. In the same way too, if it used movements that are too rapid, do this by cutting them back moderately; if it used relaxed movements, do this by increasing the vigor. On the same basis, after more vigorous movements, undertake more gentle exercises, while after weaker movements, undertake those that are stronger; substitute the violent for the weak and the opposite for the violent. In short, correct every imbalance, keeping the person healthy through the opposite excess. In regard to not erring in the corrections, it is first necessary to recognize accurately the conditions of the body, then to recall all those things that occurred on the previous day. The conditions indicate what is wrong, while the memory of what previously occurred dictates by how much it is necessary to change what is customary. Now in the conditions, the following are imbalances: a body that appears thinner   315K

15

ρότερον ἢ μαλακώτερον ἢ ὑγρότερον ἢ ξηρότερον ἢ
ἀραιότερον ἢ πυκνότερον ἑαυτοῦ φαίνεσθαι τὸ σῶμα
πρὸς τῷ μηδὲ τὴν κατὰ φύσιν ἀκριβῶς ἀποσῴζειν
εὔχροιαν. ἡ μνήμη δὲ τῶν προγεγονότων αὐτό τε τὸ
ἁμαρτηθὲν ἐνδείξεταί σοι καὶ διδάξει τὴν ἐπανόρθωσιν ἐκ τοῦ παραβάλλεσθαι τοῖς ἐνεστῶσιν.

εἰ μὲν γὰρ ἰσχνότερον τὸ σῶμα φαίνοιτο, σκοπεῖσθαι χρὴ καὶ ἀναμιμνήσκεσθαι, πότερον πλείω τοῦ
προσήκοντος ἐπόνησεν ἢ ὀξυτέραις ἐχρήσατο ταῖς
κινήσεσιν ἢ περὶ τὴν τρῖψιν ἐπλεόνασεν ἢ τὰ λουτρά·
καὶ μετὰ ταῦθ' ἑξῆς ἐπισκοπεῖσθαι, πότερον ἐφρόντι-
σεν ἢ ἠγρύπνησεν ἢ ἐξέκρινε κατὰ γαστέρα πολὺ
πλείω τοῦ προσήκοντος· ἐπισκοπεῖσθαι δὲ καί, εἰ
οἶκος θερμότερος, ἐν ᾧ διέτριψεν, ἢ εἰ[4] ἔφαγεν ἔλαττον
ἢ ἔπιεν ἢ ἀφροδισίοις ἐχρήσατο μὴ δεόντως·[5] εἰ δ' ἐν
ὄγκῳ μείζονι τὸ σῶμα φαίνοιτο, μὴ τρῖψις[6] ἢ γυμνάσιον ἔλαττον ἢ βραδύτερον ἢ πλείων ὕπνος ἢ ἐποχὴ
γαστρὸς ἢ σιτίων πλῆθος ἀμέμπτως πεφθέντων.

316K εἰ δὲ σκληρότερον ἑαυτοῦ φανείη τὸ σῶμα, τρίψεων
μὲν ἀναμνησθῆναι χρὴ πρῶτον, εἶτα γυμνασίων εὐ-
τόνων μετ' ἀνταγωνιστοῦ σκληροῦ,[7] καὶ πρὸς τούτοις
εἰ ἐν κόνει καὶ ταύτῃ σκληρᾷ καὶ ψυχρᾷ παντάπασιν[8]
καὶ εἰ χωρὶς τῆς καλουμένης ἀποθεραπείας· εἶθ' ἑξῆς
λουτρῶν, εἰ ψυχρὰ παντάπασιν ἢ λίαν θερμά, καὶ ὁ
οἶκος, ἐν ᾧ διέτριψεν ἐγρηγορώς τε καὶ κοιμώμενος,
εἰ[9] ψυχρός· ἔτι δὲ σιτίων ξηρότητος καὶ πομάτων ἐν-

_____

4 εἰ add. Ko   5 μὴ δεόντως Ko; πλείοσιν Ku

than it should be, or fatter, or harder, or softer, or more moist, or more dry, or more loose-textured or more condensed, and in addition to this, one that does not preserve precisely its natural color. It is the recollection of what previously existed that will show you the fault and will teach the correction through the comparison with what presently exists.

Thus, if the body seems too thin, you must consider and recall whether the person labored more than was appropriate, or used overly rapid movements, or went to excess with massage or bathing. And next after these things, you must consider whether he was anxious, sleepless, or eliminated much more than was appropriate from the stomach. You must also consider whether the house in which he spent his time was too hot, or whether he ate or drank too little, or indulged in unnecessary sexual activity. If the body seems too great in mass, consider whether or not the massage or exercise was too little or too slow, or whether he slept too much, or there was retention in the stomach, or an excess of foods faultlessly concocted.

If, however, the body seems harder than it was, it is first 316K necessary to recall the massages; then, if the exercises were vigorous with an adversary pressing hard, and in addition to these things, if they were in dust that was altogether hard and cold, and if they were done without the so-called apotherapy. Then next, it is necessary to recall if the baths were altogether cold or very hot, if the house in which he spent time waking and sleeping was cold, and

---

6 *post* τρίψις: μαλακὴ Ku

7 *post* σκληροῦ: πιλοῦντος (Ku) *om.*

8 παντάπασιν *add.* Ko   9 εἰ *add.* Ko

δείας. καὶ εἰ μαλακώτερον ἑαυτοῦ γένοιτο κατὰ τὴν
ὑστεραίαν τὸ σῶμα, πρῶτον μὲν ἀναμιμνήσκεσθαι
χρὴ τῆς τρίψεως, εἰ μαλακή τε καὶ σὺν λίπει καὶ λουτροῖς ἀτρέμα χλιαρωτέροις ἐγένετο· μετὰ δὲ τὴν τούτων ἐπίσκεψιν, εἰ τὰ γυμνάσια βραδέα καὶ ὀλίγα καὶ
μετὰ τοῦ συμπαλαίοντος ἀμετρότερον ἀπαλοῦ, κἄπειτα περὶ πόματος, εἰ πλέον, εἶθ' ἑξῆς ἐδεσμάτων, εἰ
πλέον· ἢ ὑγρότερα τὴν φύσιν, εἶθ' ὕπνων, εἰ πλείους.
ἐγγὺς δὲ τῆς μαλακῆς ἐστι τοῦ σώματος διαθέσεως ἡ
ὑγρὰ καλουμένη, πλὴν ὅσον ἡ μὲν μαλακὴ τῶν σωμάτων αὐτῶν ἐστιν οἰκεία ποιότης, ἡ δ' ὑγρὰ τῶν ἐν
αὐτοῖς ὑγρῶν. διακρίνεται δ' ἁπτομένων ἡμῶν·[10] ἡ μὲν
317K   γὰρ ὑγρὰ σὺν ἰκμάσιν ἐστίν, ἡ δὲ μαλακὴ χωρὶς
τούτων, ὄντος γε δηλονότι καὶ τοῦ μαλακοῦ σώματος
ὑγροῦ τοῖς οἰκείοις μορίοις· ἀλλ' ἕνεκα σαφοῦς διδασκαλίας μαλακὸν μὲν τοῦτο καλείσθω, τὸ δ' ἕτερον
ὑγρόν.

ἡ μὲν οὖν ἀμέτρως ξηρὰ διάθεσις εὐθὺς σκληρύνει
τὴν ἕξιν, οὐκ ἐξ ἀνάγκης δὲ μετὰ μαλακότητός ἐστιν
ἡ ὑγρά. δύναται γὰρ ἐσκληρύνθαι μὲν ἡ σάρξ, ἀνα-
φέρεσθαι δ' ἰκμὰς ἐκ τοῦ σώματος ἢ ἱδρώς. ἐπὶ μὲν
οὖν τῶν ὑγροτήτων ἤτοι γ' ἀφροδισίων χρῆσιν ἄκαι-
ρον ἢ ἀπό τινος ἑτέρας αἰτίας ἀρρωστίαν τῆς δυνάμεως ὑποπτευτέον ἢ ἀραιότητα τοῦ σώματος ἐπὶ μαλακαῖς ἀμέτρως τρίψεσιν ἢ λουτροῖς πλέοσιν ἢ ἀέρι
τῷ κατὰ τὸν οἶκον, ἐν ᾧ διέτριψε, θερμοτέρῳ παρὰ τὸ

10 ἡμῶν add. Ko

further, if foods were dry and drinks lacking. And if on the following day the body becomes softer than it was, the first thing that must be recalled is the massage—if it was gentle and with oil, and if the baths were gentle and lukewarm. After consideration of these things, it must be recalled if the exercises were slow and few, and with an excessively soft adversary, and then regarding drink, if it was much and then next foods, if they were too much or too moist in nature, and then if sleep was too much. The so-called moist [condition] is close to the soft condition of the body except to the extent that softness is a proper quality of bodies themselves, whereas moistness is a quality of the fluids in them. This is judged by our touching them, for the moistness is with fluids (humors) whereas the softness 317K is apart from these, although obviously the soft body is also moist in the appropriate parts. But for the sake of clarity in teaching, let the one be called soft and the other moist.

Thus, the excessively dry condition immediately hardens the state[2] (of the body), but a moist [condition] is not necessarily associated with softness. It is possible for the flesh to be hardened, but moisture or sweat can be carried up from the body. Therefore, in those who are moist, one must suspect untimely sexual activity or weakness of the capacity from some other cause, or rarefaction of the body due to excessively soft massages, or too many baths, or from the air in the house in which he spends his time be-

---

[2] Galen uses the term ἕξις in contrast to διάθεσις. The difference is recognized in the translation by using "state" for the former, implying relative permanence, and "condition" for the latter, which implies relative transience and susceptibility to change.

δέον. ἐπισκεπτέον δὲ καὶ περὶ πόματος εἰ πλέον, εἰ ὕπνοι πολὺ πλείους τῶν κατὰ φύσιν, ἢ εἰ τὸ περιέχον ἀθρόως μετέβαλεν εἰς ὑγρότητα καὶ θερμότητα, καὶ περὶ τροφῶν ὡσαύτως. ἐπὶ δὲ τῆς ἀπαλότητος ἢ μαλακότητος (ἑκατέρως γὰρ ὀνομάζειν ἔθος ἐστίν), ὅταν ποτὲ χωρὶς ὑγρότητος ᾖ, πεπέφθαι μὲν τὴν τροφὴν ὀρθῶς καὶ τεθράφθαι τὸ σῶμα, γεγυμνάσθαι δὲ ἐνδεέστερον· ἔμπαλιν δὲ ἐπὶ τῆς σκληρότητος ἢ τετρίφθαι σκληρῶς ἢ γεγυμνάσθαι πλεῖον μετὰ σκληροῦ σώματος ἐν κόνει.

318K

ξηρότης δὲ τῆς ἕξεως ἔνδειαν πόματος ἢ τροφῆς ἢ ἀγρυπνίαν ἢ μέριμναν βιωτικὴν ἢ πολλὴν τρίψιν ἢ γυμνάσιον ἄμετρον ἐνδείκνυται.[11] ταῦτ᾽ οὖν ἐπισκεπτόμενος ἐπανορθοῦσθαι δυνήσῃ καθ᾽ ἑκάστην ἡμέραν τὸ σφάλμα, πρὶν αὐξηθὲν δυσίατον γενέσθαι. μέμνησο δ᾽ ἀεὶ τοῦ πᾶσαν ἀμετρίαν εἰς ἐπανόρθωσιν ἄγεσθαι διὰ τῆς ἐναντίας ἀμετρίας. ὁ γάρ τοι σκοπὸς οὗτος ἁπάντων ἐστὶ κοινὸς τῶν παρὰ φύσιν. ὥστε χρὴ προσθέντα σε τῷδε τήν τε τῶν σωμάτων αὐτῶν διάγνωσιν, ὧν προνοεῖσθαι μέλλεις, ἑκάστου τε τῶν βοηθημάτων τὴν δύναμιν ἁπάσης ἐπιστήμονα τῆς περὶ τὸ σῶμα τέχνης γενέσθαι, καθ᾽ ἣν οὐ μόνον τοὺς ὑγιαίνοντας ἐν ὑγείᾳ διαφυλάξεις, ἀλλὰ καὶ τοῖς νοσοῦσιν ἀναλήψῃ τὴν ἀρχαίαν ἕξιν.

319K

3. Ὁποῖον μὲν οὖν ἐστι τὸ ἄριστον σῶμα, πρόσθεν εἴρηται. τὸ δ᾽ ἀπολειπόμενον τοῦδε γινώσκειν μέν σε

11 post ἐνδείκνυται.: ταῦτ᾽ οὖν ἐπισκεπτόμενος Ko; ταῦτά τε οὖν βίον σκεπτόμενος Ku

ing hotter than it should be. Also one must give consideration to drinks, whether there were too many, or if sleep was much longer than accords with nature, or if the ambient air changed suddenly toward moisture and heat, and similarly in regard to nutriments. In the case of tenderness or softness (for it is customary to use either term),[3] whenever at any time this exists apart from moisture, [one must consider whether] the nutriment has been concocted properly and the body nourished but exercised too little. 318K Conversely, in the case of hardness, one must consider whether there has been hard massage or too much exercise with a hard body in dust.

Dryness of the state [of the body] indicates a lack of drink or food, or sleeplessness, anxious thoughts about life, much massage or excessive exercise. Considering these things, then, will enable you to correct the error each day before it is increased to become hard to cure. Always remember to bring every excess to correction through the opposite excess. This certainly is the objective common to all the things contrary to nature. As a result, it is necessary for you to apply yourself to the diagnosis of the actual bodies which you are going to care for, and to the potency of each of the remedies, to become knowledgeable about the whole art concerning the body, through which you will not only maintain the healthy in good health (the health of the healthy) but also restore the original state to those who are sick.

3. What kind of body is best was stated previously. 319K What you must know is that what departs from this does

---

3 The two terms are ἀπαλότης (softness, tenderness) and μαλακός (soft, tender).

χρὴ διὰ τρεῖς αἰτίας ἀπολειπόμενον, ἢ ὅτι κακῶς ἐξ
ἀρχῆς κατεσκευάσθη κυούμενον, ἢ ὅτι μετὰ ταῦτα
πρός τινος αἰτίας εἰς τὴν παρὰ φύσιν ἤχθη διάθεσιν,
ἢ τῷ τῆς ἡλικίας λόγῳ. πειρᾶσθαι δ᾽ ἐπανορθοῦσθαι
πάντη διὰ τῆς ἀντικειμένης ἀμετρίας· οἷον αὐτίκα[12] τὸ
γῆράς ἐστι μὲν ψυχρὸν καὶ ξηρόν, ὡς ἐν τοῖς Περὶ
κράσεων ἀποδέδεικται λόγοις. ἐπανόρθωσις δ᾽ αὐτοῦ
διὰ τῶν ὑγραινόντων καὶ θερμαινόντων γίνεται. τοιαῦτα δ᾽ ἐστὶ θερμὰ λουτρὰ γλυκέων ὑδάτων καὶ οἴνου
πόσις ὅσαι τε τῶν τροφῶν ὑγραίνειν τε ἅμα καὶ θερμαίνειν πεφύκασι. περὶ δὲ γυμνασίων ἢ τρίψεων ἀπά-
σης τε κινήσεως (ἀπὸ τούτων γὰρ ἄρξασθαι βέλτιον,
ἐπειδὴ καλῶς εἴρηται "τοὺς πόνους τῶν σιτίων ἡγεῖσθαι") τάδε χρὴ γινώσκειν, ὡς ἐκ μέρους μέν τινος
ὀρθῶς εἶπεν ὁ ποιητής·

ἐπὴν λούσαιτο φάγοι τε,
εὐδέμεναι μαλακῶς· ἡ γὰρ δίκη ἐστὶ γερόντων.

320K   οὐ μὴν τό γε σύμπαν ἐν τούτῳ τέτακται. δέονται
γὰρ οἱ γέροντες οὐδὲν ἧττον τῶν νέων εἰς κίνησιν
ἄγειν τὸ σῶμα, κινδυνευούσης αὐτοῖς σβεσθῆναι τῆς
ἐμφύτου θερμασίας. ἐπὶ μέν γε τῶν ἀκμαζόντων σωμάτων εὑρίσκονταί τινες φύσεις ἡσυχίας δεόμεναι,
περὶ ὧν εἴρηταί που καὶ Ἱπποκράτει· γέρων δ᾽ οὐδεὶς
χρῄζει παντελοῦς ἡσυχίας, ὥσπερ οὐδὲ γυμνασίων
σφοδρῶν· ῥιπίζεσθαι μὲν γὰρ αὐτῶν δεῖται τὸ θερ-

---

[12] post οἷον αὐτίκα: περὶ γήρως. (Ku) add.

so for three reasons; that it was badly constituted from the beginning at the time of pregnancy; that after this for some reason it was brought to a condition contrary to nature; and by virtue of the stage of life (age). One must attempt to restore all these through the opposing excess. To begin with the example of old age, this is cold and dry, as was shown in the writings *On Mixtures*.[4] Correction of this occurs through moistening and heating agents. Such things are hot baths of sweet waters, drinking wine, and those nutriments that are naturally moistening and heating at the same time. Regarding exercises or massage and every movement (for it is better to begin from these since it was well said, "let exertions precede food"),[5] one must realize that what the poet said was right in part:

> when he has bathed and eaten,
> let him rest gently, for this is right for the old.[6]

But the whole matter is not so ordained. For old men need to bring the body to movement no less than young men do, since there is a danger of the innate heat being quenched in them. In the case of bodies in the prime of life, some natures discover they need rest. Concerning these, Hippocrates also said somewhere: "none of those who are old need complete rest, just as they do not need overly violent exercise."[7] For their heat needs to be fanned

320K

---

[4] *Mixt.,* I.509–694K (English trans., Singer, *Galen: Selected Works*). See particularly Book 2, chap. 2, I.580K.

[5] Hippocrates, *Epidemics* 6.4(23), *Hippocrates* VII, LCL 477, 254–55.   [6] This quote is also found in *Marc.* 5, VII.682K, where it is attributed to Homer.

[7] Hippocrates, VI.84, 10.19 Littré (Koch).

μόν, ἐξελέγχεται δὲ κατὰ τὰς σφοδροτέρας κινήσεις.
αἱ μὲν οὖν μεγάλαι φλόγες οὐδὲν ἔτι χρῄζουσι τοῦ
ῥιπίζοντος, ἀλλ' ἑαυταῖς εἰσιν ἱκαναὶ πρὸς τὸ διασῴ-
ζεσθαί τε καὶ κρατεῖν τῆς ὕλης. οὔκουν οὐδ' ἀνα-
τρίβεσθαι μετὰ τοὺς ὕπνους ἕωθεν[13] οἱ ἀκμάζοντες
δέονται, καθάπερ οἱ γέροντες. ὁ γάρ τοι σκοπὸς τῆς
τοιαύτης ἀνατρίψεως ἅμα λίπει γινομένης διττός
ἐστιν, ἢ κοπώδη διάθεσιν ἰάσασθαι, πρὶν αὐξηθεῖσαν
ἀνάψαι πυρετόν, ἢ ἀρρωστοῦσαν ἀνάδοσιν ἐπεγεῖραι.
πολλοὺς γὰρ ἀτροφοῦντας ἐκ μακροῦ χρόνου ῥᾳδίως
ἐν ὀλίγαις ἡμέραις εὐσαρκώσαμεν, ἐπὶ τὴν τοιαύτην
321K ἀνάγοντες τρίψιν. ἀλλ' ὅπερ τοῖς ἄλλοις κατὰ πάθος
ἐν χρόνῳ τινὶ γίνεται, τοῦτ' ἀεὶ τοῖς γέρουσιν ὑπάρχει. ψυχρὸν γὰρ ὅλον αὐτοῖς ἐστι τὸ σῶμα καὶ ἀδύνατον ἕλκειν τὴν τροφὴν ἐφ' ἑαυτὸ καὶ κατεργάζεσθαι
καλῶς καὶ τρέφεσθαι πρὸς αὐτῆς.

ἀλλ' ἡ τρίψις ἐπεγείρουσα τὸν ζωτικὸν τόνον αὐτῶν
καὶ θερμαίνουσα τὰ μέτρια τήν τ' ἀνάδοσιν εὐπετε-
στέραν ἐργάζεται καὶ τὴν θρέψιν ἑτοιμοτέραν. οὕτω
τοι καὶ διὰ τὸ πιττοῦσθαι νέοι[14] πολλοὶ τῶν ἀτροφούν-
των ἐσαρκώθησαν οἵ τε γέροντες ὀνίνανται πάντες. ἐν
μὲν δή σοι τοῦτο καθάπερ τι γυμνάσιον ἔστω τοῖς
γέρουσιν ἕωθεν γινόμενον, ἡ μετ' ἐλαίου τρίψις, ἐφεξῆς δὲ περίπατοί τε καὶ αἰωρήσεις ἄκοποι, στοχαζο-
μένῳ τῆς τοῦ γέροντος δυνάμεως.[15] οὐ γὰρ μικρά τις

13 post ἕωθεν: ἅπαντες (Ku) om.
14 διὰ τὸ πιττοῦσθαι νέοι Ko; διατώμενοι νέοι τε Ku

but is counteracted by violent movements. Great fires do
not still need fanning but are sufficient to maintain themselves and overcome the material. Therefore, all those in
the prime of life do not need to be massaged excessively
in the early morning after sleep, as old men do. For certainly the objective of such massage occurring with oil is
twofold: to cure the fatigue condition before it is increased
to kindle a fever, and to activate a weakened distribution.
I have easily restored flesh in a few days, in many who have
been emaciated for a long time, by resorting to such mas-   321K
sage. But what occurs to others as an incidental affection
at a particular time is always in existence in the aged. Thus,
in them, the whole body is cold and unable to draw the
nutriment to itself, process it properly and be nourished
from it.

But massage, since it stimulates their (i.e., old people)
vital tone and heats moderately, makes distribution easier
and nourishing more effective. Certainly, in this way too,
many young men who are atrophic become enfleshed
when following a regimen,[8] while all old men are benefited. There is the one thing you should do for old people
in the early morning as an exercise: after massage with oil,
next get them to walk about and carry out passive exercises
without becoming fatigued, taking into account the capacity of the old person. For there is no small difference be-

[8] The Kühn text is followed here—that is, "following a regimen" rather than "have pitch applied," since it seems unlikely that
this would be a measure used in old people.

---

[15] post δυνάμεως. A substantial section of the Kühn text,
comprising 321K, lines 13–17, 322–28K complete, and 329K, lines
1–8 (part), is transferred to follow Koch, section 10, line 13.

ἐν αὐτοῖς ἐστιν διαφορά, τινῶν μὲν ἐσχάτως ἀρρώστων εἰς τὰς κινήσεις ὄντων, εἰ καὶ μηδέπω τύχοιεν
ἑβδομηκοντοῦται τὴν ἡλικίαν ὄντες, ἐνίων δὲ πολὺ
ῥωμαλεωτέρων ἢ κατὰ τούσδε γεγονότων ἔτη πλείω ἢ
ὀγδοήκοντα. τοὺς μὲν ἀσθενεστέρους αἰωρεῖν μᾶλλον
ἢ περιπατεῖν κελεύων κινήσεις, τοὺς δ' ἰσχυροτέρους
δι' ἀμφότερα[16] γυμνάσια· ἄξεις δὲ πρὸς τὴν δευτέραν
τρίψιν οὐχ ὁμοίως ἀμφοτέρους, ἀλλ' ἀεὶ τὸν ἀρρω-
στότερον θᾶττον. ἔστω γάρ σοι καὶ τοῦτο τῶν κοινο-
330K τάτων παραγγελμάτων, ἐπὶ μὲν ἀσθενοῦς δυνάμεως
ἀνατρέφειν τὸ σῶμα πυκναῖς καὶ βραχείαις τροφαῖς,
ἐπὶ δὲ ἰσχυρᾶς ἀραιαῖς καὶ πολλαῖς.

4. Εἰπεῖν μὲν οὖν καὶ ταῦτα καὶ τἄλλα ῥᾶστον,
ἐπιστατῆσαι δὲ γέροντι διαφυλάττοντα τὴν ὑγείαν
αὐτοῦ τῶν χαλεπωτάτων, ὥσπερ γε καὶ τῶν ἀνακομι-
ζομένων ἐκ νόσου. καλεῖται δὲ ὑπὸ τῶν νεωτέρων ἰατρῶν τουτὶ μὲν τὸ μέρος τῆς τέχνης ἀναληπτικόν, τὸ
δὲ ἐπὶ τῶν γερόντων γηροκομικόν. καὶ δοκοῦσιν αἱ
διαθέσεις ἀμφοῖν οὐ κατὰ τὴν ἀκριβεστάτην ὑπάρχειν ὑγείαν, ἀλλ' ἤτοι μέσαι τινὲς εἶναι νόσου τε καὶ
ὑγείας ἢ πάντως μὴ τῆς καθ' ἕξιν, ἀλλὰ τῆς κατὰ
σχέσιν ὀνομαζομένης ὑπ' αὐτῶν ὑγείας. εἴτ' οὖν νόσον εἴτε νοσώδη διάθεσιν εἴτε μέσην ὑγείας τε καὶ
νόσου διάθεσιν εἴτε κατὰ σχέσιν ὑγείαν ὀνομάζειν

---

[16] post ἀμφότερα: γυμνάσια· ἄξεις δὲ Ko; γυμνάσαντα,
ἄξαντά τε Ku

tween those who are exceedingly weak in movements, even if they do not yet happen to be seventy, and those who are much stronger than them who are more than eighty years old. Direct those who are weaker to passive exercises more than walking around, whereas exercise those who are stronger in both ways. You will bring both toward the second massage but not similarly; always bring the weaker quicker. Take this as one of your most general precepts: in the case of a weak capacity, nourish the body 330K with thick and few nutriments, while in the case of a strong capacity with thin and many.

4. It is very easy, then, to say these and other things, but taking care of an old person and maintaining his health is one of the most difficult things, just as restoring someone to health from disease also is. This part of the art is called by younger doctors "analeptic" (restorative), while that which pertains to old people is called "geriatric." And both the conditions seem not to be in accord with the most perfect health, but to be midway between disease and health in some cases or altogether not of health in terms of a stable state, but what is called by them health in terms of an unstable state.[9] Therefore, whether we ought to call old age a disease, or a morbid condition, or a condition between health and disease, or an unstable state of health,

[9] There is a distinction between ἕξις, taken as a stable or permanent state, and σχέσις, which is taken as an unstable or impermanent state, somewhat similar to διάθεσις. On these terms in relation to fevers, see Galen's *MM*, X.533K (Johnston and Horsley, *Galen: Method of Medicine*, 2.533), and for a more general discussion, Johnston, *Galen: On Diseases and Symptoms*, 27–28. See also note 2 above.

χρὴ τὸ γῆρας, οὐ πάνυ τι τῶν τοιούτων[17] φροντίζοντας ζητήσεων μέχρι λόγου προερχομένων ἐπίστασθαι δὲ χρὴ τὴν κατάστασιν τοῦ τῶν γερόντων σώματος, ὅτι ἐπὶ σμικροῖς αἰτίοις εἰς νόσον μεθισταμένην

331K ὁμοίως τοῖς ἀναλαμβάνουσιν ἐκ νόσου τὴν ἀρχαίαν ὑγείαν διαιτᾶν χρή.

διόπερ εἰπεῖν μὲν ῥᾴδιον, ἄμεινον εἶναι χρίεσθαι λίπει μετ' ἀνατρίψεως ἔωθεν τὸν γέροντα, προσηκόντως δὲ ἐργάσασθαι τοὖργον ἁπάντων χαλεπώτατον. ἥ τε γὰρ σκληροτέρα βραχὺ τρίψις κοπώδης, ἥ τ' ἄγαν μαλακὴ πλέον οὐδὲν ἐργάζεται, καθάπερ οὐδ' ἡ βραχεῖα παντάπασιν, ἡ δέ γε πολλὴ διαφορεῖ μᾶλλον ἢ ἀνατρέφει. καὶ μὴν καὶ τὸ χωρίον, ἐν ᾧ γυμνοῦται τὸ τοῦ γέροντος σῶμα, ψυχρότερον μὲν ὂν οὐ μόνον οὐδὲν ἐργάσεται χρηστόν, ἀλλὰ καὶ πυκνώσει καὶ καταψύξει, θερμότερον δ' εἴπερ εἴη τοῦ δέοντος, ἐν μὲν χειμῶνι πλέον ἢ προσήκει τὸ σῶμα τοῦ γέροντος ἀραιότερον ἀποτελέσαν εὔψυκτον ἐργάσεται, θέρους δὲ διαφορήσει τε καὶ καταλύσει τὴν δύναμιν αὐτοῦ. τοῖς μὲν γὰρ καθ' ἕξιν ὑγιαίνουσιν οὐδὲ τῶν ἰσχυροτέρων αἰτίων οὐδὲν ἀλλοιοῖ τὸ σῶμα, τοῖς γέρουσι δὲ καὶ τὰ σμικρότατα μεγίστην ἐργάζεται μεταβολήν. οὕτω τοίνυν ἔχει κἀπὶ τῆς τῶν σιτίων ποσότητός τε καὶ ποιότητος. καὶ γὰρ κἂν τούτοις ἂν βραχύ τι τοῦ προσήκοντος ὑπερβῶσιν οἱ γέροντες, οὐ σμικρὰ βλάπτονται, τῶν νεανίσκων ἐπὶ μεγίστοις

332K ἁμαρτήμασι βραχέα βλαπτομένων.

ἀσφαλέστερον οὖν ἐστι τοῖς ἀσθενέσι γέρουσιν

not giving very much consideration to any such inquiries as far as the argument of our predecessors are concerned, we must know the condition of the body of those who are old, in that it can change to disease following minor causes and must be restored to its original healthy way of life, like 331K those recovering from disease.

This is why it is easy to say it is better to anoint the old person with oil after massage early in the morning, but to carry out the task properly is the most difficult thing of all. For a slightly harder massage is fatiguing, whereas that which is very soft accomplishes nothing more, just like that which is altogether brief, while a large amount of massage disperses more than it nourishes. Furthermore, if the place in which the old man's body is exposed is too cold, the massage not only accomplishes nothing useful, but also condenses and chills. If, on the other hand, the place is hotter than it needs to be, in the winter it makes the body of the old man more loose-textured than is appropriate and causes him to be easily chilled, whereas, in summer, it disperses and dissipates his capacity. Thus, in those who are healthy in terms of a stable state, none of the stronger causes change the body, whereas in old people even the smallest causes bring about great change. Moreover, it is like this in the case of the quantity and quality of the foods, for even in these, if old people overstep what is appropriate by just a little, they are harmed a lot, whereas young men are harmed very little by the largest 332K errors.

Therefore, it is safer to give weak old people a little

---

17 *post* τῶν τοιούτων: φροντίζοντας ζητήσεων μέχρι λόγου Ko; ζητήσεων φροντίζοντας, βωμολόχων Ku

ὀλίγα διδόναι τρὶς τῆς ἡμέρας, ὡς Ἀντίοχος ὁ ἰατρὸς
ἑαυτὸν διῆτα, γεγονὼς μὲν ἐτῶν πλείω τῶν ὀγδοήκοντα, προϊὼν δὲ καθ᾽ ἑκάστην ἡμέραν εἰς τὸ χωρίον,
ἐν ᾧ τὸ συνέδριον ἦν αὐτῷ τῶν πολιτῶν, ἔστι δ᾽
ὅτε μακρὰν ὁδὸν ἀπιὼν ἐπισκέψεως ἀρρώστων ἕνεκα.
ἀλλ᾽ εἰς μὲν τὴν ἀγορὰν ἀπὸ τῆς οἰκίας ἐβάδιζεν,
ὁδὸν ἕως τριῶν σταδίων· οὕτω δὲ καὶ τοὺς πλησίον
ἀρρώστους ἑώρα· πορρώτερον δ᾽ εἴ ποτ᾽ ἠναγκάσθη
πορευθῆναι, τὸ μὲν ἐν δίφρῳ βασταζόμενος ἐκομί-
ζετο, τὸ δ᾽ ἐπ᾽ ὀχήματος αἰωρούμενος. ἦν δ᾽ αὐτῷ τι
κατὰ τὴν οἰκίαν οἴκημα διὰ καμίνου θερμαινόμενον ἔν
γε τῷ χειμῶνι, θέρους δὲ εὔκρατον ἔχον ἀέρα καὶ χωρὶς τοῦ πυρός. ἐν τούτῳ πάντως ἀνετρίβετο καὶ χειμῶνος καὶ θέρους ἔωθεν ἀποπατήσας δηλονότι πρότερον. ἐν δὲ τῷ κατὰ τὴν ἀγορὰν χωρίῳ περὶ τρίτην
ὥραν ἢ τὸ μακρότερον περὶ τετάρτην ἤσθιεν ἄρτον
μετὰ μέλιτος Ἀττικοῦ, πλειστάκις μὲν ἑφθοῦ, σπα-
νιώτερον δ᾽ ὠμοῦ. καὶ μετὰ ταῦτα τὸ μέν τι συγγιγνό-
μενος ἑτέροις διὰ λόγων, τὸ δέ τι καθ᾽ ἑαυτὸν ἀναγινώσκων εἰς ἑβδόμην ὥραν παρέτεινε, μεθ᾽ ἣν ἐτρίβετό
τε κατὰ τὸ δημόσιον βαλανεῖον ἐγυμνάζετό τε τὰ
πρέποντα γέροντι γυμνάσια, περὶ ὧν τῆς ἰδέας ὀλίγον ὕστερον εἰρήσεται. κἄπειτα λουσάμενος ἠρίστα
σύμμετρον, πρῶτα μὲν ὅσα λαπάττει τὴν γαστέρα
προσφερόμενος, ἐφεξῆς δὲ ἰχθύων τὸ πλεῖστον, ὅσοι
πετραῖοί τε καὶ πελάγιοι. κἄπειτ᾽ αὖθις ἐπὶ τοῦ δεί-
πνου τῆς μὲν τῶν ἰχθύων ἐδωδῆς ἀπείχετο, τῶν δ᾽
εὐχυμοτάτων τε καὶ δυσφθάρτων ἐλάμβανεν, ἤτοι

food three times a day, as Antiochus the doctor[10] prescribed for himself. When he was more than eighty years
old, he went out every day to the place in which the coun-
cil of citizens was held, and sometimes took a long route
for the purpose of visiting the sick. But he walked to the
Agora from his house—a distance of some three *stadia*—
and in this way also saw the sick nearby. If, at any time, he
was forced to be taken further, he was lifted into a chair
and carried or transported in a chariot. In his house, there
was a chamber heated by a fire in the winter, while in the
summer this had *eukratic* air without the fire. In this above
all, he spent time in the early morning, both winter and
summer, obviously excreting beforehand. In his place at
the Agora, around the third hour or later around the
fourth hour, he used to eat bread with Attic honey. Mostly
this was toasted; more rarely it was uncooked. After these
things, either mingling with others in discussions or read-   333K
ing by himself, he continued to the seventh hour, after
which he was massaged in the public baths and performed
the exercises suitable for an old man—I shall speak about
what kind these were a little later. Then, having bathed,
he lunched moderately, first taking those things that
empty the stomach and next mostly fish—those from
around the rocks and those from the deep sea. Then again,
at dinner, he abstained from eating fish and took what was
most *euchymous* and not easily spoiled—either gruel with

10 It is not entirely clear to whom Galen is referring here.
Elsewhere he refers to Antiochus the philosopher. It may be Antiochus Paccius (fl. 20 BC–AD 14), who is credited with a number
of compound medications (EANS, 95).

χόνδρον μετ' οἰνομέλιτος ἢ ὄρνιν ἐξ ἁπλοῦ ζωμοῦ. τούτῳ μὲν οὖν τῷ τρόπῳ γηροκομῶν ἑαυτὸν ὁ Ἀντίοχος ἕως ἐσχάτου διετέλεσεν ἀπήρωτος ταῖς αἰσθήσεσι καὶ τοῖς μέλεσιν ἄρτιος ἅπασι.

Τήλεφος δὲ ὁ γραμματικὸς ἐπὶ πλείονας μὲν ἐξίκετο χρόνους Ἀντιόχου σχεδὸν ἑκατὸν ἔτη βιούς. ἐλούετο μὲν τοῦ μηνὸς δὶς ἐν τῷ χειμῶνι, τετράκις δὲ ἐν τῷ θέρει, τρὶς δὲ ἐν ταῖς μεταξὺ τούτων ὥραις. ἐν αἷς δ' ἡμέραις οὐκ ἐλούετο, περὶ τρίτην ὥραν ἠλείφετο μετὰ βραχείας ἀνατρίψεως. εἶτα χόνδρον ἡψημένον ἐν ὕδατι, μέλι μιγνὺς ὠμὸν ὅτι κάλλιστον, 334K ἤσθιε, καὶ τοῦτ' ἤρκει μόνον αὐτῷ τήν γε πρώτην. ἠρίστα δὲ καὶ οὗτος ὥρας ἑβδόμης ἢ βραχεῖ τινι θᾶττον, λάχανα μὲν πρῶτον προσφερόμενος, εἶθ' ἑξῆς ὀρνίθων ἢ ἰχθύων γευόμενος. ἑσπέραν δὲ μόνον ἄρτον ἤσθιε διαβρέχων ἐν οἴνῳ κεκραμένῳ.

5. Ὥσπερ δὲ τοῖς παισὶν ὁ οἶνος ἐναντιώτατόν ἐστιν, οὕτω τοῖς γέρουσι χρησιμώτατον. ἔστω δὲ τῶν φύσει θερμοτέρων, ὁποῖοι τῶν Ἑλληνικῶν ὁ Ἀριούσιός ἐστι καὶ ὁ Λέσβιος καὶ ὁ καλούμενος Μύσιος, οὐκ ἐκ τῆς παρὰ τὸν Ἴστρον Μυσίας, ἀλλ' ἐκ τῆς Ἑλλησποντίας ὀνομαζομένης, ἥτις ἐστὶ κατὰ τὴν ἡμετέραν Ἀσίαν ὁμοροῦσα Περγάμῳ, τῶν δ' ἐκ τῆς Ἰταλίας ὅ τε Φαλερῖνος καὶ ὁ Σουρεντῖνος. ἐφεξῆς αὐτῶν εἰσι κατὰ μὲν τὴν Ἰταλίαν ὅ τε Τιβουρτῖνος καὶ ὁ Σιγνῖνος, ἀμφότεροι παλαιωθέντες, ὡς νέοι γε ὄντες οὔτ' εἰς ἀνάδοσιν ὁρμῶσιν οὔτ' οὖρα κινοῦσιν, ἀλλ' ἐπὶ πολὺ κατὰ τὴν γαστέρα μένουσι κλύδωνας

oxymel or a bird with a simple sauce. Looking after himself in old age in this way, Antiochus continued on until the very end, unimpaired in his senses and sound in all his limbs.

However, Telephus the grammarian[11] reached an even greater age than Antiochus, living almost a hundred years. He was in the habit of bathing twice a month in winter and four times a month in summer. In the seasons between these, he bathed three times a month. On the days he didn't bathe, he was anointed around the third hour with a brief massage. Then he used to eat gruel boiled in water mixed with raw honey of the best quality, and this alone was enough for him at the first meal. He also dined at the seventh hour or a little sooner, taking vegetables first and next tasting fish or birds In the evening, he used to eat only bread, moistened in wine that had been mixed.

5. Just as wine is most inimical for children, so too is it most useful for the aged. It should be warmer in nature such as the Ariusian and Lesbian of the Greek wines are and the so-called Mysian—although this is not from Mysia around the Ister but from the so-called Hellespontine Mysia, which is in our own Asia, bordering on Pergamum. Among the wines from Italy, there are the Falernian and Surrentine. Second [in quality] to these in Italy are the Tiburtine and Signine, both aged, because when they are new, they do not stimulate distribution or promote urination but remain for a long time in the stomach creating

334K

---

[11] Telephus of Pergamum (2nd c. AD) was a Stoic grammarian who is credited with a number of works, including one on Attic syntax in five books. All these works are lost.

ἐργαζόμενοι· δεύτεροι δ᾽ ἐπ᾽ αὐτοῖς Ἀδριανός τε καὶ
Σαβῖνος καὶ Ἀλβανὸς καὶ Γαυριανὸς καὶ Τριφυλῖνος
335K   ὅσοι τ᾽ Ἀμιναῖοι κατὰ τὴν Ἰταλίαν γεννῶνται περί τε
Νεάπολιν καὶ κατὰ τὴν Θούσκων γῆν. κατὰ μέν γε
τὴν τοῦ Σουρεντίνου δύναμιν ὁ Μύσιός ἐστι, κατὰ δὲ
τὴν τοῦ Φαλερίνου τῶν Τμωλιτῶν οἱ κάλλιστοι, Σα-
βίνῳ δὲ καὶ Ἀδριανῷ Τιτακαζηνός τε καὶ Ἀρσυηνὸς
ἐοίκασι.

τούτων οὖν μετρίως παλαιωθέντων πίνειν χρὴ
τοὺς πρεσβύτας, ὅσοι μὴ πάνυ τὴν κεφαλὴν ἰσχυρὰν
ἔχουσιν· οἷς δ᾽ ἰσχυρά, τούτοις Φαλερῖνός τε καὶ Σουρεντῖνος Ἀριούσιός τε καὶ Λέσβιος καὶ Μύσιος καὶ
Τμωλίτης ἐπιτήδειοι. δῆλον οὖν, ὡς καθ᾽ ἕκαστον
ἔθνος ἐκ τῶν εἰρημένων παραδειγμάτων αἱρεῖσθαι τὸν
ἐπιτηδειότατον ἕκαστος τῶν γερόντων[18] δυνήσεται
σκοπὸν ἔχων ἐπὶ τῇ τῶν οἴνων δοκιμασίᾳ κατὰ μὲν
τὴν σύστασιν ἀεὶ τὸν λεπτότατον αἱρεῖσθαι, κατὰ δὲ
τὴν χρόαν, ὃν ὁ Ἱπποκράτης εἴωθε "κιρρὸν" καλεῖν·
δύναιτο δ᾽ ἄν τις καὶ ξανθὸν ὀνομάζειν αὐτόν. ἀγαθὸς
δὲ καὶ ὁ ὠχρός, ἐν τῷ μέσῳ καθεστὼς ξανθοῦ τε καὶ
λευκοῦ. καὶ γὰρ εἰ βουληθείης μῖξαι τὸν ξανθὸν οἶνον
τῷ λευκῷ, τὸν μικτὸν ἐξ ἀμφοῖν ὠχρὸν ἐργάσῃ, καὶ
336K   κραθεὶς μεθ᾽ ὕδατος ὁ ξανθὸς τοιοῦτος γίνεται. παρὰ
δὲ τὸ πλέον ἢ ἔλαττον ὕδωρ ἐπιβάλλειν ἐνίοτε μὲν
ὠχρός, ἐνίοτε δὲ οἷον ὠχρόλευκος ἢ ὠχρόξανθος φαίνεται.

θερμότατος μέν, ὅσον ἐπὶ τῇ χρόᾳ, τῶν εἰρημένων
οἴνων ὁ ξανθός, ἥκιστα δὲ θερμὸς ὁ λευκός· οἱ δ᾽ ἐν

"splashings." Second again to these are Adrian, Sabine, Albanian, Gaurian and Tryphiline, and those Aminaean   335K wines produced in Italy around Naples and in Tuscany. In respect of potency, Mysian wine is similar to the Surrentine and the best of the Tmolitan to the Falernian, while the Titacazene and Arsynian are like the Sabine and the Adrian.

It is necessary for old men who do not have a very strong head to drink these when they are moderately aged. For those who do have a strong head, the Falernian, Surrentine, Ariusian, Lesbian, Mysian and Tmolitan are suitable. It is clear, therefore, from the stated examples, that in each region, each old man will be able to choose the most suitable wine, with the objective in the case of those in old age being, by examination of the wines, to always choose the thinnest in consistency, and in color what Hippocrates was wont to call "orange-tawny," although someone might also be able to call it "yellow." Good also is the pale yellow, sitting midway between yellow and white. For if it is preferred to mix the yellow wine with the white, the mixture from both makes pale yellow, and if the yellow is mixed with water, it becomes such a wine. Depending on   336K whether more or less water is added, it sometimes appears pale yellow, sometimes like whitish yellow, and sometimes pale yellow.

The yellow is the hottest of the wines mentioned in terms of the color, while the white is the least hot. Those

---

[18] τῶν γερόντων *add.* Ko

τῷ μεταξύ, καθ᾽ ὅσον ἂν ἑκατέρῳ πλησιάζωσι, τῆς
ἐκείνου μετέχουσι δυνάμεως. ἐν μὲν οὖν τοῦτο μέγιστον ἀγαθὸν ἐξ οἴνου τοῖς γέρουσι περιγίνεται, τὸ
θερμαίνεσθαι πάντ᾽ αὐτῶν τὰ μόρια, δεύτερον δὲ τὸ
δι᾽ οὔρων καθαίρεσθαι τὸν ὀρρὸν τοῦ αἵματος. διὸ καὶ
κάλλιστος ἐπὶ γερόντων ὁ οἶνος, ὃς ἂν ταῦτ᾽ ἐργάζη-
ται μάλιστα. τοιοῦτος δ᾽ ἐστὶν ὁ τῇ συστάσει μὲν
λεπτός οὖρα γὰρ οὗτοι κινοῦσι, τῇ χρόᾳ δὲ ξανθός
ἴδιον γὰρ τοῦτο τῶν θερμῶν ἱκανῶς οἴνων τὸ χρῶμα.
διὸ κἂν πάνυ λευκοὶ κατ᾽ ἀρχὰς ὦσι, παλαιούμενοι
προσλαμβάνουσί τινα ξανθότητα, δι᾽ ἣν ὕπωχροι
μὲν τὸ πρῶτον, ὕστερον δὲ τελέως ὠχροὶ γίνονται,
κἂν ἐπὶ πλεῖστον ἥκωσι χρόνου, τελευτῶντες ὠχρό-
ξανθοι φαίνονται· τελέως γὰρ οὐχ οἷόν τε τοῖς λευ-
337K   κοῖς οἴνοις γενέσθαι ξανθοῖς. ὅσοι δὲ τῶν ὠχρῶν ἢ
ξανθῶν οἴνων παχεῖς εἰσιν, αἷμα γεννῶσιν οὗτοι καὶ
τρέφουσι τὸ σῶμα· διὸ γένοιτ᾽ ἄν ποτε καὶ αὐτοὶ
χρήσιμοι τοῖς γέρουσι, καθ᾽ ὃν δηλονότι χρόνον οὔτ᾽
ὀρρῶδες ὑγρὸν ἔχουσιν ἐν ταῖς φλεψὶ καὶ δέονται θρέ-
ψεως περιττοτέρας. ὡς ἐπὶ τὸ πολὺ δὲ τῶν οὐρητικῶν
οἴνων χρῄζουσιν οἱ τὴν ἡλικίαν ταύτην ἄγοντες διὰ
τὸ πλεονάζειν ἐν αὐτοῖς ὑδατῶδες περίττωμα. τῶν δ᾽
ἐπὶ πλέον ἐν τῇ γαστρὶ μενόντων οἴνων οὐδέποτ᾽ οὐδεὶς οἰκεῖος γέροντι. τοιοῦτοι δ᾽ εἰσὶν οἱ ἀπὸ τῆς Βι-
θυνίας Ἀμιναῖοι καὶ τῶν ἀπὸ τῆς Ἰταλίας ὁ Μάρσος
καὶ Σιγνῖνος καὶ Τιβουρτῖνος, ἔστ᾽ ἂν ὦσι νέοι.

ἀλλ᾽ οὗτοι μὲν πάντες λευκοί, μέλανες δὲ ἄλλοι δὲ
παχεῖς, ὅσοι στύφουσιν, ἐν τῇ γαστρὶ μένουσι χρόνῳ

in between, to the extent that they approach either, partake of the potency of that. This one thing, then, is superior as the greatest good from wine for old men—that it heats all their parts. Second, however, is that through urination it purifies the serum of the blood. On this account, the best wine in those who are old is that which most brings these things about. Such a wine is thin in consistency, for such wines are diuretic, and yellow in color, for this color is specific to wines that are sufficiently hot. So even if they are very white at first, as they age they take on a certain yellowness, through which they are first pale and later become completely pale yellow. If they go on for a longer time, they finally appear pale yellow, for it is not possible for white wines to become completely yellow. 337K Those pale or yellow wines that are thick, generate blood and nourish the body. For this reason, they are also useful for old people at the time, obviously when they have no serous fluid in the veins and require more nourishment. For the most part, those who have reached this age need the urine-producing wines because there is also an excess of watery superfluity in them. However, since these wines remain longer in the stomach, none are ever suitable for an old man. Such are the Aminaian wines from Bithynia, and of those from Italy, the Marsian, Signine and Tiburtine, if they are new.

But these are all white wines; others are dark and thick and these are astringent, remaining a long time in the

πολλῷ καὶ κλύδωνας ἐργάζονται κατ' αὐτήν, ὥσπερ ὁ
ἐν Κιλικίᾳ μὲν Ἀβάτης,[19] ἐν Ἀσίᾳ δὲ Αἰγεάτης τε καὶ
Περπερίνιος. ὅσοι δὲ χωρὶς τοῦ στύφειν μέλανές τέ
εἰσι καὶ παχεῖς, οἷος ὁ Σκυβελίτης τε καὶ ὁ Θηραῖος,
ἧττον μὲν ἐν τῇ γαστρὶ μένουσιν, οὔρησιν δὲ οὐδ'
338K αὐτοὶ κινοῦσιν, ἀλλ' ὑπέρχονται κάτω· διὸ καὶ προπί-
νουσιν αὐτοὺς ἐδωδῆς σιτίων. ἀλλ' οὐδ' οὗτοι χρήσιμοι τοῖς γέρουσιν οὔτε προπίνειν οὔτε πολὺ μᾶλλον
ἐν ἑτέρῳ καιρῷ προσφέρεσθαι, καθότι μηδ' ἄλλο μηδὲν τῷ γέροντι παχύχυμον· ἐμφράττεται γὰρ ἐξ αὐτῶν
ἧπάρ τε καὶ σπλὴν καὶ νεφροί· κἀντεῦθεν οἱ μὲν ὑδε-
ριῶσιν, οἱ δὲ λιθιῶσι τῶν ἐπὶ πλέον αὐτοῖς χρη-
σαμένων γερόντων.

εἴπερ οὖν ἐθέλοι μετὰ τὸ λουτρὸν οἴνῳ γλυκεῖ χρήσασθαι τῶν πρεσβυτέρων τις, ἐδηδοκότι μὲν ἔωθεν
αὐτῷ, καθότι τὸν Ἀντίοχον ἔφην πράττειν, ἄριστος ὁ
Φαυστιανὸς Φαλερῖνος, ἐκείνου δ' ἀποροῦντι τῶν
ὁμοίων τις. ὅμοιοι δ' ἂν εἶεν οἱ γλυκεῖς τε ἅμα καὶ
ὠχροὶ κατὰ χρόαν, ἐφεξῆς δὲ τούτοις ὅ τε Θηρῆνος
καὶ ὁ Κυριῆνος. οὐ κωλύω δὲ οὐδὲ τοῖς ἐσκευασμένοις
διὰ μέλιτος οἴνοις χρῆσθαι, καὶ μάλισθ' ὅσοις τῶν
γερόντων ὑποψία τίς ἐστιν ἐν νεφροῖς λιθιδίων[20] γενέσεως ἢ καὶ ποδάγρα τις ἢ ἀρθρῖτις ἐνοχλεῖ. τὸν δ'
οἶνον ἐπὶ τῆς τοιαύτης συνθέσεως ἄριστον εἶναι Σα-
βῖνον ἤ τινα τῶν ὁμοίων. ἐπεμβάλλεται δὲ αὐτῷ καὶ
πετροσέλινον, καὶ μόνον ἀρκεῖ τοῦτο τοῖς ἀρθριτικοῖς.
339K ἐπὶ δὲ τῶν λιθιώντων καὶ τῆς βετονίκης τι πόας

stomach and bringing about "splashings" in it, just as the Abatan wine in Cilicia and the Aegeatic and Perperine wines in Asia. Those that, apart from being astringent, are dark and thick, like the Scybiline and Theraean, remain less time in the stomach and do not stimulate urination but move on downward. This is why people drink them 338K before eating food. But these are not beneficial for old people, either to drink before eating, or much more to take at another time, inasmuch as they do nothing else but create a thick humor in an old person. The liver, spleen and kidneys are obstructed by these wines, and when old men use them more, they develop either dropsy or stones from that source.

If, then, some old man should wish to use a sweet wine after his bath, and he has eaten in the early morning, as I said Antiochus used to do, the Faustian Falernian is best, and if that is not available, one of those like it. Similar would be those that are sweet and at the same time pale in color. Next in order to these are the Therene and Cyriene. Nor would I prevent the use of those wines prepared with honey, and particularly for those old men in whom there is a suspicion of the genesis of stones in the kidneys, or who are troubled with some gout or arthritis. In the case of such a combination, the best wine is the Sabine or one of those like it. If some parsley is also added to this, it is alone sufficient for those with arthritis. In the case of those with stones, some grass of betony is also 339K

---

19 Ἀβάτης Ko; Συβάτης Ku
20 λιθιδίων Ko; λίθων Ku

μίγνυται καὶ κέστρου τοῦ παρὰ τοῖς Κελτοῖς γεννω-
μένου· καλοῦσι δὲ τὴν βοτάνην ταύτην σαξίφραγον.
ἔνιοι δὲ περιεργότερον σκευάζοντες τὸ φάρμακον
ἐπεμβάλλουσι καὶ ναρδοστάχυος, εἰσὶ δ' οἱ καὶ ἄλλων τινῶν οὔρησιν κινεῖν δυναμένων. ἀλλὰ τό γ'
ἁπλοῦν πόμα δι' οἴνου καὶ μέλιτος οἱ πολλοὶ συντι-
θέασι, πηγάνου τε καὶ πεπέρεως ἔχον ὀλίγον. εἰ δὲ
καὶ προεδηδοκὼς εἴη πρὶν λούσασθαι καὶ ἡ γαστὴρ
αὐτοῦ μηδεμιᾶς βοηθείας χρῄζοι, τῶν λευκῶν καὶ
ὀλιγοφόρων οἴνων πίνειν χρὴ[21] μετὰ τὸ λουτρόν. ὅσοι
δὲ παχεῖς καὶ γλυκεῖς καὶ μέλανες, ὡς ἐμφράττοντας
τὰ σπλάγχνα τοὺς τοιούτους ἅπαντας φεύγειν προσήκει.

6. Ἀλλ' αἱ μὲν ἀπὸ τῶν οἴνων ἐμφράξεις μέτριαι,
τὰς δ' ἀπὸ τῶν ἐδεσμάτων, ὅσα γλίσχρον ἢ παχὺν
ἐργάζεται χυμόν, οὐ ῥάδιον ἰᾶσθαι. διόπερ οὐ χρὴ
πλεονάζειν τοὺς γέροντας οὔτε ‹ἐν› χόνδρων ἢ τυρῶν
ἢ ὠῶν ἢ κοχλιῶν ἢ φακῆς ἢ βολβῶν ἢ χοιρείων
κρεῶν ἐδωδῇ, πολὺ δὴ μᾶλλον ἐν ταῖς τῶν ἐγχελύων
ἢ ὀστρέων ἢ ὅλως τῶν σκληρὰν καὶ δυσκατέργαστον
ἐχόντων τὴν σάρκα. διὰ τοῦτο οὖν οὐδὲ τῶν ὀστρακο-
δέρμων οὐδὲν ἢ σελαχίων ἢ θύνων ἢ ὅλως τῶν κητωδῶν[22] ἐπιτήδειον, ὥσπερ οὖν οὐδὲ τῶν ἐκ τῆς γῆς φυομένων μυκήτων ἢ τῶν κρεῶν τῶν ἐλαφείων ἢ αἰγείων
ἢ βοείων. καὶ ταῦτα μὲν οὐδ' ἄλλῳ τινὶ χρήσιμα·
προβάτεια δὲ νέοις μέν ἐστιν οὐ φαῦλον ἔδεσμα,
γέρουσι δὲ οὐδὲ ταῦτα, καὶ πολὺ δὴ μᾶλλον ἔτι τὰ

mixed in and *cestron*[12] which grows among the Celts; they call this herb "saxifrage." Some, when they prepare the medication more carefully, also put in spikenard. And there are certain other things that are able to stimulate urine formation. But there are many who put together a simple drink made from wine and honey, having in it a small amount of rue and pepper. Also, if it is taken before bathing, and the stomach derives no benefit from it, it is necessary also to drink, after the bath, one of the white wines that bears little water. It is appropriate to avoid those wines that are thick, sweet and dark, as all such wines obstruct the internal organs.

6. But the obstructions from wines are moderate whereas those from foods create a viscid and thick humor not easy to cure. This is why old people must not go to excess in eating gruel, cheese, boiled eggs, snails, lentil soup, purse-tassels or the flesh of swine, and much more than these, the flesh of eels or oysters, or in general those things which have flesh that is hard and difficult to digest. Because of this then, they should not eat any of those things that are hard shelled—cartilaginous fish, tunny fish or in general coarse fish, just as they should not eat mushrooms grown in the earth, or the flesh of stags, goats or oxen. Also these are not useful for anyone else. For those who are young, the flesh of sheep is not a bad food, but for those who are old, these are not good, and much more still,

340K

---

[12] LSJ has two different plants listed under κέστρον: betony and saxifrage, referring to this passage regarding the latter.

---

[21] χρὴ Ko; καὶ Ku   [22] post τῶν κητωδῶν: ἐπιτήδειον, ὥσπερ οὖν οὐδὲ τῶν ἐκ τῆς γῆς φυομένων μυκήτων add. Ko

τῶν ἀρνῶν· ὑγρὰ γάρ ἐστι καὶ βλεννώδη καὶ γλίσχρα καὶ φλεγματώδη. τά γε μὴν τῶν ἐρίφων οὐκ
ἀνεπιτήδεια γέροντι καὶ τῶν πτηνῶν, ὅσα μὴ καθ᾽ ἕλη
καὶ ποταμοὺς ἢ λίμνας διαιτᾶται. τὰ δὲ ταριχευθέντα
πάντα τῶν προσφάτων ἀμείνω.

χρὴ τοίνυν, ὥσπερ εἴρηται, μάλιστα μὲν ἀπέχεσθαι τῶν ἐμφραττόντων ἐδεσμάτων, εἰ δ᾽ ὑπ᾽ ἀνάγκης
ποτὲ χρήσασθαι συμβαίη πλείοσιν αὐτοῖς, αὐτίκα
προσφέρεσθαι τὸ διὰ τῆς καλαμίνθης φάρμακον, οὗ
τὴν σύνθεσιν ἔμπροσθεν εἴρηκα κατὰ τὸ τέταρτον
γράμμα. μὴ παρόντος δὲ αὐτοῦ τῷ διὰ τῶν τριῶν πε-
341K πέρεων χρηστέον. εἰ δὲ μηδὲ τοῦτο παρείη, κόπτοντα
καὶ διαττῶντα ἀκριβῶς, ὡς χνοῶδες γενέσθαι, πέπερι
λευκὸν ἅμα τε τοῖς ὄψοις ἐσθίειν ἐπιπάττειν τε τῷ
ποτῷ. καὶ κρόμμυον δὲ ἐσθίειν τηνικαῦτα συμφέρει,
κἂν τούτῳ συνήθης ᾖ,[23] σκόροδον. καὶ τῇ διὰ τῶν
ἐχιδνῶν δ᾽ ἄν τις ἣν καλοῦσι θηριακὴν ἀντίδοτον ἐπὶ
τῶν γερόντων οὐ κακῶς χρῷτο, καὶ μάλισθ᾽ ὅταν ἐν
τοῖς ἐμφράττουσιν ἐδέσμασιν αὖθις τῶν ἐκφραττόντων φθάσῃ προσενηνέχθαι τι· καλλίστη γὰρ ἡ παρὰ
τοῦ θηριακοῦ φαρμάκου βοήθεια τηνικαῦτα γίνεται.
καὶ γαστρὸς δ᾽ ἐπὶ τοῖς τοιούτοις ὑπαχθείσης χρησιμώτατον ἂν εἴη διδόναι τῇ ὑστεραίᾳ τὸ διὰ τῶν ἐχιδνῶν. οὐδὲν δὲ ἧττον αὐτοῦ τὴν ἀμβροσίαν τε καὶ ἀθα-
νασίαν ὀνομαζομένην ὅσα τ᾽ ἄλλα διὰ τῶν εἰρημένων
ἀρωμάτων σύγκειται φάρμακα. τῷ γε μὴν ἀκριβῶς

23 ᾖ Ko; εἴης καὶ Ku

the flesh of lambs, for this is moist, slimy, viscid and phlegmatous. However, the flesh of kids is not unsuitable for an old man, nor is that of birds that do not live in swamps, rivers or marshes. All those that are preserved are better than those that are fresh.

Accordingly, it is necessary, as I said, to stay away particularly from obstructing foods, although if at some time, of necessity, one happens to use more of these, immediately take the medication made from catmint—I have spoken previously about the composition of this in the fourth book. However, if this is not available, one must use the medication made from the three peppers. If this is not 341K available, crush and sift white pepper thoroughly so it becomes a fine powder and eat it with cooked foods or sprinkled on a drink. And it helps to eat onions under such circumstances, and if one is accustomed to it, garlic. Also, in the case of old people, what they call "theriac antidote"[13] made from vipers is not bad to use, and especially when, with the obstructing foods, it is administered before the obstruction occurs. Under these circumstances, the best benefit from the theriac medication occurs. And when the stomach is emptied below by these things, it would be very useful to give the medication from vipers on the following day. Not less [useful] than this are the so-called ambrosia and athanasia and those other medications compounded from the aromatics mentioned. For the

13 There are two works on this in Kühn—*Ther.* (XIV.210–94K) and *Ther. Pamph.* (XIV.295–310K). The latter may be spurious.

ἑαυτῷ προσέχοντι γέροντι φαρμάκου μὲν οὐκ ἄν ποτε
γένοιτο χρεία τοιούτου· δεηθέντι δὲ ἐνίοτε λεπτυνούσης διαίτης ἀρκέσει τὰ λελεγμένα δι' ἑτέρου γράμματος ἰδίᾳ, καθ' ὃ Περὶ τῆς λεπτυνούσης διαίτης ὁ
λόγος ἡμῖν ἐγένετο.

342K 7. Πρόδηλον δ', ὅτι καὶ τῶν ἄρτων τοὺς μήτ' ἐνδεῶς
ἔχοντας ἁλῶν ἢ ζύμης ἢ φυράσεως ἢ ὀπτήσεως ἐσθίειν χρὴ μήτε τὴν ἐπαινουμένην ὑπὸ πάντων σεμίδα-
λιν ἢ τὰ δι' αὐτῆς πέμματα· καὶ γὰρ δύσπεπτα πάντα
καὶ παχύχυμα καὶ σπλάγχνων ἐμφρακτικά. καὶ εἴ γε
μὴ τοῖς διὰ βουτύρου καὶ σεμιδάλεως σκευαζομένοις
πλακοῦσιν ἐμίγνυτο μέλι δαψιλές, οὐδὲν εἴη ἂν ἔδεσμα πολεμιώτερον ἀνθρώποις πᾶσιν, οὐ μόνον τοῖς
πρεσβύταις. ἀλλὰ τί δεῖ τοῖς βλαβεροῖς τὸ χρήσιμον
μιγνύειν, ἐνὸν αὐτῷ μόνῳ χρῆσθαι τῷ μέλιτι, τοῦτο
μὲν ἀφεψῶντα, τοῦτο δὲ σὺν ἄρτῳ λαμβάνοντα πρὶν
ἑψηθῆναι; προνοητέον δὲ τῆς κατὰ τὸν ἄρτον ἀρετῆς
μᾶλλον ἢ τῆς κατὰ τὸ μέλι. βέλτιον γάρ, εἰ τὸ μέλι
τοιοῦτον εἴη τήν τε ἰδέαν καὶ τὴν δύναμιν οἷον τὸ
Ἀττικόν· εἰ δὲ καὶ μὴ τοιοῦτον ἔχοιμεν, ἀλλὰ παντὶ
μέλιτι χρηστέον ἐστὶ πλὴν τῶν δυσωδῶν ἢ ἐν ὅσοις
αἰσθητῶς κηροῦ ποιότης ἐμφαίνεται, καὶ πολὺ δὴ
μᾶλλον, εἴ τινος ἑτέρας ἀλλοκότου. τῶν δ' ἄρτων ὁ
μὲν τοιοῦτος, οἷον ἀρτίως εἶπον, οὐ μόνον οὐδὲν ἀγα-
343K θὸν ἐργάζεται κατὰ τὸ σῶμα τοῦ γέροντος, ἀλλὰ καὶ
βλαβερώτατός ἐστι, καὶ μάλισθ' ὅσῳπερ ἂν ᾖ καθ-
αρώτερος.

ὁρῶ δὲ τοῖς ἀθληταῖς αὐτὸν ἐπίτηδες σκευαζόμε-

old man who carefully looks after himself, there should not, at any time, be a need for such a medication. However, if sometimes there is a need for a thinning diet, those things that have been said in another specific work will suffice—the work is my book entitled *On the Thinning Diet*.[14]

7. It is clear, however, that of the breads, it is necessary not to eat those lacking in salt or leaven, or mixing or baking, or the finest wheaten flour praised by all, or the pastries made from it, for these are all difficult to digest, thick-humored and obstruct the internal organs. And if, in fact, an abundance of honey is not mixed with the flat cakes prepared using butter and the finest wheaten flour, there is no food more adverse for all people and not only for those who are old. But why is there need to mix the useful with those things that are harmful when it is possible to use honey itself alone, in part cooked and in part taken with bread before it is cooked? However, we must give thought to the goodness of the bread more than to that of the honey. It is better if the honey is such that, in terms of kind and potency, it is like the Attic. If we do not have such honey, we must use any honey apart from those that are malodorous or in which a waxy quality is perceptibly displayed, and much more, if there is anything else unusual. Such breads, as I said just now, not only bring about nothing good in the body of the old person, but are also very harmful, especially to the extent it is more pure.

I see this purposely prepared for athletes. But for them

---

[14] *Vict. Att.*, CMG V.4.2.

342K

343K

νον. ἀλλ᾽ ἐκείνοις μὲν εἰς ὅσον ἐπιτηδεύουσιν ἁρμόττει, γέροντι δέ, εἰ μὴ πολὺ μὲν ἁλῶν ἔχοι, πολὺ δὲ ζύμης ὠπτημένος τε ἀκριβῶς εἴη, παχὺν ἐργάζεται καὶ γλίσχρον χυμόν, ὃς οὐδ᾽ ἄλλῳ μὲν ἀγαθός ἐστιν ἐπὶ πλέον αὐξανόμενος. ἀτὰρ οὖν[24] καὶ τὰς καθ᾽ ἧπάρ τε καὶ σπλῆνα καὶ νεφροὺς ἐμφράξεις ἐργάζεται, καὶ μάλισθ᾽ ὅσοις φύσει στενότερα τῶν ἐπὶ τούτοις τοῖς σπλάγχνοις ἀγγείων ἐστὶ τὰ πέρατα. καθάπερ γὰρ ἐπὶ τῶν ἐκτὸς τούτων φλεβῶν, ἃς ἐναργῶς ὁρῶμεν, οὐ σμικρὰ διαφορὰ φαίνεται κατὰ τὸ εὖρος ἄλλου τε πρὸς ἄλλον ἄνθρωπον καὶ καθ᾽ ἕκαστον ἐν τοῖς μορίοις, οὕτως εἰκὸς ἔχειν κἀπὶ τῶν ἔνδον. ἀλλ᾽ οὐχ οἷόν τε γνῶναι τὴν διαφορὰν πρὸ τῆς πείρας· λέγω δὲ πεῖραν, ἣν ἐφ᾽ ἑκάστῳ τῶν προσφερομένων ἔνεστι ποιεῖσθαι.

γέροντα γοῦν τινα γεωργικὸν ἔγνωμεν, ἔτη πλείω τῶν ἑκατὸν βιώσαντα κατ᾽ ἀγρόν, ᾧ τὸ πλεῖον τῆς τροφῆς αἴγειον ἦν γάλα, ποτὲ μὲν αὐτίκα λαμβανόμενον ἄρτου θρυμμάτων ἐν αὐτῷ διαβρεχομένων, ἔστι δ᾽ ὅτε καὶ μέλιτος ἐμίγνυε, καί ποτε καὶ ἤψει ἐμβάλλων ἀκρέμονας θύμων ἅμα τοῖς ἄρτοις. ἀλλὰ τοῦτόν γέ τις μιμησάμενος ᾤετο γὰρ αἴτιον αὐτῷ τὸ γάλα τῆς πολυχρονίου ζωῆς γεγονέναι διὰ παντὸς ἐβλάβη καὶ κατὰ πάντα τρόπον προσφορᾶς. ἐβαρύνετο γὰρ αὐτῷ τὸ στόμα τῆς γαστρὸς καὶ μετὰ ταῦτα τὸ δεξιὸν ὑποχόνδριον ἐτείνετο. καί τις ἕτερος, ὁμοίως ἐπιχειρήσας χρήσασθαι τῷ γάλακτι, τῶν μὲν ἄλλων οὐδὲν ἐμέμφετο καὶ γὰρ ἔπεττε καλῶς αὐτὸ καὶ οὔτε ὀξυρε-

it is sufficient as far as the purposes are concerned, whereas for an old man, if it doesn't have much salt or much leaven and is baked completely, it creates a thick and viscid humor, which is not good for anyone else, and more so when it is increased. Moreover, this brings about obstructions in the liver, spleen and kidneys, and particularly when the terminal parts of the vessels in such internal organs are narrower in nature. For as in the case of those external veins which we see clearly, no small difference is apparent in the width [of these vessels] between one person and another, and in respect of each person, in the parts, so also is it likely in the veins within. But it is not possible to know the difference prior to experience. I call "experience" what it is possible to bring about by each of the things given.

Anyway, I knew a certain old man, a farmer, who had lived on the land for more than a hundred years. His main nutriment was goat's milk, which on occasion he took immediately after soaking pieces of bread in it. Sometimes he also mixed honey [with this] and sometimes he boiled it, throwing in twigs of thyme along with the bread. But someone who imitated this man, thinking the milk was the reason for his longevity, was continually harmed, regardless of how it was taken. For the opening of the stomach was weighed down by this, and afterward the right hypocondrium was distended. And someone else, who similarly tried to use the milk, found fault with none of the other things, for he digested it well and no sour indigestion, ir-

344K

---

24 *post* οὖν: καὶ τὰς Ko; οὗτος Ku

γμία τις ἢ ἐρυγὴ κνισσώδης ἢ πνευμάτωσις ἢ βάρος
ἐπεγίνετο καθ' ὑποχόνδριον, ἑβδόμῃ δ' ἡμέρᾳ μετὰ
τὴν ἀρχὴν ἐναργῶς ἔφη τοῦ ἥπατος αἰσθάνεσθαι βα-
ρυνομένου· δοκεῖν γὰρ ἑαυτῷ κατὰ τὸ δεξιὸν ὑποχόνδριον ἐγκεῖσθαί τι καθάπερ λίθον, ὑφ' οὗ κατασπᾶ-
σθαί τε τὰ ὑπερκείμενα καὶ μέχρι κλειδὸς ἀνήκειν τὴν
τάσιν. εὔδηλον οὖν, ὅτι τούτῳ μὲν ἐνεφράττετο τὸ
ἧπαρ, ἐπνευματοῦτο δὲ θατέρῳ.

καὶ μὲν δὴ καὶ ἄλλον²⁵ ἐπὶ γάλακτος χρήσει πολυ-
χρονίῳ λίθον γεννήσαντα κατὰ τοὺς νεφροὺς οἶδα,
345K καί τινα ἕτερον ἀπολέσαντα²⁶ πάντας τοὺς ὀδόντας.
τοῦτο μὲν οὖν καὶ ἄλλοις ἐγένετο πολλοῖς τῶν ἐπὶ
γάλακτι μακρῶς διαιτηθέντων. ἀλύπως δ' ἕτεροι διὰ
παντὸς ἐχρήσαντο τῷ γάλακτι καὶ μετ' ὠφελείας με-
γίστης, ἄλλοι δὲ παραπλησίως τῷ κατ' ἀγρὸν βιώ-
σαντι πλείω τῶν ἑκατὸν ἐτῶν, ὡς ἔφην. ὅταν γὰρ ἥ
τε ποιότης αὐτοῦ τῇ τοῦ χρωμένου φύσει μηδὲν ὑπ-
εναντίον ἔχῃ, τῶν τε σπλάγχνων εὐπετεῖς αἱ διέξοδοι
διὰ τὴν τῶν ἀγγείων εὐρύτητα ὦσι, τῶν μὲν ὠφελί-
μων ἐκ γάλακτος ἀπολαύουσιν οὗτοι, μοχθηροῦ δ'
οὐδενὸς πειρῶνται.

τὰ δ' ἐκ τοῦ γάλακτος ἀγαθὰ λέλεκται καὶ τοῖς
ἔμπροσθεν ἰατροῖς, ὑπαγωγὴ μετρία γαστρὸς εὐχυ-
μία τε καὶ θρέψις, οὐ βραχέα συντελούσης εἰς ταῦτα
καὶ τῆς νομῆς τῶν ζῴων, ὧν μέλλεις χρήσασθαι τῷ
γάλακτι. καίτοι γ' ἀμελοῦσιν ἔνιοι παντάπασι τῆς
νομῆς, ὡς ἤτοι μηδὲν ἢ βραχύτατον εἰς ἀρετὴν γάλακτος συντελούσης. ἀλλ' ἐναργῶς γε θεώμεθα τὰ

ritating eructations, flatulence or heaviness in the hypocondrium occurred. But on the seventh day after he began, he said he was clearly aware of a heaviness of the liver,
for it seemed to him that something was pressing on the
right hypocondrium like a stone, and that those things
lying above were drawn down by it, and the tension extended right to the clavicle. It was clear, then, that in this
man the liver was blocked up, but in the other there was
a filling with wind.

Furthermore, I know another person who formed a
stone in the kidneys through using milk for a long time,
and someone else who lost all his teeth. This has also occurred in many others who used a milk diet over a long
period. And yet others have used milk continually without
trouble and with very great benefit, just like the man who
lived for more than a hundred years in the country, as I
said. For when its quality has nothing inimical in nature
to its use, and the passages through the wide vessels of
the internal organs are favorable, these people enjoy the
benefits [of milk] and experience nothing bad.

The good aspects of milk have also been spoken of by
many previous doctors—moderate downward evacuation
of the stomach, *euchymia* and nourishment. Not a little is
contributed to these things by the pasturage of the animals, whose milk is going to be used. However, some pay
no attention at all to the pasturage, as if it contributes
nothing or very little to the goodness of the milk. But we

345K

νεμηθέντα τῶν ζῴων, ὧν μέλλομεν χρήσασθαι τῷ γάλακτι, ἢ σκαμμωνίαν ἢ τῶν τιθυμάλλων τινάς, καθ-
αρτικὸν ἴσχοντα τὸ γάλα. δῆλον οὖν, ὡς καὶ δριμὺ
346K καὶ ὀξὺ καὶ αὐστηρὸν ἐπὶ ταῖς μοχθηραῖς ἔσται νο-
μαῖς, ἐξομοιούμενον ἀεὶ τῇ φύσει τῆς πόας. καὶ διὰ
τοῦτο καὶ τοῖς πρὸ ἡμῶν ἰατροῖς ἐκ τῆς πείρας διδα-
χθεῖσιν εἴρηνταί τινες εὐγάλακτοι²⁷ νομαί προσαγο-
ρεύουσι γὰρ οὕτως αὐτάς, ὑπὲρ ὧν καὶ ἡμεῖς ἑτέρωθι
διερχόμεθα.

νῦν δ' ἀρκεῖ τό γε τοσοῦτον ἐπίστασθαι περὶ αὐτῶν, ὡς οὔτε δριμείας οὔτ' ὀξείας οὔτ' αὐστηρὰς πάνυ
χρὴ τὰς τροφὰς εἶναι τῶν ζῴων, ὧν τῷ γάλακτι μέλλομεν ὡς εὐχυμοτάτῳ χρῆσθαι. καὶ μὴν καὶ ὅτι κατὰ
τὴν ἡλικίαν ἀκμάζον εἶναι χρὴ καὶ κατὰ τὴν ἕξιν τοῦ
σώματος ἄμεμπτον τὸ ζῷον, εὔδηλον δήπου, κἂν ἐγὼ
μὴ λέγω. καὶ κάλλιόν γε τὸ μὲν αἶγα, τὸ δ' ὄνον εἶναι,
χρῆσθαι δὲ τῷ γάλακτι παρὰ μέρος ἑκατέρου· λεπτότερον μὲν γὰρ²⁸ καὶ ὀρρωδέστερον τὸ τῆς ὄνου, σύμμετρον δὲ τῷ πάχει τὸ τῆς αἰγός. ὥστε τοῦτο μὲν
ἀνατρέφει μᾶλλον, εἰ τούτου χρεία, τὸ δ' ὄνειον
ἀσφαλέστερον πάντη. καὶ γὰρ εἰ μόνον ποτὲ λαμβά-
νοιτο χωρὶς ἄρτου, καὶ ὑπέρχεται θᾶττον καὶ ἥκιστά
ἐστι φυσῶδες οὐ τυροῦταί τε κατὰ τὴν γαστέρα, καὶ
μάλισθ' ὅταν ἁλῶν καὶ μέλιτος προσλάβῃ.

347K χρὴ δὲ καὶ τούτου καὶ τῆς ἄλλης ἁπάσης ὕλης τῶν
βοηθημάτων τὰς δυνάμεις ἰδίᾳ προμεμαθηκέναι τὸν
μέλλοντα χρήσασθαι καλῶς, ἵνα μὴ πολλάκις ἀναγ-

clearly see that, when those things used as pasturage of the animals whose milk we are going to use are either scammony or one of the spurges, the milk is strongly cathartic. It is clear, then, that sharpness, acidity, and bitterness will   346K
exist with bad pasturages, [the milk] always becoming like the nature of the grass. And because of this also, some pastures were said to be good for milk production by doctors before us who learned from experience, and they named them accordingly—I also go over these elsewhere.[15]

For the present, however, that it is enough to know this much about these—the nutriment of the animals whose milk we are going to use as most *euchymous* must not be very sharp, acidic or bitter, and further, that the animal must be in the prime of life and faultless in the state of its body. This, I presume, is also clear, even if I do not say so. In fact, it is better to use goat's and ass's milk alternately, for that of the ass is thinner and more serous while that of the goat is moderate in thickness. Consequently, the latter nourishes more, if there is need of this, whereas that of the ass is altogether safer. If, on occasion, one takes it alone without bread, it passes through quicker, is least flatulence-producing and doesn't curdle in the stomach, particularly when one takes salt and honey in addition.

It is also necessary to have learned beforehand the   347K
potencies of both this and every other material of the remedies individually, if one is going to use them properly,

[15] See Galen's *MM*, 5.12 (X.363K); Johnston and Horsley, *Galen: Method of Medicine*, 2.86–89.

---

[27] εὐγάλακτοι Ko; ἀγάλακτοι Ku
[28] post μὲν γὰρ: add. ἐστι Ku

καζώμεθα περὶ τῶν αὐτῶν λέγειν τὰ αὐτά. καὶ νῦν γέ
μοι δοκῶ μακρότερον ἢ²⁹ δεῖ τοῖς ἐνεστῶσι διεληλυ-
θέναι περί τε γάλακτος καὶ οἴνων. ἄμεινον γὰρ ἦν
εἰπόντα τὴν ἐξ αὐτῶν ὠφέλειαν τοῖς γέρουσι γινομένην ἐπὶ τὴν τῆς ὕλης ἐκλογὴν ἀποπέμψαι τὸν ἤδη
μεμαθηκότα τάς τε κοινὰς δυνάμεις καθ' ἑκάτερον
αὐτῶν καὶ τὰς ἐν μέρει διαφοράς, ἐπὶ μὲν τῶν οἴνων
εἰπόντα τὰς διαφορὰς τοὺς θερμοτέρους τε καὶ οὐρη-
τικωτέρους ἀμείνους εἶναι τοῖς γέρουσι, ἐπὶ δὲ τοῦ
γάλακτος, ὡς οὐδὲ πᾶσι δοτέον, ἀλλὰ μόνοις ὅσοι γε
πέττουσιν αὐτὸ καλῶς καὶ συμπτώματος οὐδενὸς αἰ-
σθάνονται κατὰ τὸ δεξιὸν ὑποχόνδριον. ἐπεὶ δ' ἔστιν
ὅτε διὰ τὴν πολλῶν ὀλιγωρίαν οὐχ ὑπομενόντων ἀνα-
γινώσκειν τὰ βιβλία, δι' ὧν ἐπὶ πλέον ὑπὲρ τῆς τῶν
βοηθημάτων ὕλης λέλεκται, μηκύνειν ἀναγκαζόμεθα
πολλάκις, εἰκότως ἄν τις ἡμῖν καὶ νῦν συγγνοίη τοῦ
348K   τρόπου τῆς διδασκαλίας, οὐ κατὰ τὴν ἀκριβῆ βραχυ-
λογίαν ἐπὶ ταῖς καθόλου μεθόδοις προερχομένοις.

ἰστέον γε μήν, ὡς ἀδύνατόν ἐστιν ἀμέμπτως χρήσασθαι ταῖς γραφομέναις ὕλαις ἐκ μόνων τῶν τοιούτων λόγων, οἵους καὶ νῦν εἶπον ὑπὲρ οἴνου τε καὶ
γάλακτος. ἀλλὰ χρὴ τὸν ἄριστα μεταχειριούμενον
αὐτὰς ἰδίᾳ πρότερον ὑπὲρ ἑκάστης μεμαθηκέναι τὸν
οἰκεῖον λόγον, ἐν ᾧ τήν τε κοινὴν τῆς ὕλης δύναμιν
ἐπισκεπτόμεθα καὶ τὰς κατὰ μέρος ἐν αὐτῇ διαφορὰς
μέχρις ἐσχάτων εἰδῶν. οὐδὲ γὰρ περὶ τῶν ἄλλων
ἁπάντων, ὅσα χρὴ γινώσκεσθαι τῷ μέλλοντι προνοήσασθαι γέροντος, ἄμεινον ἐν τῇδε τῇ πραγματείᾳ δι-

so we are not frequently compelled to say the same things about them. And I seem now to have gone on longer than is appropriate for the present circumstances about milk and wine. For it is better, having stated what benefit arises from these for old people, to defer to the selection of material what has already been taught with regard to the general potencies in each of them, and the individual differences, having stated in respect of the differences of wines that the hotter and more urine-producing wines are better for old people, and in the case of milk, that one must not give this to everyone, but only to those who digest it well and are not aware of any symptom in the right hypocondrium. But since sometimes, due to the negligence of many who do not have the patience to read the books of those who have said still more about the material of remedies, we are often compelled to delay, so it is likely that someone might now excuse us for our way of teaching insofar as we are not completely succint in the methods previously gone through generally.

348K

In fact, one must know that is impossible to use faultlessly the materials written about from such words alone, as I said just now about wine and milk. But it is necessary, for one to handle these in the best way, to have learned specifically beforehand the relevant discussion about each, in which we consider the general potency of the material and the differences in this individually up to the ultimate kinds (*infima species*). It is better in a work such as this not to go over at length all the other things which it is necessary to know for one who is going to care for the

---

29 *post* μακρότερον ἦ: δεῖ τοῖς ἐνεστῶσι Κο; ὡς τοῖς ἐνεστῶσι προσήκει Κu

GALEN

ἔρχεσθαι μακρῶς, ἀλλ᾽ ἀρκεῖ περὶ μὲν ἐνίων, ὅσα
μάλιστά ἐστι χρήσιμα τοῖς πρεσβύταις, οὕτως εἰπεῖν
ὡς νῦν εἴρηται περί τε γάλακτος καὶ οἴνου, περὶ δὲ
ἐνίων ἔτι βραχύτερον ἢ κατὰ ταῦτα διελθεῖν, ὥσπερ
γε καὶ περὶ ἄλλων ὅλως μηδὲν εἰπεῖν. καίτοι τῷ γε
μηκύνειν βουλομένῳ δυνατόν ἐστιν ἁπάντων μνημονεύειν, ὅσα πλεονάζει τοῖς γέρουσιν, ὧδέ πως λέγοντι.

349K   8. Κάλλιον δὲ καὶ περὶ τῶν ἄλλων διελθεῖν, ὅσα
τοῖς πλείστοις τῶν γερόντων γίνεται, βράγχοι καὶ
κόρυζαι καὶ λίθων γενέσεις ἐν νεφροῖς ἀρθρίτιδές τε
καὶ ποδάγραι καὶ ἄσθματα καὶ τἆλλα ὅσα τοιαῦτα.
πρὸς μὲν οὖν τοὺς βράγχους τε καὶ τὰς κορύζας δίαιταν μὲν τήνδε, φάρμακα δὲ τάδε προσφέρειν χρή,
πρὸς δὲ τὰς τῶν λίθων γενέσεις ταῦτα, καθ᾽ ἕκαστόν
τε τῶν ἄλλων οὕτως ἐπερχόμενος ἐπιμελής τε εἶναι
δόξεις καὶ περιττὸς εἰς ἐπιστήμην γηροκομικήν. ἀλλ᾽
ὥσπερ τούτων οὐδὲν χρὴ γράφειν ἐν συγγράμματι
γηροκομικῷ μεταφέρειν γε δυναμένων αὐτὰ τῶν γεγυμνασμένων ἐν ταῖς τῶν νοσημάτων θεραπείαις,
οὕτως αὖ πάλιν οὐδ᾽ ἐπὶ μόνοις τοῖς κοινοῖς παύσα-
σθαι, προειπόντα δ᾽ αὐτὰ προστιθέναι τινὰ τῶν κατὰ
μέρος, ἃ μάλιστ᾽ ἐστὶν οἰκεῖα τῷ προκειμένῳ σκέμματι, καθάπερ ἡμεῖς ἐποιήσαμεν. ἐπειδὴ γὰρ ἐδείχθη
τὸ γῆρας ἐν τοῖς Περὶ κράσεων ὑπομνήμασιν ὁμολογουμένως μὲν εἶναι ψυχρόν, οὐχ ὁμολογουμένως δὲ
350K   ξηρόν, ἐνίων ὑγρὸν αὐτὸ φάντων εἶναι, προσήκει δήπου, λαβόντας ὑπόθεσιν εἰς τὸ γηροκομικὸν μέρος

54

aged; it is sufficient to go over some that are most useful for the aged, to speak in the way I spoke just now about milk and wine, and to go over some even more briefly than these, and to say nothing at all about others. And yet, for someone who wishes to speak at length, it is possible to mention all the things that are too much for old people, speaking as follows.

8. It is better to go over the other things which arise in   349K
the majority of those who are old—sore throats, colds, the genesis of stones in the kidneys, arthritides, gout, asthmas[16] and other such things. Thus, for the sore throats and colds it is necessary to follow this particular regimen and to apply these particular medications, while for the genesis of kidney stones, apply these, [and so on]. To go over the care in each of the other things in this way, you will think is redundant for the science of geriatrics. But just as it is unnecessary to write any of these things in a geriatric treatise, since we are able to carry them over having practiced them in the treatments of diseases, contrariwise, it is necessary not to stop at the common things alone, but having previously spoken about these, to add something to them individually which is particularly relevant to the matter before us, as I did. Since old age was shown in the treatise *On Mixtures*[17] to be agreed upon as being cold but not agreed upon as being dry, some saying it is moist, it is appropriate, I presume, to extend as an hypothesis for   350K

[16] Here this term probably means shortness of breath generally, rather than the specific condition known today as asthma, although it may, of course, include this—see Hippocrates, *Aphorisms* 3.22, and Grmek, *Diseases in the Ancient World,* 34.

[17] See Galen's *Mixt.,* I.582K.

τῆς τέχνης, ὃ νῦν ἡμῖν πρόκειται, τὰ δειχθέντα περὶ
τῆς κράσεως αὐτοῦ, τὸν μὲν σκοπὸν αὐτῶν τῶν πραγμάτων ἀπ᾽ ἐκείνης λαβεῖν, ἐπελθεῖν δ᾽ ἔνια τῶν κατὰ
μέρος, τὰ μὲν ἕνεκα τοῦ γυμνάσαι τὸν μαθητήν, τὰ δ᾽
ὡς παραδείγματα πρὸς τὴν τῶν οὐκ εἰρημένων εὕρεσιν γενησόμενα, διὰ τὸ μὴ πάντας οὕτως εἶναι συν-
ετούς, ὡς ἐκ μόνου τοῦ καθόλου γνωσθέντος εὑρίσκειν
τὰ κατὰ μέρος, ἀλλὰ τοῦ ποδηγήσοντος ἐπ᾽ αὐτὰ
προσδεῖσθαι.

γεγυμνασμένος γάρ τις τὸν λογισμὸν ἀκούσας τὸ
τοῦ γέροντος σῶμα κατὰ μὲν αὐτὰ τὰ μόρια ψυχρὸν
εἶναι[30] καὶ ξηρόν, ἐμπίπλασθαι δὲ ῥᾳδίως ὀρρωδῶν τε
καὶ φλεγματωδῶν περιττωμάτων δι᾽ ἀρρωστίαν τῆς
δυνάμεως,[31] ἐξοχετεύειν μὲν ταῦτα πειράσεται, τὰ
στερεὰ δ᾽ αὐτὰ τοῦ σώματος μόρια θερμαίνειν τε
καὶ ὑγραίνειν. ὅσοι δ᾽ ἐκ τοῦ τῶν περιττωμάτων
πλήθους ἀπατηθέντες ἀδιορίστως ἀπεφήναντο τὴν
τῶν γερόντων κρᾶσιν ὑπάρχειν ὑγράν, οὗτοι περὶ
351K τὸν πρῶτον εὐθέως ἐσφάλησαν σκοπόν, ἡγούμενοι
χρῆναι ξηραίνειν τὰ πρεσβυτικὰ σώματα. τῶν τε οὖν
ἐδεσμάτων ὅσα ξηραντικὰ διδόασι μᾶλλον, οἷον ἐν
μὲν τοῖς λαχάνοις τὴν κράμβην πρὸ τῆς μαλάχης καὶ
τοῦ βλίτου καὶ τοῦ λαπάθου καὶ τῆς ἀτραφάξυος καὶ
τῆς θριδακίνης, ἐν δὲ τοῖς ὀσπρίοις τὴν φακὴν πρὸ
τῆς πτισάνης καὶ τὸν κέγχρον πρὸ τοῦ κυάμου καὶ
τὸν ἔλυμον πρὸ τῆς ζειᾶς, ἐν δὲ τοῖς καρποῖς ἀμύ-
γδαλα καὶ τερμίνθου σπέρμα μᾶλλον ἢ κολοκύνθης
τε καὶ σικύων πέπονα καὶ κοκκύμηλα καὶ μόρα, κρεῶν

the geriatric part of the art, which now lies before us, what has been shown about the *krasis* of this, and to take from that the objective of these same matters, going through some of these individually. This is in part for the sake of training the student, and in part to provide examples toward the discovery of those things not said that will come about, due to the fact that not everyone is wise enough to know how to discover the individual matters from the general alone, but needs as well a guide for these.

Thus someone practiced in reasoning, on hearing that the body of an old person is cold and dry in the parts themselves and is easily filled with serous and phlegmatous superfluities due to weakness of the capacity, will try to draw these off, heating and moistening the actual solid parts of the body. But those who are deceived by the excess of the superfluities and proclaim loosely that the *krasis* of old people is moist are immediately mistaken regarding the primary objective, thinking that aged bodies need to be dried. Therefore, they give more of the drying foods, such as, among vegetables, cabbage in preference to mallow, blite, and monk's rhubarb, orach and lettuce; among the pulses, they give lentils in preference to ptisane, millet (*kenchros*) in preference to beans, and millet (*elumos*) in preference to wheat; among the fruits, almonds and seeds of terebinth more than colocynth, ripe bottle gourd, plums and blackberries; among the meats, those of wild animals more than those of domesticated animals, and those that

351K

---

30 εἶναι *add.* Ko
31 τῆς δυνάμεως *add.* Ku

δὲ τὰ τῶν ἀγρίων ζῴων μᾶλλον ἢ τὰ τῶν ἡμέρων καὶ
τὰ ταριχηρὰ τῶν προσφάτων, ἐν ὅλῃ τε τῇ διαίτῃ
φεύγουσι μὲν ὧν ἡ δύναμις ὑγρά, προαιροῦνται δὲ
διδόναι τὰ ξηραίνοντα.

καίτοι γε τοὐναντίον ἅπαν ἐστὶν ἀληθὲς ὠφε-
λουμένων ἐναργῶς τῶν γερόντων ὑπὸ τῶν ὑγραινόντων ἐδεσμάτων. ἐπεὶ δ' ἐξ αὐτῶν ἔνια ψυχρὰ ταῖς
κράσεσίν ἐστι διὰ τοῦτο φλέγμα γεννῶσιν εὐθὺς ἐν
τῇ γαστρὶ καὶ κατὰ τὰς πρώτας φλέβας, ἐντεῦθεν οὖν
ἔδοξεν εἶναι βλαβερὰ τοῖς μήτε τὴν ὅλην κρᾶσιν ἐπι-
σταμένοις τῶν γερόντων μήτ' εἰς τὰ κεφάλαια τῆς
352K   διαίτης ὁρῶσιν. οὔτε γὰρ ἔλαιον αὐτοὺς οὔτε λουτρὰ
θερμὰ ποτίμων ὑδάτων οὔτ' οἴνου πόσις ὠφελεῖ μὴ
δεομένους ὑγραίνεσθαι. καὶ μὴν καὶ τὸ μέτρια κινεῖσθαι καὶ τὸ χρῄζειν ὕπνων ἱκανὰ μαρτύρια τοῖς
σκοποῖς τῆς διαίτης. ὥστ' εἰ καί ποτε φλέγματος ἐν
γαστρὶ γεννηθέντος ἐξ ἀνάγκης ἑλοίμεθά τι τῶν
τεμνόντων, μετιέναι χρὴ ταχέως ἐπὶ τὴν ὑγραίνουσαν
δίαιταν. οὕτω δὲ κἂν δι' ὑποψίαν ἐμφράξεως ἐπί τι
τῶν ἐκφραττόντων ἐδεσμάτων ἢ φαρμάκων ἀφικώ-
μεθα, οὐδὲ τὰ θρέψοντα τῶν σιτίων ὑγρᾶς εἶναι κράσεως οὐ χρὴ παντάπασιν οὐδὲ κατ' ἐκείνην αὐτὴν
ἀποστῆναι τὴν ἡμέραν πολὺ δὲ μᾶλλον ἐπὶ τῆς ὑστε-
ραίας ἔχεσθαι προσήκει τοῦ σκοποῦ, τόν τε χόνδρον
ἅμα τῷ μέλιτι διδόντας ἢ ὄξους αὐτῷ μιγνύντας, ὅταν
ὡς πτισάνην σκευάζωμεν, δι' οἰνομέλιτος ἢ οἴνου Φαλερίνου προσφέρεσθαι κελεύοντας ἢ τῷ γάλακτι χρωμένους, ὡς προεῖπον, ἢ πτισάνην καλῶς ἑφθήν, πε-

are preserved more than those that are fresh. In the whole diet they avoid those things whose capacity is moist, choosing instead to give those things that are drying.

And yet the complete opposite is true, since old people are clearly benefited by moistening foods. However, since some of these are cold in their *krasias*, and because this immediately generates phlegm in the stomach and primary veins, here then they seemed to be harmful to those who do not know the whole *krasis* of old people, nor see the chief points of the regimen. For neither olive oil nor hot baths of potable waters, nor the drinking of wine help them, if they do not need to be moistened. And furthermore, moderate movement and the need for sleep are sufficient testimony to the objectives of the regimen. As a consequence, if also at any time phlegm is generated in the stomach, we choose of necessity one of the things that cuts this, and must cease quickly from the moistening diet. Similarly, even if, due to a suspicion of obstruction, we come to the foods or medications that clear obstructions, it is not necessary for the nourishing foods to be altogether of a moist *krasis*, nor to abstain on that same day; rather it is much more appropriate on the next day to hold to the objective, giving gruel along with honey, or mixing some vinegar with it, as when we prepare ptisane, directing the giving of it with honey and wine mixture (mead) or Falernian wine, or using milk, as I said before, or ptisane well cooked, putting in completely ground pepper. In this way

352K

πέρεως ἀκριβῶς χνοώδους ἐπεμβάλλοντας. οὕτω δὲ
καὶ πέπειρα σῦκα πρὸ τῆς ἄλλης ὀπώρας αἱρετέον
ἐστὶν καὶ κατὰ τὸν χειμῶνα τὰς ἰσχάδας ἄρτον τε
353K παρεσκευασμένον, ὡς εἶπον, ἢ μετὰ μέλιτος ἢ οἰνομέλιτος ἢ μετά τινος οἴνου τῶν ἐπιτηδείων. ὡς παραδείγματα γὰρ ἕνεκα σαφηνείας λέγεται ταῦτα πρὸς
τὸ καὶ τοὺς ἀναγνόντας αὐτὰ δύνασθαι κρίνειν τὴν
ὕλην ὁμοίως ἀποβλέποντας εἰς τὸν καθόλου σκοπόν,
ὃν ἐν τῷ θερμαίνειν τε καὶ ὑγραίνειν ἔφην κεῖσθαι.

9. Ὅτι δὲ τά τε φλεγματώδη καὶ ὀρρώδη περιττώματα κατὰ τὸ τῶν πρεσβυτέρων ἀθροίζεται σῶμα,
τήν τε οὔρησιν ἐφ' ἡμέρᾳ προτρέπειν χρή, μὴ διὰ τῶν
φαρμακωδῶν, ἀλλὰ σελίνῳ καὶ μέλιτι καὶ οἴνοις οὐ-
ρητικοῖς, ὑπάγειν τε τὴν γαστέρα δι' ἐλαίου μάλιστα,
καταρροφοῦντας αὐτὸ πρὸ τῶν σιτίων. εὔδηλον δέ,
ὅτι καὶ τὰ λαχανώδη πάντα πρὸ τῶν ἄλλων σιτίων
ἐσθίειν χρὴ δι' ἐλαίου καὶ γάρου. τὸ μὲν οὖν ἐφ'
ἡμέρᾳ διά τε τούτων ἱκανῶς ἐνίοις ἡ γαστὴρ λαπάτ-
τεται καὶ σύκων, ὁπότ' εἴη, καὶ κοκκυμήλων ὅσα τ'
ἄλλα κατὰ θέρος καὶ φθινόπωρον ἀκμάζει, χειμῶνος
δὲ διά τε τῶν ἰσχάδων καὶ τῶν Δαμασκηνῶν κοκκυμήλων ἤτοι γ' ἑψημένων ἢ ἁπλῶς βεβρεγμένων ἐν
μελικράτῳ τὸ πλέον ἔχοντι μέλιτος. ἔσται δὲ τοῦτο
κάλλιον, ἐὰν Ἀττικὸν ᾖ τὸ μέλι. πολύ γε μὴν τῶν
354K Δαμασκηνῶν ὑπακτικώτερα τὰ ἀπὸ τῆς Ἰσπανίας
ἐστί. καὶ τῶν ἐλαιῶν δὲ τῶν ἐκ τῆς ἅλμης ἐγχωρεῖ
ποτε λαμβάνειν. οὐ μὴν ἀλόην γε συμβουλεύω προσφέρεσθαι, καθάπερ ὁρῶ πολλοὺς τῶν γερόντων, ὅσοις

too, we must choose ripe figs before other fruit and, in the
winter, dried figs and bread prepared, as I said, with either   353K
honey or honey wine mixture (mead) or one of the wines
that are suitable. I have spoken of these things as examples
for the sake of clarity, so those who read them are able to
judge the materials similarly, keeping a focus on the general objective, which I said lies in heating and moistening.

9. Because the phlegmatous and serous superfluities
collect in the body of old people, it is necessary to provoke
urination daily, not through medications, but with parsley,
honey and diuretic wines, and to empty the stomach
downward through oil particularly, swallowing this down
before food. It is clear, however, that it is also necessary
to eat all the vegetables in preference to the other foods,
with oil and fish sauce. It is then enough in some for the
stomach to be emptied every day through these things and
figs, whenever they are available, and plums and those
other fruits that ripen in summer and autumn, whereas in
winter, through dried figs and Damascene plums, either
boiled or simply soaked in melikraton having a preponder-
ance of honey. This will be better if the honey is Attic. The   354K
plums from Spain are much more aperient than the Damascene. And sometimes it is permissible to take the oils
from brine. I do not advise using aloes, as I see many old

ἡ γαστὴρ ξηρά, τινὰς μὲν ἀναπλάττοντας εἰς κατα-
πότια καὶ χυλῷ κράμβης, ἐνίους δὲ καὶ μόνης αὐτῆς
λείας ἐπιπάττοντας ὑγρῷ τινι, καὶ γὰρ καὶ τοῦτο τισὶ
μὲν ὕδωρ ἐστί, τισὶ δὲ μελίκρατον· ὅσοι δ' ἐξ αὐτῶν
πλούσιοι, τὴν μετὰ κινναμώμου σκευαζομένην ἀλόην
λαμβάνουσιν. ὀνομάζουσι δὲ τὸ φάρμακον ἔνιοι μὲν
ἱερὰν τὴν διὰ τῆς ἀλόης, ἔνιοι δέ τινες πικράν. ἔχουσι
δὲ τινὲς μὲν ξηρόν, ὡς ἐπιπάττειν τῷ ποτῷ, τινὲς δὲ
μετὰ μέλιτος ἀνέφθου μετρίως ἀνειλημμένου.

ἀλλ' οὐδενὸς τούτων ἐστὶ χρεία τοῖς γέρουσιν, ὅτι
μὴ μεγάλης ἀνάγκης καταλαβούσης. ὑπακουούσης
μὲν γὰρ τοῖς εἰρημένοις τῆς γαστρός, εἰ καὶ μὴ καθ'
ἑκάστην ἡμέραν, ἀλλὰ παρὰ μίαν, οὐδὲν ὅλως χρὴ
φαρμακῶδες προσφέρειν· εἰ δὲ μὴ διὰ τρίτης ὑπέλθοι
δυοῖν ἡμερῶν ἐπισχεθεῖσα, τηνικαῦτα καὶ λινόζωστις
αὐτάρκης ἐστὶ καὶ ἡ καλουμένη θαλασσοκράμβη καὶ
κνίκος ἐν πτισάνῃ διδόμενος ὅσα τ' ἄλλα μετρίως
φαρμακώδη, καθάπερ οὖν καὶ ἡ τερμινθίνη ῥητίνη.

355K λαμβάνουσι δὲ αὐτῆς ἐνίοτε μὲν καρύου Ποντικοῦ
μέγεθος, ἐνίοτε δὲ καὶ δυοῖν ἢ τριῶν· οὐ μόνον γὰρ
ἀλύπως λαπάττειν πέφυκεν, ἀλλὰ καὶ τὰ σπλάγχνα
πάντα διαρρύπτειν, ἧπαρ καὶ σπλῆνα καὶ νεφροὺς
καὶ πνεύμονα. ποικίλως δὲ χρῆσθαι δεῖ τοῖς εἰρημένοις, οὐχ ἓν ἐπιλεξάμενον ἐκεῖνο μόνον προσφέρε-
σθαι· συνήθης γὰρ ἐν τῷ χρόνῳ τοῦ λαμβάνοντος ἡ
φύσις αὐτῷ γινομένη καταφρονεῖ τῆς τοῦ φαρμάκου
δυνάμεως.

ὑπαλλάττειν οὖν χρὴ τὰ εἰρημένα καὶ πρὸς αὐτοῖς

62

people, in whom the stomach is dry, do, some making it into little pills with the juice of cabbage. Others, however, take this ground fine by itself alone, sprinkled over some liquid; for some, the liquid is water and for some melikraton. Those who are rich take the aloes prepared with cinnamon. Some call the medication "sacred aloes" whereas some call it "bitter aloes." Some have it dry, so as to sprinkle it over a beverage, and some with honey, taking this moderately unboiled.

But there is no need of any of these things for old people, unless they are seized by a great necessity. For when the stomach responds to the things mentioned, if not every day, at least every other day, there is no need to administer medications at all. If, however, the bowels do not move on the third day, retaining for two days, under these circumstances a mercurial is sufficient, as are the so called sea cabbage (sea kale) and safflower given in ptisane, and those other things that are moderately medicinal, such as terebinth resin. Sometimes they take this in the size of a Pontic nut, and sometimes two or three, for not only is it of a nature to evacuate painlessly, but also to thoroughly cleanse all the internal organs—liver, spleen, kidneys and lungs. You need to use the things spoken of variously, not choosing one and administering that alone, because in time the nature of the one taking this becomes accustomed to it and "disdains" the potency of the medication.

355K

Therefore, it is necessary to change a little the things

τὸ διὰ τῶν ἰσχάδων σκευαζόμενον· ἔστωσαν δὲ λιπα-
ραὶ καὶ ἀφαιρείσθω τὸ περικείμενον ἔξωθεν αὐταῖς
οἱονεὶ³² δέρμα· κατὰ δὲ τὸν αὐτὸν τρόπον καὶ τοὺς
κνίκους, εἶτ᾽ ἄμφω μιγνύμενα κοπτέσθω. τῷ σταθμῷ
δ᾽ ἔστω πολλαπλάσιος τοῦ κνίκου ἡ ἰσχάς. ἔξεστι δὲ
κἀνταῦθα πειρωμένους τοῦ φαρμάκου, ὡς ἔχει πρὸς
τὴν τοῦ λαμβάνοντος φύσιν, ἐνίοτε μὲν ἔλαττον, ἐνίοτε δὲ πλέον μιγνύναι τοῦ κνίκου. λαμβάνειν δὲ τὸ
μέγεθος ἰσχάδων δυοῖν ἢ τριῶν. ἐξ αὐτῆς δέ, ὡς εἶπον, τῆς πείρας ἑαυτῷ τις εὑρήσει τὸ μέτρον ἐπὶ
τῶν τοιούτων ἁπάντων· ἐνίοις μὲν γὰρ μᾶλλον, ἐνίοις
356K δ᾽ ἧττον ὑπακούουσα τοῖς αὐτοῖς φαρμάκοις ἡ γαστήρ. ἀλλ᾽ ἱκανῶς γε δαψιλὲς οὐδὲν οὐδέποτε χρὴ
τῶν εἰρημένων³³ τι λαμβάνειν φαρμάκων. ἐν μὲν γὰρ
τῷ παραχρῆμα χαίρουσιν ἔνιοι σφοδρῶς κενωθέντες,
ὅσον δ᾽ ἂν μᾶλλον ἐκκενωθῶσι, τοσοῦτον μᾶλλον αὐτοῖς ἡ γαστὴρ ἴσχεται κατὰ τὰς ἐφεξῆς ἡμέρας. διὰ
τοῦτο δ᾽ ἐγὼ καὶ τοὺς ἐν νόσοις χρονίαις τὴν κοιλίαν
ἐπεχομένους, ὁποῖαι μάλιστα κατὰ τὸν χειμῶνα γίνονται πολλοῖς, καὶ μετὰ μακρὰν ἀρρωστίαν ἐν ταῖς
ἀναλήψεσιν ὁμοίως ἐνοχλουμένους οὐ κλύζω δριμέσι
κλύσμασιν, ἀλλ᾽ ἔλαιον ἐνίημι μόνον. ὅπερ οὐδὲν
κωλύει καὶ τοῖς ὑγιαίνουσιν γέρουσιν ἐγχεῖν ἐνίοτε
τῆς γαστρὸς ἐπισχεθείσης· καὶ γὰρ διαβρέχεται τὰ
σκληρὰ τῶν περιττωμάτων, καί τις ὄλισθος ἐν τῇ δι-
όδῳ γίνεται, καὶ τὸ σῶμα τῶν γερόντων αὐτὸ μαλάτ-
τεται ταῖς σκληραῖς ὁμοίως διφθέραις ἐσκληρυμμέ-
νον. ἀλλ᾽ οὐδὲ ταῦτα τοῦ γέροντος, ᾗ γέρων ἐστίν,

spoken of, and besides these, [give] the preparation from
dried figs. Let the figs be oily and remove the kind of skin
lying around them. In the same way also, pound the safflower, then pound both mixed together. Let the dried figs
be much more than the safflower in weight. And even here
it is possible to test how the medication is in relation to
the nature of the one taking it, mixing sometimes less
and sometimes more of the safflower. Take the amount of
two or three dried figs. From actual experience, as I said,
someone will discover for himself the measure in the case
of all such things; for in some people the stomach re-   356K
sponds more to these medications and in some less. But it
is never at any time necessary to take any of the medications mentioned in excessive amounts. For some, when
they are purged strongly, are pleased for the moment, but
the more they are evacuated, the more the stomach retains
in them over the following days. Because of this, I do not
purge those who are retaining in the belly in the chronic
diseases, of the kind that particularly arise in many people
in the winter, and similarly those troubled with a long ill-
ness in the recovery phase, with sharp clysters but insert
oil alone. There is also nothing to prevent instilling this in
old people who are healthy, when sometimes there is retention in the stomach, for it soaks the hard superfluities
and a certain slipperiness occurs in the passage. And in old
men, the body itself, which has been hardened like tough
hides, is softened. But these things are not specifically for

---

32 οἱονεὶ Ko; ἔξω Ku
33 post τῶν εἰρημένων: τι λαμβάνειν add. Ko

ἴδια· καὶ γὰρ καὶ τοῖς ἐκ νοσημάτων μακρῶν ἀνακο-
μιζομένοις συμφέρει.

τί τοίνυν ἴδιόν ἐστι τοῦ γέροντος, ᾗ γέρων ἐστί; τὸ
τῇ κράσει σύμφορον. αὕτη γάρ ἐστι, δι' ἣν καὶ γη-
ρῶμεν, ἄλλος ἐν ἄλλῳ χρόνῳ, πρωϊαίτερον ἢ ὀψιαί-
357K τερον, ὡς ἂν καὶ φύσεως ἔχωμεν ἐξ ἀρχῆς ἢ ⟨δι'⟩
ἐπιτηδεύματα ἢ δίαιταν ἢ νόσους ἢ φροντίδα ἤ τι
τοιοῦτον ξηρανθέντες ἀμετρότερον τύχωμεν. ἔστι μὲν
γάρ, ὅ γε κυρίως ἅπαντες ἄνθρωποι γῆρας ὀνομάζουσιν, ἡ ξηρὰ καὶ ψυχρὰ κρᾶσις τοῦ σώματος ἐκ πολυ-
ετίας γινομένη. συμβαίνει δέ ποτε καὶ διὰ πυρετῶδες
νόσημα, καὶ καλοῦμεν αὐτὴν ἐκ νόσου γῆρας, ὡς ἐν
τῷ Περὶ μαρασμοῦ λέλεκται γράμματι· μαρασμὸς
γὰρ καὶ ἡ τοιαύτη διάθεσίς ἐστιν, οὐκ ἐν ζῴοις μόνον,
ἀλλὰ καὶ ἐν φυτοῖς γινομένη. γέγραπται δὲ καὶ κατὰ
τὸ πρῶτον βιβλίον τῆσδε τῆς πραγματείας ἡ ἀναγ-
καία τοῦ γήρως γένεσις. ἔκ τε οὖν ἐκείνων καὶ ἐξ ὧν
ἐν τοῖς Περὶ κράσεων εἴρηται καὶ προσέτι τοῦ Περὶ
μαρασμοῦ γράμματος εὐπορώτερος ἄν τις εἰς τὸ γηροκομικὸν μέρος τῆς τέχνης γενηθείη.

10. Σύγκειται γὰρ ἅπασα πρόνοια σωματικῆς διαθέσεως ἔκ τε τοῦ γινώσκειν αὐτῆς τὴν οἰκείαν οὐσίαν
καὶ τῆς τῶν βοηθημάτων ὕλης τὰς δυνάμεις· οἷον
358K εὐθέως ἐπὶ τοῦ προκειμένου νῦν ἡμῖν, τοῦ γήρως, ὁ
γνοὺς τὴν διάθεσιν ἐπιστημονικῶς, ὅτι ξηρότης ἐστὶ
μετὰ ψύξεως, ἐὰν ἐκμάθῃ τὰς ὑγραινούσας τε καὶ θερ-
μαινούσας ὕλας τῶν βοηθημάτων, ἀγαθὸς ἂν ἰατρὸς
εἴη γερόντων. οὐσῶν δὴ τεττάρων ὑλῶν κατὰ γένος,

the old person because he is old; they are also helpful for those recovering from prolonged illnesses.

What then is specific for the old person by virtue of his being old? It is what benefits the *krasis*. For the *krasis* is why we grow old, one at one time and one at another, either sooner or later, because we happen to be dried   357K
to such an excessive degree from our original nature, or through ways of life, or regimen, disease or anxiety, or some other such thing. For all people legitimately call old age the dry and cold *krasis* of the body arising from many years of life. However, sometimes it happens due to a febrile disease, and we call this old age from disease, as I have said in the work *On Marasmus*.[18] There is certainly such a condition as marasmus, which occurs not only in animals but also in plants. I have written about the inevitable development of old age in the first book of this particular treatise. From what has been said, as well as what was said in the work *On Mixtures* and besides from the work *On Marasmus,* someone would be well provided for regarding the geriatric part of the art.

10. It is agreed that all foreknowledge of a bodily condition comes from knowing the specific essence of this condition and the potencies of the material of the remedies. For example, in the case of the matter immediately   358K
before us now—old age—the one who knows the condition scientifically, that it is dryness along with coldness, is a good doctor for the aged, if he thoroughly learns the moistening and heating materials of remedies. Since there are four materials in terms of class, which they term those

18 Galen, *Marc.*, VII.666–703K; see particularly 685K.

ἃς ὀνομάζουσι, προσφερομένων ποιουμένων κενου-
μένων καὶ προσπιπτόντων ἔξωθεν, ἐκλέγεσθαι προσήκει καθ᾽ ἑκάστην αὐτῶν τὰ θερμαίνειν τε καὶ ὑγραίνειν δυνάμενα. πρὸς δὲ τὴν οἰκείαν ἑκάστου χρῆσιν ἡ
ἐπὶ τῶν παραδειγμάτων ἄσκησις ὠφελιμωτάτη.[34]

321K   διόπερ ἡμῖν εἴρηται πρόσθεν περὶ ἐδεσμάτων τε
καὶ πομάτων οὐκ ὀλίγα παραδείγματος ἕκενα· καὶ νῦν
εἰρήσεται περὶ τῶν γυμνασίων, ὑπὲρ ὧν ἐμάθομεν ἔμπροσθεν, ὡς τὰ μὲν ὀξέα (προσαγορεύουσιν δ᾽ οὕτως,
ὧν αἱ κινήσεις ταχεῖαι) λεπτύνει τὸ σῶμα, τὰ δ᾽ ἐναν-
322K τία παχύνει, καὶ τὰ πολλὰ μὲν ξηραίνει, τὰ δὲ μέτρια
σαρκοῖ. λέλεκται δὲ καὶ περὶ τῶν ἄλλων ἐν αὐτοῖς
διαφορῶν ἁπασῶν, ὥσπερ γε καὶ περὶ τῶν τρίψεων.
αἱ δὲ τῶν γυμνασίων ἰδέαι τοῖς γέρουσιν, ὅσαι τ᾽
ὠφέλιμοι καὶ ὅσαι βλαβεραί, κριθήσονται τῇ τε τοῦ
σώματος ὅλου διαθέσει[35] καὶ τοῖς ἔθεσι καὶ τοῖς ἐν-
οχλοῦσι παθήμασι. τῇ μὲν ὅλου τοῦ σώματος διαθέσει κατὰ τάδε. τὸ μὲν ἄριστον τῇ κατασκευῇ σῶμα,
περὶ οὗ μέχρι δεῦρο πεποίημαι τὸν λόγον ἐξ ἀρχῆς,
ὥσπερ ἐν νεότητι πρὸς ἅπαντας τοὺς σφοδροτάτους
πόνους ἐπιτηδειότατόν ἐστιν, οὕτως ἐν γήρᾳ πρὸς
ἅπαντας τοὺς μετρίους. τὸ δ᾽ ἤτοι παχυσκελὲς ἢ εὐρύ-
στερνον ἢ περαιτέρω τοῦ προσήκοντος ἰσχνοσκελὲς
ἢ ὅσοις ὁ θώραξ μικρός ἐστιν καὶ κομιδῇ στενὸς ἢ
τὸ βλαισὸν ἢ τὸ ῥαιβὸν ἢ ὁπωσοῦν ἄλλως ἀσύμμε-
τρον εἰς πολλὰ τῶν γυμνασίων οὐκ ἐπιτήδειον. ἐν μὲν
γὰρ τοῖς διὰ τῆς φωνῆς γινομένοις ὁ κακῶς κατ-

that are applied, those that are done, those that evacuate
and those that befall externally, it is appropriate to chose
from each of these the heating and the moistening potencies. In regard to the proper use of each, the practice with
reference to the examples is most useful.

This is why I said quite a lot about foods and drinks   321K
earlier. I shall now speak by way of example about exercises. What I taught before about these is that rapid exercises (they term thus those in which the movements are
quick) thin the body, whereas the opposites thicken it; and   322K
much exercise dries the body whereas moderate exercise
enfleshes. I have spoken about all the other differences
in these, just as I also have about massages. The specific
kinds of exercises for old people—those that are beneficial
and those that are harmful—will be determined by the
condition of the whole body, by the customs and by the
troublesome affections. Determination by the condition
of the whole body is as follows: with respect to the best
body in terms of the constitution, which I have made the
discussion about from the beginning to this point, that in
the young person is most suitable for all the most violent
exertions, while that in the old is most suitable for all those
that are moderate. But the body that is thick-legged, or
broad-chested, or more thin-legged than is appropriate, or
in which the chest is small or quite narrow, or in those who
are splay-footed or bandy-legged, or is in any other way
whatsoever asymmetrical is not suitable for many of the
exercises. Thus, the badly constituted chest is not suitable

---

34 *This marks the start of the passage found in chapter 3 of the
Kühn text as detailed in n. 15 above. Kühn page numbers are given
in the present text.*   35 *post* διαθέσει: καὶ τοῖς ἔθεσι *add.* Ko

εσκευασμένος θώραξ, ἐν δὲ τοῖς διὰ περιπάτων τὰ
σκέλη βλάπτεται. κατὰ δὲ τὸν αὐτὸν τρόπον ἐπί τε
χειρῶν εἰρῆσθαί μοι νόει καὶ τραχήλου καὶ νώτου καὶ
323K ὀσφύος ἰσχίων τε καὶ τῆς ῥάχεως ὅλης. ὅ τι γὰρ
ἂν αὐτῶν κακῶς ᾖ κατεσκευασμένον, ἐξελέγχεται
μᾶλλον ἢ ῥώννυται γυμναζόμενον, εἰ μὴ ἄρα τὰς
συμμέτρους τις κινήσεις ὀνομάζειν ἐθέλοι γυμνάσια
σύμμετρα τοῖς ἀσθενέσι μορίοις· ἀλλ' οὕτω γε περὶ
ὀνόματος μᾶλλον ἢ περὶ πράγματος ἡ ἀμφισβήτησις
ἔσται.

τὰ δ' εἰς ὑγείαν διαφέροντα τῷ γέροντι γυμνάσια
διὰ τῶν ἰσχυροτέρων χρὴ ποιεῖσθαι μορίων· συγκι-
νεῖται γὰρ ἐκείνοις καὶ συγγυμνάζεται τὰ λοιπά. καὶ
μὲν δὴ καὶ τὰ ἔθη μεγίστην ἔχει μοῖραν εἰς εὕρεσιν[36]
ἰδέας γυμνασίων. ἄκοποί τε γὰρ αὐτοῖς αἱ εἰθισμέναι
γίνονται κινήσεις ἥδονταί τε κατ' αὐτὰς ἐνεργοῦντες,
ὥσπερ αὖ πάλιν ἐν τοῖς ἀήθεσιν ἄχθονταί τε καὶ κο-
πώδεις γίνονται. τῶν γε μὴν τεχνικῶν ἐνεργειῶν οὐδ'
ἅψασθαι δυνατόν ἐστιν οἷον[37] αὐλεῖν ἢ σαλπίζειν ἢ
κιθαρίζειν, ὥσπερ γε τῶν κατὰ παλαίστραν, ὅσοι παλαισμάτων ἀμαθεῖς. ἕκαστον οὖν τῶν γερόντων ἐν
τοῖς συνήθεσι γυμνάζειν, ἀνιέντας αὐτῶν τὴν σφοδρότητα.

324K τρίτος δὲ σκοπὸς ἰδέας γυμνασίων ἀπὸ τῶν παθημάτων λαμβάνεται, κοινὸς μὲν ἁπάσης ἡλικίας ὑπάρχων, οὐ μὴν ἴσην γε τὴν δύναμιν ἐν ἁπάσαις ἔχων,
ὅτι μηδὲ τὴν βλάβην ἴσην ἐργάζεται παροφθείς. ὅσοι
μὲν γὰρ ἑτοίμως ἁλίσκονται σκοτώμασιν ἢ ἐπιλη-

for those things arising through the voice, while legs that
are damaged are not suitable for walking. And assume I
have spoken in the same way in regard to the arms, neck,
lower back, hips and the whole spine. Whatever of these   323K
is badly constituted from the beginning is weakened
more than strengthened when exercised, unless someone
should wish to designate moderate movements for the
weak parts. But in this way the dispute will be more about
a name than a matter.

The exercises that make a difference to health in the
old person must be done through the stronger parts, for
the rest are moved and exercised together with those. Furthermore, the customs have the greatest share in discovering the kinds of exercises. Thus, customary movements
occur without fatigue in people and they find pleasure in
doing them, just as conversely, with those that are not
customary, they are burdened and fatigues occur. It is
impossible for them to engage in the technical activities,
like playing the pipes, trumpet or cithara, just as it is for
those untutored in the holds of wrestling to take part in
the activities of the wrestling school. Thus, each old man
should take exercise in those things that are customary
while keeping away from what is violent among these.

The third objective of the kind of exercises is taken   324K
from the affections, and is general for every age, although
it doesn't have equal power in all ages in that if it is disre-
garded, it doesn't bring about equal harm. For those who
are readily seized by dizzy spells, epilepsies, severe opthal-

---

36 εὕρεσιν Ko; αἵρεσιν Ku
37 οἷον Ko; ἢ Ku

ψίαις ἢ ὀφθαλμίαις σφοδραῖς ἢ ὠταλγίαις, οὐ χρὴ
τούτους ἐπινεύοντας ἢ κατακύπτοντας ἢ καλινδουμέ-
νους χαμαὶ γυμνάζεσθαι, περιπάτοις δὲ πολλοῖς καὶ
δρόμοις μετρίοις αἰωρήσεσί τε ἐπὶ τῶν ὀχημάτων
ἀκόπως χρῆσθαι. παραπλήσιον δὲ τρόπον καὶ ὅσοις
παρίσθμια ῥᾳδίως ἢ ἀντιάδες ἢ συνάγχαι γίνονται
καὶ ὅσοις γαργαρεὼν ἑτοίμως ῥευματίζεται καὶ ὅσοις
οὖλα καὶ ὅσοις ὀδόντες ἢ ὅλως τι τῶν κατὰ τὸν τρά-
χηλόν τε καὶ τὴν ὅλην κεφαλὴν μορίων. ἡμικρανίᾳ
γοῦν ἐνοχλοῦνται πολλοί, καὶ τένοντας ἄλλοι συνεχῶς ἀλγοῦσιν ἐπὶ μικραῖς προφάσεσιν. ὧν οὐδεὶς
ἀνέχεται γυμνασίου πληροῦντος τὴν κεφαλήν, ἀλλ'
ἔστιν ἅπασιν αὐτοῖς τὰ διὰ τῶν σκελῶν ὠφέλιμα,
καθάπερ γε τοῖς ἀσθενέσι φύσει τὰ σκέλη βελτίω τὰ
διὰ τῶν ἄνω μερῶν γυμνάσια, χειρονομίαι καὶ ἀκρο-
325K χειρισμοὶ καὶ δίσκων βολαὶ καὶ ἁλτήρων χρήσεις ὅσα
τε κατὰ τὴν παλαίστραν γυμνάζεται χαμαὶ πάντα.
τοῖς γε μὴν τὰ μέσα μόρια χειρῶν καὶ σκελῶν
πάσχουσι ῥᾳδίως ἅπαν μὲν εἶδος ἐπιτήδειόν ἐστι
γυμνασίων, εἰ μὴ[38] τῶν ἄλλων τις ἀπείργοι σκοπῶν.

ἤδη δὲ τὰ μὲν κατὰ θώρακα τοῖς κάτω μᾶλλον
χαίρει, τὰ δὲ κατὰ κύστιν ἢ νεφροὺς τοῖς ἄνω, σπλὴν
δὲ καὶ γαστὴρ καὶ ἧπαρ ἔντερά τε καὶ κῶλον, ὥσπερ
ἐν τῷ μέσῳ τῶν ἄνω τε καὶ[39] κάτω μερῶν ἐστιν, οὕτω
καὶ τοῖς γυμνασίοις ἀμφοτέροις[40] ὁμοίως χαίρει. τάς

[38] εἰ μὴ Ko; ἐπεὶ Ku   [39] ἄνω τε καὶ add. Ko
[40] ἀμφοτέροις add. Ko

mias, or otalgias must not exercise by nodding their heads, bending down or rolling around on the ground. Rather, they should walk a lot, run a moderate amount, and use passive movements in chariots that are not fatiguing. The same applies to those in whom sore throat, tonsillitis or synanche occur easily, to those in whom there are readily fluxes of the uvula, and to the gums and teeth, or generally any of the parts in the neck and the whole head. Anyway, many are troubled by hemicrania[19] and others suffer pains continuously in the tendons for minor reasons. None of these are relieved of pain by an exercise that congests the head, but in all of them (exercises) with the legs are helpful, just as in those who are naturally weak in the legs, the exercises with the upper parts are better—shadowboxing, *acrocheirism* (hand wrestling), discus throwing, the use of handheld weights when jumping, and all those exercises done on the ground in the wrestling school. In those who are easily affected in the middle parts of the arms and legs, every kind of exercise is suitable, unless it prevents one of the other objectives.

325K

Now the parts in relation to the chest take more pleasure in exercises than those parts below, while those in relation to the bladder and kidneys take more pleasure in exercises than those above. However, those parts in relation to the spleen, stomach, liver, intestines and colon, just as parts that are between the parts above and below, find pleasure in both [groups of] exercises similarly. It is ap-

[19] The term "hemicrania" together with "hemicephalgia" is now essentially a synonym for migraine.

γε μὴν τρίψεις τῶν ἀσθενῶν μορίων, ἐν οἷς μὲν χρόνοις νοσεῖ, φυλάττεσθαι προσῆκεν, ἐν οἷς δὲ ὑγιαίνει,
προσφέρειν μᾶλλον ἢ τοῖς ἄλλοις καὶ μάλιστα τὰς
ξηράς, ἐπὶ πλεῖστον ἀνατρίβοντα διά τε τῶν ὀθονίων
καὶ μόναις ταῖς χερσί. καὶ μὲν δὴ καὶ ὅσαι κατὰ
περίοδον ὀδύναι γίνονται μορίοις τισί, οὐ σμικρὸν
γίνεται κώλυμα τοῦ μὴ⁴¹ γίνεσθαι τοιαύτη τρίψις ἐν
τῷ μεταξὺ χρόνῳ παραλαμβανομένη, καὶ μάλιστα
πρὸ δυοῖν ἢ τριῶν ὡρῶν τοῦ παροξυσμοῦ· ῥώννυται
γὰρ ὑπ᾽ αὐτῆς τὰ μόρια καὶ ἧττον δέχεται τὰ συνήθη
κατασκήπτοντα ῥεύματα.

326K   ταυτὶ μὲν οὖν ἄπαντα κοινὰ γερόντων ἐστὶ καὶ τῶν
ἄλλην ἡλικίαν ἐχόντων ἡντινοῦν. τὸ δὲ μὴ γυμνάζεσθαι τοῖς ἀσθενέσι μορίοις ἐπὶ γερόντων μόνον συμ-
βουλεύω· τῶν δ᾽ ἄλλων ὅσον τις ἂν ᾖ τοῦ γήρως
ἀπωτέρω, τοσούτῳ μᾶλλον αὐτῷ γυμναστέον ἐστὶ τὸ
ἀσθενές. ὅπως δὲ⁴² καὶ διὰ παραδείγματος⁴³ ὁ λόγος
ᾖ σαφής, οὐκ ὀκνήσω διελθεῖν. ὑποκείσθω τι σῶμα
τῶν ἔτ᾽ αὐξανομένων ἰσχνὸν τοῖς σκέλεσι. τούτῳ συμφέρει πιττοῦσθαί τε τὰ σκέλη καὶ τρίβεσθαι μετρίως
καὶ δρόμῳ χρῆσθαι μᾶλλον ἢ ἄλλῳ τινὶ γυμνασίῳ.
χρὴ δ᾽ ἐπιστατεῖν αὐτῷ τινα τὸ σύμμετρον ὁρίζοντα
τῶν κινήσεων, ὡς μήτε τοῦ προσήκοντος ἐνδεέστερον
γυμνάζοιτο μήθ᾽ ὑπερβάλλοι τοσοῦτον ὡς ἁλῶναι
κόπῳ. τοῦτο δὴ καίτοι δύσγνωστον εἶναι δοκοῦν οὐκ
ἐν γυμνασίοις μόνον, ἀλλὰ καὶ τοῖς ἄλλοις ἅπασιν
ἡμεῖς ἐπεδείξαμεν εὔγνωστον ἔν γε τοῖς κατὰ τὴν ὑγιεινὴν δίαιταν. οὐ γάρ ἐστιν ἐπ᾽ αὐτοῖς ὀξὺς ὁ καιρός,

propriate to avoid massages of the weak parts during the times they are diseased, but during the times they are healthy, apply massages more than at other times, and especially the dry massages, rubbing more with linen cloths and with the hands only. And furthermore, those pains that occur periodically in certain parts are no small impediment to such massage being done. Such massage should be undertaken in the time between, and particularly two or three hours before the paroxysm. For the parts are strengthened by this and receive less of the customary fluxes which pass down.

In these matters, then, everything is common to old people and those of any other age whatsoever. However, I recommend not exercising with the weak parts in the case of old people only. For the others, the more someone is remote from old age, the more he must exercise what is weak. Because the argument is made clear through an example, I shall not hesitate to provide one. Let us assume the body of someone who is still growing is thin in the legs. In this person, it is beneficial to apply pitch to the legs, to massage moderately and to use running more than any other exercise. However, it is necessary for someone to be in charge of him to determine the right measure of the movements, so he may exercise neither less than is appropriate nor excessively to such a degree that he is seized by fatigue. And although this seems difficult to understand, not only in exercises but in all other things, I shall show it to be easy to understand in those matters related to a healthy regimen. For in these, time is not of the es-

326K

---

41 μὴ add. Ko   42 ὅπως δὲ Ko; ὅπερ δὴ Ku
43 post παραδείγματος: ἵνα μᾶλλον (Ku) om.

ὡς ἐν ταῖς νόσοις, ἀλλ᾽ ἔνεστιν ἄρξασθαι μὲν ἀπὸ
327K τῶν ἀσφαλεστάτων ἐν ἑκάστῳ μέτρῳ, ἐπισκοπούμε-
νον δὲ τὸ ἀποβαῖνον ἤτοι γ᾽ ἀφαιρεῖν ἢ προστιθέναι
τι, καθ᾽ ἑκάστην ἡμέραν ἐπανορθούμενον τὸ παρ-
οφθέν.

οἷον εὐθέως ἐπὶ τῶν ἰσχνοσκελῶν, ἐγώ ποτε παιδάριον ἐτῶν τρισκαίδεκα παραλαβών, εἶτα παντὶ τῷ
μετὰ ταῦτα χρόνῳ τῆς αὐξητικῆς ἡλικίας προνοησά-
μενος αὐτοῦ κατὰ τοὺς ὑπογεγραμμένους σκοπούς,
ἀπέφηνα νεανίσκον εὔρυθμόν τε καὶ σύμμετρον. ἐν
μὲν γὰρ τῇ πρώτῃ τῶν ἡμερῶν αὐτὸν κατέχρισα τῇ
πίττῃ δὶς ἐφεξῆς, ὡς εἴωθα, καὶ δραμεῖν ἐκέλευσα
μήτ᾽ ὀξέως μήτε πολύ. κατὰ δὲ τὴν δευτέραν τρίψει
συμμέτρῳ μὲν κατὰ σκληρότητά τε καὶ μαλακότητα,
βραχείᾳ δὲ κατὰ τὸ πλῆθος ἅμα λίπει προμαλάξας,
ἐκέλευσα βραχὺ πλείω δραμεῖν, οὐ μὴν ὀξύτερόν γε
τῶν πρόσθεν. ἐπὶ δὲ τῷ δρόμῳ δηλονότι ταῖς ἀπο-
θεραπευτικαῖς ὀνομαζομέναις ἐχρώμην τρίψεσιν. ἐκέλευον δὲ αὐτῷ καὶ περιπατεῖν καθ᾽ ἑκάστην ἡμέραν
ἀπὸ μετρίων ἀρξάμενος ἀεὶ πλείω, κατασκεπτόμενος
τά τ᾽ ἄλλα τῶν σκελῶν καὶ μάλιστα τὰς μεγάλας
φλέβας, εἰ μὴ περιττότερον εὐρύνοιντο τῆς τῶν σκελῶν εὐτροφίας. χαλεπὸν γὰρ τοῦτο, ῥευματικὰ καὶ
328K βλαισώδη τῷ χρόνῳ παρασκευάζον, οὐκ εὔτροφα τὰ
κῶλα.

τούτῳ τε οὖν προσέχειν σε χρὴ τῷ γνωρίσματι,
καὶ εἰ θερμότερα περαιτέρω τοῦ προσήκοντος γίνοιτο
τὰ σκέλη, καὶ εἰ κοπώδη τινὰ αἴσθησιν ἔχοντα. μη-

sence, as it is in diseases, so it is possible to begin from what is safest in each measure, while, by giving consideration to what results, it is possible to either take away or add something every day, correcting what was overlooked. 327K

An immediate example is the case of those with thin legs. On one occasion, when I undertook the care of a thirteen-year-old boy, and subsequently provided for him according to the previously written objectives for the whole time of the age of growth, I produced a young man well proportioned and symmetrical. On the first day, I smeared him with pitch twice in succession, as is my custom, and directed him to run neither fast nor much. On the second day, having made him supple beforehand with massage intermediate between hard and soft, and with a small amount of oil, I directed him to run a little more but not faster than before. After the running, I obviously used what are termed apotherapeutic massages. I directed him to walk around every day, always starting moderately and increasing. I observed closely the other things in the legs, and particularly the large veins, lest they were dilated more excessively than accords with proper nourishment of the legs. For this is a problem that, being subject to fluxes, in time produces limbs that are distorted,[20] as they are not 328K well nourished.

You must, therefore, direct your attention to this sign, and whether the legs become warmer than is fitting, and whether there is a sense of fatigue. If none of these signs

---

[20] The term βλαισώδης is listed in LSJ as equivalent to βλαισός, attested by the present passage. It is taken in the general sense rather than the more specific meanings of bandy and splay-footed.

δενὸς μὲν γὰρ τούτων τῶν σημείων φανέντος αὐξάνειν τε χρὴ τὸ πλῆθος τῶν περιπάτων καὶ τῶν δρόμων
καὶ πιττοῦν διὰ τρίτης. ὀφθέντος δέ τινος αὐτῶν,
ἀνάρροπά τε σχηματίζειν ἐν τῇ κοίτῃ τὰ κῶλα καὶ
τοῦ πλήθους ἁπάντων ὧν διῆλθον ἀφαιρεῖν, ἐλάττω
μὲν περιπατεῖν, ἐλάττω δὲ τρέχειν κελεύοντα καὶ τῇ
τρίψει κάτωθεν ἄνω. ὅταν δέ σοι ταῦτα πράξαντι
κατὰ φύσιν ἀκριβῶς ἔχῃ τὰ σκέλη, πάλιν ἐπὶ τὴν ἐξ
ἀρχῆς εἰρημένην χρῆσιν τῶν βοηθημάτων ἰέναι,
βραχυτάταις αὐξήσεσι χρώμενον ἢ καὶ διὰ παντὸς
ἐπὶ τῶν αὐτῶν καταμένοντα. παρεμβάλλειν δὲ χρὴ
καὶ τὰς στρογγύλας καλουμένας τρίψεις μετὰ σκλη-
ρότητός τε καὶ μαλακότητος, ἔσθ᾽ ὅτε μὲν ἄνωθεν,
ἐνίοτε δὲ ἐκ τῶν κάτω μερῶν ἀρχομένους. καί που
κεκονισμένοις τοῖς σκέλεσι παραχέοντας ἔλαιον οὕτω
τρίβειν. ἔστω δὲ κόνις ἡ καλουμένη λιπαρά· προσ-
αγορεύουσι δ᾽ οὕτως, ἐν ᾗ μήτε τραχὺ μήτε δριμύ.
λεπτύνουσι γὰρ μᾶλλον ἢ σαρκοῦσιν, ὅσαι τραχύτη-
τος κισσηρώδους ἢ δριμύτητος λιτρώδους ἢ ἁλμώ-
δους μετέχουσιν. ὅτῳ δὲ παχύτερα μὲν τὰ σκέλη,
βραχίονες δὲ καὶ πήχεις ἰσχνοί, τούτῳ τὰ μὲν διὰ
χειρῶν γυμνάσια πάντ᾽ ἐπιτηδευτέον ἐστί, φευκτέον[44]
δὲ τὰ διὰ τῶν σκελῶν. αἵ γε μὴν τρίψεις τε καὶ πιτ-
τώσεις ὁμοίως γινέσθωσαν, ὅσα τ᾽ ἄλλα περὶ τῶν
κατὰ τὰ σκέλη λέλεκται γυμνάσια.[45]

329K

---

[44] φευκτέον Ko; φυλακτέον Ku   [45] *This marks the end
of the transferred passage—see n. 15 above.*

are apparent, you must increase the amount of walking around and running, and apply pitch on the third day. If, however, any of these signs are observed, you must arrange the legs in an elevated position in bed, reduce the amount of all the things I went over, directing the boy to walk around and run less, and use massage from below upward. When you have done these things and the legs are entirely in accord with nature, you must go back again to the use of the remedies described at the beginning, using very small increases or continuing throughout with the same amounts. You must also add the so-called circular massage midway between hard and soft, sometimes beginning from the parts above and sometimes from the parts below. And on legs that have been sprinkled over with powder to some degree, pour on oil and massage in this way. Let the powder be the so called fatty powder. They term thus what is neither rough nor sharp. Those powders   329K that partake of the roughness of pumice stone, or the sharpness of niter or salts, thin more than they enflesh. In the person whose legs are thicker but whose upper arms and forearms are thin, one must practice all the exercises with the arms, and avoid those with the legs. The massages and applications of pitch should be done in the same way, as well as those other exercises I have spoken of in relation to the legs.[21]

[21] This marks the end of the transposed section from the Kühn text.

358K   οἶδα δέ ποτε θώρακα παιδὸς ἀπολειπόμενον οὐκ
ὀλίγῳ τῆς τῶν ἄλλων μορίων συμμετρίας αὐξήσας
εἰς τὸ μέτριον ἐκ τοιαύτης ἰδέας βοηθημάτων. ἐζών-
νυον μὲν αὐτοῦ τὰ κάτω τοῦ θώρακος ἅπαντα μέχρι
τῶν ἰσχίων ζώνῃ συμμέτρως πλατείᾳ, οὕτω παρα-
βάλλων, ὡς ἐρηρεῖσθαι μὲν ἀλύπως, μήτε δὲ χαλαρὸν
ἀπολείπειν τι μήτε πεπιέσθαι, γυμνάσια δὲ τά τε δι'
ὅλων τῶν χειρῶν ἐπιτηδευόμενα καὶ τὰς καλουμένας
ὑπὸ τῶν φωνασκῶν ἀναφωνήσεις παρελάμβανον, ἐφ'
ἑκάτερα δὲ κατάληψιν πνεύματος ἐκέλευον ποιεῖσθαι.
359K γίνεται δ' αὕτη σφοδρῶς μὲν θλιβόντων ἡμῶν ἐκ
παντὸς μέρους τὸν θώρακα, τὴν δ' ἐκπνοὴν κατεχόν-
των, ὡς ἐπέχεσθαι πᾶν ἔνδον τὸ πνεῦμα τὸ διὰ τῆς
εἰσπνοῆς φθάνον εἱλκύσθαι. διὸ καὶ πλέον εἰσπνευ-
στέον ἐστὶ τῷ μέλλοντι καλῶς τοῦτο δράσειν· ὅσῳ
γὰρ ἂν ᾖ πλείων ὁ θλιβόμενος ἀήρ, τοσούτῳ μᾶλλον
ὁ θώραξ εὐρύνεται διατεινόμενος. ὅτι δὲ καὶ τὰς ἀναφωνήσεις ἐν μεγέθει τε καὶ τῇ κατ' ὀξύτητα τάσει τῆς
φωνῆς ποιητέον ἐστὶν ἐπὶ τούτων, οὐκ ἄδηλον, εἴγε
δὴ πρόκειται γυμνάζειν ἰσχυρῶς πάντα τοῦ θώρακος
τὰ μόρια. ταῦτα μὲν οὖν ἐπὶ τῶν αὐξανομένων σωμάτων εἰς συμμετρίαν ἄξει τὰ φύσει κακῶς διαπε-
πλασμένα· μέτρια δὲ καὶ τοὺς ἤδη τελείους ὠφελήσει.

    οὐ μὴν ἐπί γε τῶν πρεσβυτῶν ἐπιχειρήσεις ἔργῳ
τοιούτῳ· βέλτιον γὰρ ἐπ' ἐκείνων ἡσυχάζειν τοῖς
ἀσθενέσι μορίοις, ὡς εἴρηται, βέλτιον δὲ καὶ τὸ διὰ
τῶν συνήθων ἐπιτηδευμάτων γυμνάζειν, εἰ καὶ βλαβερὰ μετρίως εἴη. τοῖς γε μὴν νέοις ὑπαλλάττειν πει-

I have seen on one occasion the chest of a child, that   358K
departed not a little from the symmetry of the other parts.
I increased this in due measure from such a kind of remedies. In him, I bound all the parts below the chest as far
as the hips with a symmetrically wide girdle applied in
such a way that it was firmly fixed without pain, allowing
neither relaxation nor compression, and employed suitable exercises with the whole arms and the so-called voice
exercises of singers, and with each I directed him to hold
his breath. This occurs when we strongly compress the   359K
chest in every part, while withholding expiration so as to
retain within all the breath which was drawn in beforehand through inspiration. On this account, someone must
inhale more, if he is going to do this well. For the more
the air is compressed, the more the chest, being stretched
to the uttermost, is distended. And it is clear that the vocal
exercises, in magnitude and the tension in relation to
sharpness of the voice must be made in the case of these,
if in fact the purpose is to exercise all the parts of the chest
strongly. These things, then, will bring to proportion growing bodies malformed in nature, and will also benefit moderately those parts already complete.

In the case of old people, you will not attempt such an
action, for as was said, it is better for old people to rest
the weak parts, and better also to exercise through those
things that are customary and suitable, even if they are
moderately harmful. In young people, attempt to change

ρᾶσθαι τὰ βλάπτοντα, κἂν ἐκ παίδων ᾖ συνηθέστατα·
360K δύναται γὰρ ἡ δύναμις αὐτῶν ὑπομεῖναι τὴν μετα-
βολὴν ἐν μέτρῳ γινομένην, ἐλπίς τ᾽ ἐστὶ τῷ λοιπῷ
χρόνῳ τῆς ζωῆς ὀνήσασθαί τι πρὸς τῶν ἀμεινόνων
ἐπιτηδευμάτων. ὁ δὲ γέρων, εἰ δυνηθείη χρόνῳ συχνῷ
κατὰ βραχὺ μεταστῆναι τῶν φαύλων ἐθῶν, οὐδ᾽ ἐν ᾧ
χρόνῳ χρήσεται τοῖς βελτίοσιν ἕξει, συμβήσεται δ᾽
αὐτῷ πονῆσαι μάτην, ὥσπερ εἰ καὶ τέχνην ἄρχοιτο
μανθάνειν ὀγδοηκοντούτης ὤν. ὥσπερ δὲ τῶν γερόντων αὐτῶν οὐ σμικρά τίς ἐστι διαφορὰ πρὸς ἀλλήλους, εἴτε κατὰ τὴν ἡλικίαν αὐτὴν εἴτε κατὰ τὴν
διάθεσιν τοῦ σώματος, οὕτω καὶ τῶν φύσει ξηρο-
τέρων τε ἅμα καὶ ψυχροτέρων, ὡς πρὸς τὸν σύμμετρον παραβάλλειν, οὐκ ὀλίγον τὸ ἐν μέσῳ τῆς δυσκρασίας ἐστὶ πρὸς ἀλλήλους τε καὶ τοὺς γέροντας.
ὑγραίνειν μὲν οὖν αὐτοὺς δηλονότι καὶ θερμαίνειν
προσήκει, θαρραλεώτερον δὲ πράττειν ταῦτα, διὰ τὴν
ἡλικίαν ἀνεχομένοις ἰσχυρῶν γυμνασίων, ὡς ἂν ἐρ-
ρωμένης ἔτι τῆς δυνάμεως. ὁ γάρ τοι πρῶτος σκοπός,
ἀφ᾽ οὗ τὴν ἔνδειξιν τῶν ἰαμάτων λαμβάνομεν, ἡ τοι-
άδε δυσκρασία τοῦ σώματός ἐστιν, οὐ τὸ γῆρας ἢ
ὅλως ἡλικία τις.

361K   ἀλλ᾽ ἐπειδὴ στοχαστικῶς ἡμῖν τὸ τῶν διαθέσεων
μέτρον λαμβάνεται, διὰ τοῦτο καὶ τὴν ἡλικίαν ἐπι-
σκοποῦμεν. ἡ δ᾽ αὐτὴ χρεία καὶ τῆς τῶν ἐθῶν γνώ-
σεώς ἐστι καὶ τῆς τῶν προκαταρκτικῶν ὀνομαζομένων
αἰτίων· εἰς γὰρ τὴν τῆς διαθέσεως ἀκριβεστέραν γνῶσιν ἐκ τῶν τοιούτων ἁπάντων ὠφελούμεθα, τῶν δ᾽

82

what has been harmed, even if they are very accustomed to it from childhood, for their strength is able to endure the change which occurs in moderation, and there is hope for the remaining time of life that they will be helped toward better practices. However, the old person, if he is enabled over a long time to change slightly his bad habits, will not have the time to use the better habits, so what will happen to him is that he will labor in vain, like someone who is eighty beginning to learn a craft. But just as old people themselves differ from one another to no small extent, whether in age itself or in the condition of the body, so too, those who are drier and colder in nature as compared to the median state have no little difference in the *dyskrasia* between each other and between them and old people. Obviously, then, it is appropriate to moisten and heat them, and to do these things with greater confidence if, due to their age, you keep away from strong exercises, such as you would use in those whose capacity is still strong. For certainly the first objective, from which we take the indication of the cures, is this *dyskrasia* of the body and not old age or any age generally.

360K

But since our measure of the conditions is taken conjecturally, we also, because of this, take into consideration the age. There is the same use also of the knowledge of the customs and of the so-called *prokatarktic* causes. We benefit from all these things when it comes to the more accurate knowledge of the condition, although the indica-

361K

ἰαμάτων ἡ ἔνδειξις οὐκ ἐκ τούτων γίνεται. τῷ γε μὴν ἐμπειρικῷ μέρος τῆς ὅλης συνδρομῆς ἐστι καὶ τὰ τοιαῦτα, τετηρημένης ἐπ᾽ αὐτοῖς τῆς θεραπείας, οὐκ ἐνδεικτικῶς εὑρισκομένης.

εἰκότως τοιγαροῦν Ἱπποκράτης τὰ πλεῖστα περὶ διαγνώσεών τε καὶ προγνώσεων ἔγραψε γυμνάζων ἡμᾶς ἐν τοῖς κατὰ μέρος, οἵ τ᾽ ἄριστοι τῶν μετ᾽ αὐτὸν ἰατρῶν ὡσαύτως ἔπραξαν, εὖ εἰδότες, ὡς τῷ μὴ[46] γνόντι τὴν διάθεσιν ἀκριβῶς τοῦ σώματος, ὃ μεταχειριζόμεθα[47] νοσοῦν [ὑγιαῖνον], χαλεπῶς εὑρίσκεται τὰ βοηθήματα. χρὴ γάρ, ὡς εἴρηται πολλάκις, ἄρχεσθαι μὲν ἀπὸ θεραπευτικοῦ παραγγέλματος τοῦ "τὰ ἐναντία τῶν ἐναντίων ἰάματα," τῆς δ᾽ ὕλης τῶν βοηθημάτων ἐπιστήμονα γενόμενον, ὡς ἐκμαθεῖν αὐτῆς τὰς δυνάμεις, ἀεὶ προσφέρειν ἅπαντι σώματι, τῷ μὲν

362K φαύλως διακειμένῳ τἀναντία, τῷ δ᾽ ἄριστα κατεσκευασμένῳ τὰ παραπλήσια ταῖς δυνάμεσι. καὶ τοίνυν καὶ τὰς δυσκρασίας αὐτάς, ὅσαι κατὰ τὸ τῆς ὑγείας ἔτι πλάτος εἰσίν, ἐπανορθοῦσθαι μὲν βουλόμενος ὑπεναντίως διαιτήσεις. ὅπερ ἐπὶ πολλῆς γίνεται σχολῆς. ἐν ἀσχολίᾳ δὲ οὖσι διὰ τῶν ὁμοίων ὑπηρετήσεις, καὶ μάλισθ᾽ ὅταν οὕτως ὦσιν εἰθισμένοι. συμβαίνει δὲ ὡς τὰ πολλὰ θάτερον τούτων μᾶλλον, ὡς ἂν τῶν φύσεων ἀδιδάκτως αἱρουμένων τὰ οἰκεῖα, καὶ μέχρι γε τοσούτου προσφερομένων αὐτά, μέχριπερ ἂν εἰς τὴν ἀρχαίαν ἐπανέλθωσι κατάστασιν.

[46] μὴ add. Ko [47] post μεταχειριζόμεθα: νοσοῦν [ὑγιαῖνον], χαλεπῶς Ko; ὑγιαῖνον ἢ νοσοῦν, οὐ χαλεπῶς Ku

tion of the cures does not arise from them. For the Empiric,[22] such things are a part of the whole syndrome while the actual treatment is based by them on observation and is not discovered indicatively.

So, for example, Hippocrates reasonably wrote most about diagnosis and prognosis, training us in these individually, and the best doctors after him acted in similar fashion, being well aware that by not knowing accurately the condition of the body which we are dealing with in health or disease, the remedies are difficult to discover. For, as is often said, it is necessary to begin from a therapeutic precept—opposites are the cures of opposites—to become knowledgeable about the material of remedies, so you may learn thoroughly the potencies of this, and always apply to the whole body, if it is in a bad state, opposites, while to a body with the best constitution, apply those things that are similar in their powers. And moreover, you will also feed in an opposite way, if you wish to correct the actual *dyskrasias* that are still within the range of health, which is time-consuming. When time is short, you will minister to them with similars, and particularly when they are accustomed to this. What happens for the most part, when one of these is more, is that the natures choose what is suitable without teaching, and the application of these continues to such a point that they (i.e., the natures) are restored to their original condition.

362K

[22] For a summary of the main tenets of medical empiricism as it existed in Galen's day, and its relation to the other schools and sects, see Johnston, *Galen: On Diseases and Symptoms*, 17–20. For Galen's own position in some detail, see his *Sect.*, I.64–105K (English trans., Walzer and Frede, *Three Treatises*).

11. Ἡ δέ γ᾽ ἐπανορθουμένη τὴν δυσκρασίαν δίαιτα περαιτέρω προέρχεται τὰ πολλά. διὸ καὶ τοῖς οἰκείοις νοσήμασιν ἁλισκόμεθα πλείοσι, τοῦ μὲν φύσει θερμοτέρου ῥᾳδίως τὰ θερμὰ νοσήματα νοσοῦντος τοῦ δὲ ψυχροτέρου τὰ ψυχρά, καὶ τῶν ἄλλων ὡσαύτως. ἐπανέρχεται τοιγαροῦν θᾶττον εἰς τὴν ἰδίαν φύσιν ἕκαστον τῶν δυσκράτων σωμάτων ἤπερ ἐπὶ τὴν ἀρίστην κρᾶσιν. ἐν τῷ μεσαιτάτῳ γὰρ αὕτη πασῶν οὖσα

363K τῶν δυσκρασιῶν, ἐὰν μὲν οἰκείῳ τις ἁλῷ τῇ φύσει νοσήματι, πλέον ἀφέστηκεν, ἐὰν δὲ οὐκ οἰκείῳ, βραχύτερον. οὐκοῦν οὐδὲ τὰ ἔθη μεταβλητέον ἐστί, κἂν ᾖ μοχθηρά, δυσαρεστουμένων ἔτι τῶν σωμάτων· ἀλλ᾽ ἐν ταῖς ἀκριβεστέραις ὑγείαις οὐδὲ ἀεὶ τοῦτο πρακτέον, ἀλλ᾽ ὅταν ἀπὸ τῶν πολιτικῶν πραγμάτων ἄγῃ σχολὴν ὁ μέλλων μεταχθήσεσθαι.

ταῦτ᾽ οὖν ὄντα κοινὰ παραγγέλματα πασῶν τῶν κράσεων ὑπαλλάττεται κατὰ μέρος, εἰς ὅσον ἂν ἑκάστη δυσκρασία τῆς ἄκρας εὐκρασίας ἀπολείπηται, καὶ θαυμαστὸν οὐδέν ἐστι, τὰς δυσκράτους φύσεις ἐν τῷ μέσῳ καθεστηκυίας ἀκριβοῦς τε ὑγείας αἰσθητοῦ τε νοσήματος ἐπαμφοτερίζειν καὶ τῷ τρόπῳ τῆς διαίτης. οὐ μὴν οὐδὲ τὸ φαίνεσθαι τοὺς ἀνθρώπους ὑπὸ τῶν αὐτῶν ὠφελουμένους τε καὶ βλαπτομένους θαυμαστόν. εἰ μὲν γὰρ ἅπαντες ὁμοίαν ἀλλήλοις εἶχον κατασκευήν, τότ᾽ ἂν ἦν θαυμαστὸν ὑπὸ τῶν ἐναντίων ἐνίους ὠφελεῖσθαι, βλαπτομένων ἐνίων ὑπὸ τῶν αὐτῶν· ἐπεὶ δ᾽ ἐναντίαι πολλῶν τῶν ἀνθρώπων εἰσὶ κατασκευαὶ τοῦ σώματος, εὔλογόν ἐστι καὶ τὴν ὠφέ-

11. The regimen that corrects the *dyskrasia* in many instances extends further. This is because we are seized by many similar diseases, the person hotter in nature easily becoming ill with hot diseases, the person colder in nature with cold diseases and similarly with the others. So too does each of the *dyskratic* bodies return more quickly to its own particular nature than to the best *krasis*. That which is most midway of all the *dyskrasias,* if it is seized by a disease of the same kind in nature, deviates more, whereas if it is not by one of the same kind, it deviates less. Therefore, one must not change the customs, even if they are bad, while bodies are still suffering malaise. The exception is those who are more perfectly healthy, but this must only be done when the person takes time away from civil matters, with the intention of being changed.

363K

These things, then, are the common precepts of all the *krasias* that are changed individually to the extent that each *dyskrasia* is wanting in regard to the highest *eukrasia*—and it is no wonder that the *dyskratic* natures, situated in the middle between perfect health and perceptible disease, are corrected by the method of regimen. Nor is it any wonder that some people are obviously benefited and some harmed by the same things. For if all people had a similar constitution to each other, then it would be surprising for some to be benefitted by opposite things and some harmed by these same things. However, since the constitutions of the body of many people are opposite, it is rea-

364K λειαν αὐτοῖς ὑπὸ τῶν ἐναντίων γίνεσθαι. διὸ καὶ θαυμάσειεν ἄν τις ἁπάντων τῶν ἰατρῶν, ὅσοι γράφειν ἐπεχείρησαν ὑγιεινὰ συγγράμματα, μὴ διελομένων τῷ λόγῳ τὰς φύσεις. ὥσπερ γὰρ ἑνὶ καλάποδι πρὸς ἅπαντας ἀνθρώπους ἀδύνατον χρῆσθαι τοῖς σκυτοτόμοις, οὕτω καὶ τοῖς ἰατροῖς ἰδέᾳ βίου συμφερούσῃ μιᾷ. διὰ τοῦτο οὖν ἐνίοις μὲν ὑγιεινότατον εἶναί φασι τὸ γυμνάζεσθαι καθ᾽ ἑκάστην ἡμέραν ἱκανῶς, ἐνίοις δὲ μηδὲν κωλύειν, εἰ καὶ παντάπασί τις ἐν ἡσυχίᾳ διατρίβει, καὶ λούεσθαι τοῖς μὲν ὑγιεινότατον εἶναι δοκεῖ, τοῖς δ᾽ οὔ, καὶ πίνειν ὕδατος καὶ οἴνου, καὶ περὶ τῶν ἄλλων ὡσαύτως, οὐ μόνον ὑγιεινῶν διαιτημάτων, ἀλλὰ καὶ τῶν τοῖς νοσοῦσι προσφερομένων ἰαμάτων, ἐναντιώτατα γράφουσιν ἀλλήλοις, ὡς σπάνιον εὑρεῖν ὁμολογούμενον ἕν τι πᾶσιν αὐτοῖς. ἥ γε μὴν πεῖρα δείκνυσιν ὑπό τε τῶν αὐτῶν ἐνίους βλαπτομένους τε καὶ ὠφελουμένους, ὑπό τε τῶν ἐναντίων ὡσαύτως. οἶδα γοῦν τινας, οἳ τρεῖς ἡμέρας ἀγύμναστοι μείναντες εὐθέως νοσοῦσιν, ἀγυμνάστους τε ἑτέρους ἀεὶ δια-

365K τελοῦντας, ὑγιαίνοντας δέ, καὶ τούτων ἐνίους μὲν ἀλούτους, ἐνίους δέ, εἰ μὴ λούσαιντο, πυρέττοντας αὐτίκα, καθάπερ ὁ Μιτυληναῖος Πριμιγένης.[48]

 ὅτι μὲν οὖν οὕτω φαίνεται ταῦτα γινόμενα, καὶ οἱ τῇ πείρᾳ μόνῃ τὴν τέχνην ἀθροίζοντες ἴσασιν. οὐ μὴν ἔγραψέ γέ τις αὐτῶν γνωρίσματα, καθάπερ ἐπὶ τῶν νοσημάτων, οἷς προσέχοντες εὑρήσομεν, ὁποίας ἕκα-

---

[48] Πριμιγένης Ko; Πρημιγένης Ku (here and subsequently).

sonable that benefit should occur to them from opposites. On which account also, someone might wonder at all the doctors who attempt to write treatises on health without differentiating the natures in the discussion. For just as it is impossible for cobblers to use one last for all people, so too is it for doctors to use a single beneficial kind of life. Because of this, then, they say it is most healthy for some to exercise sufficiently every day, whereas for others, there is nothing to prevent their passing their lives wholly in idleness. Also, for some it seems to be most healthy to bathe, whereas for others it does not, and to drink water and wine, and about the other things similarly, not only for hygienic regimes but also for the cures prescribed for diseases. And they write things completely opposite to one another, so it is rare to find one point of agreement among them all. In fact, experience shows that some people are harmed and some are benefited by the same things and similarly with opposites. Anyway, I know of some who immediately become sick, if they remain three days without exercise, and others who continue indefinitely without exercise and yet are healthy; and of some who never bathe, and others who, if they do not bathe, immediately become febrile, like Premigenes the Mitylenaean.[23]

And those who gather their skill by experience alone know that these things are manifestly so. Yet none of them has written the signs, as has been done for diseases, by which, if we pay attention to them, we shall discover what

364K

365K

---

[23] Apart from the fact that he was a Peripatetic philosopher, I have been unable to find anything of substance about Premigenes in the usual sources. He is mentioned here in the *Hygiene,* but there is no other listing for him in Ackermann's index.

στος δεῖται διαίτης. ἐπ᾽ ὀλίγων γὰρ ἔστιν εὑρεῖν τὰς καλουμένας ὑπὸ τῶν ἐμπειρικῶν συνδρομὰς ἠκριβωμένας, ὡς ἐν περιπνευμονίᾳ καὶ πλευρίτιδι· τὰ δὲ πλεῖστα νοσήματα στοχαστικὴν ἔχει τὴν διάγνωσιν, ὡς οὐκ ἐξ ἀθροίσματος ὡρισμένων συμπτωμάτων γινομένην,[49] ἀλλ᾽ ἀνδρὸς δεομένην ἀκριβῶς μὲν ἐπισταμένου τὴν διάθεσιν τοῦ σώματος, εὑρίσκειν δ᾽ ἱκανοῦ τὰ κατὰ μέρος ἅπαντα τὰ τῇ τοιαύτῃ συμφωνοῦντα. τὸ γοῦν προκείμενον αὐτῷ τῷ λόγῳ ποδηγηθέντες ἡμεῖς εὕρομεν ἔτι νέοι τὴν ἡλικίαν ὄντες. ᾧ καὶ δῆλον, ὡς ἡ μακρὰ πεῖρα χωρὶς λόγου τὰ τοιαῦθ᾽ εὑρίσκειν ἀδυνατεῖ.

ἐπὶ γοῦν τοῦ Πριμιγένους ἀκούσας, ὅτι πάντως 366K πυρέττοι μὴ λουσάμενος, ἐλογισάμην αὐτῷ καπνώδη περιττώματα γεννᾶσθαι διαπνεῖσθαι δεόμενα, πυκνοτέρου δὲ τοῦ δέρματος ὄντος ἢ ὡς ἐπιτρέπειν αὐτοῖς κενοῦσθαι πᾶσιν ἀθροιζομένοις ὑπὸ τῷ δέρματι θερμασίαν γεννᾶν. διὸ καὶ τὰ λουτρὰ χρησιμώτατα ταῖς τοιαύταις φύσεσιν, οὐ μόνον τῷ κενοῦν τὸ καπνῶδες, ἀλλὰ καὶ τῷ τέγγειν ὑγρότητι γλυκείᾳ. καταμαθεῖν οὖν ἔδοξέ μοι πλατεῖαν ἐπιβαλόντι τὴν χεῖρα τῷ θώρακι τοῦ Πριμιγένους, ὁποία τίς ἐστιν ἡ τῆς θερμασίας ποιότης. ὡς δὲ δριμεῖα καὶ δακνώδης εὑρέθη παραπλησίως τοῖς κρόμμυα προσενηνεγμένοις δαψιλῆ, πολὺ δὴ μᾶλλον ἔτι τὴν αἰτίαν τοῦ γινομένου καλῶς εὑρῆσθαί μοι πεισθείς, ἐπυθόμην, εἴ τινες ἱδρῶτες αὐτῷ γίνοιντο χωρὶς τοῦ λουτροῦ, καὶ φάντος μὴ γίνεσθαι, βεβαιοτέραν γνῶσιν ἐπείσθην ἔχειν τῆς

kind of regimen is needed for each person. It is possible
to discover, down to the smallest details, what are called
defined syndromes by the Empirics, like peripneumonia
and pleuritis. Most diseases have, however, a conjectural
diagnosis, so as not to be known from a collection of defined symptoms; what is required is a man who accurately
knows the condition of the body and is adequate to discover all the individual things in keeping with such a condition. Then, in the subject before us, being guided by
reason itself, we find we are still youngsters in age, while
from this it is also clear that long experience without reason is unable to discover such things.

Anyway, on hearing of the case of Premigenes—that if
he didn't bathe he always became febrile—I reckoned that   366K
he generated smoky superfluities, which needed to be dissipated by exhalation, but because the skin was too thick
to allow all the things collected under the skin to be evacuated, this produced heat. This is why baths are very useful for such natures; they not only evacuate the smoky
superfluities, but they also moisten with sweet moisture.
Therefore, it seemed to me that, by placing my hand flat
on Premigenes' chest, I could learn thoroughly what kind
of quality of heat this was. As it was found to be sharp and
biting, similar to those to whom an abundance of onions
had been given, and being even more persuaded that I had
correctly discovered the cause of what had occurred, I
inquired whether certain sweats occurred in him apart
from bathing. When he said they did not, I was convinced
I had a more certain knowledge of his condition. Further-

---

49 γινομένην Ko; γινωσκόμενα Ku

91

διαθέσεως αὐτοῦ. καὶ μὴν δὴ καὶ ἄλλους τινὰς εἶδον, ὁμοίως μὲν ἐκείνῳ δακνῶδες ἔχοντας τὸ θερμόν, οὐ μὴν ἐπ᾽ ἀλουσίᾳ γε μιᾷ πυρέττοντας, ἐπειδὴ δι᾽ ἱδρώτων αὐτοῖς ἐκενοῦτο τὸ περίττωμα.

τῷ Πριμιγένει δὲ πρὸς τῇ φυσικῇ κατασκευῇ καὶ
367K ὁ τρόπος τῆς διαίτης αἴτιος ἦν τοῦ πυρέττειν ἐπ᾽ ἀλουσίᾳ μιᾷ, διατρίβοντι τὸ πλεῖστον τῆς ἡμέρας⁵⁰ ἐπὶ τῆς οἰκίας, ἐν ᾗ γράφων ἢ ἀναγινώσκων διετέλει διὰ τὸ προσκεῖσθαι θεωρίᾳ Περιπατητικῇ, καθ᾽ ἣν οὐδενὸς ἦν δεύτερος τῶν κατ᾽ αὐτόν. ἴσμεν δ᾽, ὅτι καὶ τοῖς μὴ κατὰ φύσιν τοιοῦτον γεννῶσι περίττωμα φιλοπονία καὶ φροντὶς αἴτιαι γίνονται τῆς γενέσεως αὐτοῦ. διόπερ ἕτερον ἄνθρωπον ὁμοίως ἐκείνῳ τὴν θερμασίαν ἔχοντα δακνώδη συνέβαινε μὴ πυρέττειν αὐτίκα μετὰ μίαν ἡμέραν ἀλουσίας, ἐπειδὴ μήτε φροντιστὴς ἦν αἴ τε πράξεις αὐτῷ κατὰ τὴν πόλιν ἐγίνοντο περιπατοῦντι πολλὰ καὶ ὠνουμένῳ καὶ πιπράσκοντι καὶ μαχομένῳ πολλάκις, ἐν οἷς ἱδροῦν ἠναγκάζετο· καὶ διὰ τοῦτο ἕτερόν τινα τῶν νοσούντων καθ᾽ ἕκαστον θέρος ὀξὺ καὶ χολῶδες νόσημα παμπόλλων ἐτῶν ἤδη τελέως ἄνοσον ἐφύλαξα γυμνάζεσθαι κωλύσας· θερμὸς γὰρ ὢν καὶ ξηρὸς τὴν κρᾶσιν ὁμοίως τῷ Πριμιγένει γυμνασίοις τε συντόνοις ἐχρῆτο καὶ ἡλίῳ καὶ κόνει. συνέβαινεν οὖν αὐτῷ, τὴν μὲν θερμασίαν ἀμετρότερον αὐξάνεσθαι, πυκνοῦσθαι δὲ
368K τὸ δέρμα καὶ δυσδιάπνευστον γίνεσθαι· συνελθόντων δὲ εἰς ταὐτὸν ἀμφοτέρων, ἕτοιμον ἤδη πυρέξαι τῷ ταῦτα πάσχοντι σώματι.

more, I have seen some others who had a similar biting heat to Premigenes, who were not febrile after going one day without a bath, because in them the superfluity was evacuated through sweating.

For Premigenes, in addition to his natural constitution and kind of regimen, going without a bath for one day was   367K
the cause of the fever, because he spent most of the day in his house in which he continued to write and read on the Peripatetic theory he was involved with and in which he was second to none of his fellows. However, we know that, even in those who do not generate such a superfluity naturally, love of work and thought are causes of the genesis of this.[24] This is why another man, having a biting heat like him, happened not to become febrile immediately after one day without bathing, since he was not a deep thinker and his activities in the city involved a lot of walking around, buying and selling, and often wrangling, in which he was forced to perspire. And because of this, another man who was sick each summer, already having an acidic and bilious disease for many years, I finally kept free of disease when I directed him to exercise. For being hot and dry in *krasis* like Premigenes, he needed vigorous exercises in both sun and dust. Then what happened to him was that the heat was increased more immoderately, but the skin was thickened and slow to perspire. When   368K
both of these things come together to this point, they readily produce fever in the already affected body.

[24] The case of Premigenes, taken from this passage, is recounted in Samuel Tissot, *An Essay on Diseases Incident to Literary and Sedentary Persons* (1769), 59–61.

---

[50] τῆς ἡμέρας *add.* Ko

ὅπερ οὖν κατὰ τὴν ἰατρικὴν τέχνην ἅπασαν εἴωθα
λέγειν ἀεί, τοῦτο καὶ νῦν ἐρῶ· τάχιστα μὲν οὖν ὁ
λογισμὸς εὑρίσκει τὰ ζητούμενα, βεβαιοῖ δὲ τὴν πί-
στιν αὐτῶν ἡ πεῖρα. καίτοι γε θαυμασιώτερον, Ἱπποκράτους εἰρηκότος, ὡς ἐν τῷ θερμῷ φύσει ἄμεινον
ἀργεῖν ἐστιν ἢ γυμνάζεσθαι, παμπόλλους τῶν ἰατρῶν
ἔστιν ἰδεῖν μὴ διαγινώσκοντας, οἵτινές πότ᾽ εἰσὶν οἱ
τοιοῦτοι τῶν ἀνθρώπων, ἀλλ᾽ ἐξῆς ἅπασι γυμνάσια
προστάττοντας, ὥσπερ αὖ πάλιν ἄλλους οὐκ ὀλίγους
ἰατροὺς οὐδὲν ἐκ γυμνασίων πλέον εἰς ὑγείαν ἡγουμένους γίνεσθαι, καὶ πρὸς τούτοις γε τρίτους ἄλλους
τοῖς μὲν εἰθισμένοις γυμνάζεσθαι συγχωροῦντας,
εἴργοντας δὲ τοὺς ἀήθεις ἀπὸ τῶν γυμνασίων. ἅπαντες μὲν οὖν ἁμαρτάνουσιν, ἀλλ᾽ οἱ τρίτοι ῥηθέντες
ἧττον. οἱ πλείους γὰρ τῶν ἐθιζόντων ὁτιοῦν ἔθος οἰκεῖον αἱροῦνται τῇ φύσει διὰ τὸ βλαπτομένους πολλάκις ὑπὸ τῶν οὐκ οἰκείων ἀφίστασθαι. τινὲς δ᾽ ἤτοι
369K   νικηθέντες ὑφ᾽ ἡδονῆς ἢ δι᾽ ὑπερβάλλουσαν ἄνοιαν
οὐκ αἰσθανόμενοι τῶν τῆς βλάβης αἰτιῶν ἐμμένουσι
τοῖς κακοῖς ἔθεσιν. ἀλλ᾽ οὗτοι μὲν ὀλίγοι, πλείους δέ
εἰσιν οἱ μεθιστάμενοι.

διὸ καὶ τοὺς ἰατροὺς ἧττον ἁμαρτάνειν εἰκός ἐστιν,
ὅσοι διαφυλάττειν ἀξιοῦσιν ἅπαν ἔθος· ὅσοι μέντοι
νομίζουσι μηδὲν διαφέρειν εἰς ὑγείαν ἀγύμναστον ἢ
γυμνασθέντα προσαίρεσθαι σιτίον καὶ ὅσοι πάντας
ἀξιοῦσι γυμνάζεσθαι, πλείω βλάπτουσι τῶν προειρη-

Therefore, what I am always accustomed to say in relation to the medical art as a whole, I shall say again now. Reason is the quickest way to discover the things that are sought but experience gives us confidence in their belief. And indeed, what is more remarkable is that Hippocrates has said that, in a hot nature it is better to rest than to exercise,[25] and one sees very many doctors who do not know this. And not only are there such men as these, but they then order exercises for everybody, just as, on the contrary, there are not just a few other doctors who think nothing further is added to health from exercises. And, in addition to these, there is another, third group who, while agreeing with exercises for those who are accustomed to them, bar from exercises those who are not accustomed to them. They are all wrong, but the third group mentioned less so. For the majority of those accustomed to anything whatsoever choose the custom proper to their nature and stay away from what is not, having often been injured by customs that are not proper. Some however, either overcome by pleasure or due to extreme foolishness not perceiving the causes of the harm, continue on with the bad customs. But these are few; those who stay away from them are more.

369K

This is also why those doctors who think it worthwhile to maintain every custom are less likely to make a mistake. However, those who think it makes no difference to health whether a person takes food not having exercised or having exercised, and those who think it worthwhile for all to

---

[25] Hippocrates, *Epidemics* 6.4(13) has, "for a hot nature, cooling, water to drink, inactivity." *Hippocrates* VII, LCL 477, 250–51.

μένων. αὐτῶν δὲ τούτων ἀλλήλοις παραβαλλομένων
ἧττον βλάπτουσιν οἱ πρὸ τῶν σιτίων ἀεὶ γυμνάζεσθαι ἐθέλοντες. οἴονται δ' ἔνιοι καὶ Ἱπποκράτην τοῦτο
παραινεῖν "πόνους σιτίων ἡγεῖσθαι" κατὰ τοὺς Ἀφο-
ρισμοὺς εἰρηκότα, μὴ γινώσκοντες, ὡς περὶ τάξεως
νῦν ὁ λόγος αὐτῷ ἐστι γυμνασίων καὶ σιτίων, οὐ
τοῦτο λέγοντι, ὡς ἐπὶ πάντων ἀνθρώπων ἄμφω χρή-
σιμ' ἐστίν, ἀλλ' ὡς, ἐφ' ὧν συμφέρει γυμνάσια, προ-
ηγεῖσθαι χρὴ τῶν σιτίων. ὅτι δ' οὐ πᾶσι συμφέρει,
διά τε τῶν Ἐπιδημιῶν ἐδήλωσε σαφῶς, τὰς θερμὰς
φύσεις ἀξιῶν "ἐλιννύειν," ἔν τε τοῖς Ἀφορισμοῖς δυνάμει διὰ τοῦ δηλῶσαι καθόλου τὸ "τὰ ἐναντία τῶν
ἐναντίων ἰάματα" ὑπάρχειν.

370K   δύναται δέ ποτε καὶ διαφωνίας γενέσθαι φαντασία
κατὰ τὰ τοιαῦτα παραγγέλματα, τινῶν μὲν ἅπασαν
κίνησιν σύμμετρον τῷ κινουμένῳ σώματι γυμνάσιον
ὀνομαζόντων, ἐνίων δὲ τὴν σφοδροτέραν μόνην. ὥστε
κατὰ μὲν τὸ πρότερον σημαινόμενον ἅπαντας ἀνθρώπους δεῖσθαι γυμνασίων, καὶ τοῦτο εὔδηλον παντί·
κατὰ δὲ τὸ δεύτερον οὐκ ἀληθές. τῶν γὰρ ἐν ταῖς
εἱρκταῖς ὑγιαινόντων οὐκ ὀρθῶς μοι δοκοῦσιν ἔνιοι
μεμνῆσθαι· διαφθείρονται γὰρ οὗτοι πάντες ἐν τῷ
χρόνῳ, κωλυόμενοι παντάπασιν γυμνάζεσθαι[51] τὸ δὲ
πρὸς ὀλίγας ἡμέρας ἀντέχειν αὐτοὺς τῇ νοσώδει διαίτῃ θαυμαστὸν οὐδέν. εἴπερ οὖν ἅπασα κίνησις ὑπὸ

---

[51] post παντάπασιν: γυμνάζεσθαι Ko; ἀλείφεσθαί τε καὶ
λούεσθαι Ku

exercise, do more harm than those previously mentioned, whereas these same people who always wish to place exercise before food do less harm compared to others. Some also think Hippocrates recommended this: exertions should precede food—as he said in the *Aphorisms,* not recognizing that his discussion is now about the order of exercising and food. He doesn't say both are good for all people but that, in those whom exercise benefits, it is necessary for it to precede foods. That it doesn't benefit everyone, he showed clearly in the *Epidemics,* thinking it worthwhile for hot natures to take rest, and strongly in the *Aphorisms* through the general direction that "opposites are the cures of opposites."[26]

It is also possible sometimes for the appearance of discord to arise in relation to such precepts, with some calling every moderate movement by the moving body, exercise, and others only more violent movement. Consequently, according to the former signification, all people need exercises, and this is clear to everyone. However, according to the second signification, this is not true. Some seem to me not to have correctly called to mind those who are healthy in prisons. For these are altogether corrupted over time, being in every way prevented from exercising. That they would stand this disease-provoking regimen for a few days is not surprising. If, then, every movement is given

370K

[26] The three references to Hippocrates, all previously given are as follows: *Epidemics* 6.4(23), *Epidemics* 6.4(13). There is uncertainty about the reference to *Aphorisms.* Koch says this is not in *Aphorisms,* and I too have been unable to find it. For the statement, see *On Breaths* 1, *Hippocrates* II, LCL 148, 228–29.

τὴν τοῦ γυμνασίου προσηγορίαν ἄγοιτο, κἂν[52] ὁ περι-
πατήσας καὶ ὁ τριψάμενος καὶ ὁ λουσάμενος γεγυμνάσθαι λέγοιτο, ἐὰν συμμέτρως κινηθῇ τῇ παρ-
ούσῃ τοῦ σώματος καταστάσει· ἐὰν δὲ καὶ τούτων
εἴρξῃς τινά, νοσήσει πάντως. ὁρῶμεν γὰρ νῦν εἰργο-
μένους οὐ μόνον νοσοῦντας, ἀλλὰ καὶ ἀποθνῄσκοντας
ἐν ταῖς εἰρκταῖς, ἐπειδὰν κατακλεισθῶσι πλείονα χρό-
371K νον. εἰ δὲ τὰς σφοδρὰς κινήσεις μόνας ὀνομάζοι τις
γυμνάσια, γένοιτ᾽ ἂν οὕτως ἀληθὲς τὸ μὴ δεῖσθαι
πάντας ἀνθρώπους γυμνασίων.

εὐθέως μέντοι ὁ Πριμιγένης ἐκεῖνος οὐδεμιᾶς μὲν
ἐδεῖτο βιαίας κινήσεως, οὐκ ὀλίγον δὲ πρὸ τοῦ λουτροῦ περιεπάτει κατὰ τὰς πρὸ τοῦ βαλανείου στοάς.
ἀλλὰ καὶ τὸ τρίψασθαι μετ᾽ ἐλαίου καὶ τὸ μετὰ τὸ
λουτρὸν ἀπομάξασθαι τὸ ὕδωρ αὐτάρκεις εἰσὶ κινήσεις ἀνδρὶ τοιαύτης κράσεως. ὁ δὲ καὶ πρὸ τοῦ δεί-
πνου περιεπάτει τε καὶ διελέγετο τοῖς ἑταίροις, ὥσπερ
οὖν καὶ κατὰ τὴν οἰκίαν ἔωθεν. οὔκουν ἐστί τις ὑπὸ
παντελοῦς ἀργίας ὀνινάμενος. ἀλλ᾽ ὅταν "ἐλιννύειν"
ταῖς θερμαῖς φύσεσιν ὁ Ἱπποκράτης συμβουλεύῃ,
τῶν σφοδρῶν κινήσεων ἀπέχεσθαι κελεύει. κἀγὼ
πολλοὺς ὤνησα[53] νοσοῦντας οὐκ ὀλίγον ἐν ἀρχῇ φθινοπώρου διὰ τὰς ἐπιτηδευομένας αὐτοῖς κινήσεις ἐν
ὅλῳ τῷ θέρει· κωλύσας τοῦτο πράττειν ὑγιαίνοντας
ἀπέφηνα· καθάπερ ἄλλους εἶρξα τοῦ γυμνάζεσθαι

---

[52] κἂν Ko; καὶ Ku

the name of exercise, even one who walked around and
was massaged and bathed might be said to have exercised,
if he were moved moderately in the present state of the
body. If, however, you prevent someone from doing these
things also, he will be altogether diseased. Thus, we now
see those who are shut in not only being diseased, but also
dying in the prisons, whenever they are shut in for too long
a time. If, on the other hand, someone were to call only   371K
vigorous movements exercises, it would in this way be true
that not every person needs exercises.

For a start, the famous Premigenes needed no strong
movement, apart from walking around a lot on the porches
in front of the bath house before his bath. But also, to be
massaged with oil and for the water to be wiped off after
the bath are sufficient movements for a man of such a
krasis. He was also in the habit of walking around and
talking with his friends before dinner, just as he did in the
early morning at his house. He was not, therefore, a person who benefitted from complete idleness. But when
Hippocrates recommended rest for hot natures, he ordered them to abstain from violent movements.[27] I helped
many who were still quite diseased at the beginning of the
late autumn period due to the movements suitable for
them during the whole summer. I showed that preventing
this makes people healthy, just as when I prevented others
from exercising violently. Contrariwise, I kept others free

[27] This is taken to be a further general reference to Hippocrates, *Epidemics* 6.4(13).

---

[53] post ὤνησα: νοσοῦντας οὐκ ὀλίγον ἐν ἀρχῇ φθινοπώρου
Ko; ἔτι νοσοῦντας ἐν ἀρχῇ φθινοπώρου Ku

σφοδρῶς, ἑτέρους δ' ἔμπαλιν ἐξ ἀργίας εἰς γυμνάσια
καταστήσας⁵⁴ ἀνόσους διεφύλαξα, νοσοῦντας ἔμπρο-
372K σθεν οὐκ ὀλιγάκις. εἰσὶ δ' οὗτοι ψυχρότεροι τῶν εὐκράτων τε καὶ συμμέτρων, ἐναντίως διακείμενοι τοῖς
θερμοτέροις καὶ ξηροτέροις.

εἴρηται δὲ οὐ μόνον ἐν τῇ Περὶ κράσεων πραγματείᾳ τὰ γνωρίσματα τῶν κράσεων, ἀλλὰ κἀν τῇ
Τέχνῃ τῇ ἰατρικῇ· τοῦτο γάρ ἐστι τὸ ἐπίγραμμα τῶν
ἡμετέρων βιβλίων ἑνί. θαυμάσαι δ' ἔστι τῶν ἰατρῶν,
ὅσοι μηδὲ τὰ διὰ τῶν αἰσθήσεων ἐναργῶς φαινόμενα
γινώσκουσι. τίς γὰρ οὐχ ὁρᾷ τὴν διαφορὰν τῶν ἀνθρώπων ἐπὶ πλεῖστον ἐκτεταμένην, ὥσθ' ὑπὸ τῶν
αὐτῶν ἐπιτηδευμάτων καὶ σιτίων ἐνίους μὲν ὠφελεῖσθαι, βλάπτεσθαι δ' ἑτέρους, ὥσπερ γε κἀν ταῖς νόσοις; ὥσθ', ὅτι μὲν οὐ χρῄζουσι τῶν αὐτῶν ἅπαντες,
ἐναργῶς φαινόμενον ἐχρῆν αὐτοὺς εἰδέναι, γράφοντας
δὲ συγγράμματα διορίσασθαί τε καὶ διδάξαι, τίνες
ὑπὸ τίνων ὠφελοῦνται καὶ βλάπτονται. τίς οὖν ἡ αἰτία τοῦ παραλείπεσθαι τοὺς διορισμοὺς αὐτοῖς, καίτοι
γ' ὑφ' Ἱπποκράτους εἰρημένους; ἡ ἐπίτριπτος ἐπιθυμία τοῦ δόξαν ἔχειν ἐν ἀνθρώποις αἱρεσιάρχας ὀνο-
μασθέντας. διὰ ταῦτα μὲν οὖν ἐκεῖνοι μὲν ἐπεχείρη-
373K σαν οὐ τοῖς ἐναργῶς φαινομένοις ἀκολουθεῖν, ἀλλὰ
ταῖς δόξαις αἷς ὑπέθεντο. τῶν δὲ εἰς αὐτοὺς ἐκπεσόν-
των νεωτέρων ἐπείσθησαν οἱ πλεῖστοι τῷ μήτ' ἄλλο
τι μεμαθηκέναι βέλτιον ἀμαθεῖς τε εἶναι καὶ ἀγυμνά-

⁵⁴ καταστήσας Ko; μεταστήσας Ku

of disease by making them change from idleness to exercise, although previously they were diseased not infrequently. Those colder than those who are *eukratic* and balanced are oppositely affected to those who are hotter and drier.

    I have spoken about the signs of the *krasias* not only in the work *On Mixtures* but also in *The Art of Medicine,* for this is the title of one of my books.[28] Those doctors are to be wondered at who do not clearly recognize phenomena through the senses. For who does not see that the difference between people is extended to the greatest extent, so that, by the same practices and foods, some are benefited while others are harmed, just as in diseases? As a consequence, it quite obviously behooves them to know that not all people need the same things, and, when writing books, to distinguish and teach that some people are benefited by certain things and some are harmed. What then is the reason for the distinctions being omitted by them, when in fact they were stated by Hippocrates? It is the accursed desire to have the reputation that comes with being named leaders of schools. For such reasons, those men attempted not to follow the obvious phenomena, but laid down their own opinions. The majority of younger men who go astray are persuaded because they haven't learned anything better, and are ignorant and altogether unpracticed in dem-

372K

373K

----

[28] *Nat. Fac.,* II.1–204K (English trans., Brock, *On the Natural Faculties*—see particularly the opening sections). *Ars M.,* I.305–412K (French trans., Boudon, *Galien;* English trans., Johnston, *On the Constitution*—see particularly chaps. 6–13).

στους παντάπασιν ἀποδείξεων· ὡς,[55] εἴ γ' ἔποιτό τις
ἀποδείξει, ῥᾷστα διακρινεῖ τῶν τ' ἀληθῶν δογμάτων
τὰ ψευδῆ τῶν τε μετὰ διορισμοῦ γεγραμμένων τὰ μὴ
τοιαῦτα καταγνώσεταί τε πάντων ἀτεχνίας, ὅσοι χωρὶς διορισμῶν ἔγραψάν τι.

12. Πρόσχες οὖν μοι τὸν νοῦν ἄνωθεν ἐπερχομένῳ
διὰ βραχέων, ὁποίους εἶναι χρὴ τοὺς διορισμούς.
ὅσοις μὲν φύσει τὸ θερμὸν δακνῶδές ἐστιν, ὡς καπνώδη γεννᾶν περιττώματα, τούτοις λουτρὰ χρήσιμα
καὶ κινήσεις βραχεῖαί τε καὶ νωθραί, πολλαὶ δὲ καὶ
ὀξεῖαι βλαβερώτεραι. τούτοις οὖν οὐ μόνον ἅπαξ,
ἀλλὰ καὶ δὶς λούεσθαι τῆς ἡμέρας συμφέρει, καὶ
μάλιστα θέρους, ἐσθίειν τε τροφὰς εὐχύμους, μηδὲν
ἐχούσας δριμύ· πολέμιον[56] δ' αὐτοῖς καὶ τὸ ἡλιοῦσθαι
καὶ τὸ θυμοῦσθαι καὶ τὸ φροντίζειν πολλά. ταῖς δ'
374K ἐναντίαις ἢ κατὰ τούσδε φύσεσι (εἰσὶ δ' οὗτοι ψυχροὶ
καὶ ὑγροὶ τὴν κρᾶσιν) ἰσχυροτέρων τε κινήσεων
χρεία[57] καὶ διαίτης πλέον ἐχούσης τὸ λεπτόν· εἴρηται
δ', ἥτις ἐστὶν ἡ τοιαύτη, δι' ἑνὸς ὑπομνήματος· οὐ μὴν
οὐδ' ἐν ἡλίῳ γυμναζόμενοι βλάπτονται, καθάπερ οὐδ'
ἀλουτοῦντες.

μεγίστη μὲν οὖν διαφορὰ ταῖς εἰρημέναις ἐστὶ
φύσεσι πρὸς ἀλλήλας· ἐναντιωτάτη γὰρ ἡ ὑγρὰ καὶ
ψυχρὰ κρᾶσις τῇ θερμῇ καὶ ξηρᾷ· μεγίστη δὲ[58] τῇ
θερμῇ καὶ ὑγρᾷ πρὸς τὴν ψυχρὰν καὶ ξηράν. ἔστι δὲ

[55] ὡς add. Ko
[56] post πολέμιον: δ' αὐτοῖς καὶ Ko; δ' αὐτῶν ἐστι Ku

102

onstrations. So if someone were to follow demonstration, he would very easily distinguish the false doctrines from the true, and those things that have been written with discrimination from those that have not, and would pass judgment on the lack of skill of all those who wrote something without logical distinctions.

12. Give me your attention as I again go over briefly what sorts of distinctions are necessary. For those who are hot and biting in nature, so as to generate smoky superfluities, baths are beneficial, as are movements that are slow and sluggish, whereas many rapid movements are quite harmful. It is beneficial for them to bathe not only once but twice a day, and particularly in summer, and to eat *euchymous* foods having nothing acrid. To be exposed to the sun is adverse for them, and to become angry and think a lot is too. For those opposite natures in relation to these (these are cold and moist in *krasis*), there is a use for stronger movements and a diet having more of what is thin. I detailed what such a diet is in the one treatise.[29] When they exercise in the sun, they are not harmed, just as they are not harmed by not bathing.

There is, then, a very great difference from each other in the aforementioned natures. For the most opposite are the moist and cold *krasis* to the hot and dry, the greatest [opposition] being in the hot and moist in comparison to

374K

[29] *Vict. Att.*, CMG V.4.2 (English trans., Singer, *Galen: Selected Works*).

---

[57] *post* χρεία: ἔχεσθαι (Ku) *om.*   [58] *post* μεγίστη δὲ: τῇ θερμῇ καὶ ὑγρᾷ πρὸς τὴν ψυχρὰν καὶ ξηράν. Ko; ἡ ψυχρᾶς τῇ θερμῇ καὶ ὑγρᾶς πρὸς τὴν ξηράν. Ku

ἡ μὲν ψυχρὰ καὶ ξηρὰ τῇ τῶν γερόντων ὁμοία· διὸ
καὶ ταχέως γηρᾷ τὰ τοιαῦτα σώματα, καὶ λέλεκται
περὶ αὐτῶν αὐτάρκως ἔμπροσθεν. ἡ δ' ὑγρὰ καὶ
θερμὴ τοῖς ῥευματικοῖς εὐάλωτος πάθεσι. τό γε μὴν
κοινὸν ἐπὶ πασῶν τῶν δυσκράτων φύσεων ἄμεινόν
ἐστι κἀπὶ τῆσδε πράττειν. ἔστι δὲ τοῦτο κολάζειν μὲν
αὐτὴν διὰ τῶν ἐναντίων ἐπὶ πολλῇ σχολῇ, φυλάττειν
δὲ διὰ τῶν ὁμοίων, ἐπειδὰν ὑπὸ πλειόνων ἀσχολιῶν
εἴργηταί τις ἑαυτῷ σχολάζειν. περί γε μὴν τῶν ἐν
τοῖς τοιούτοις σώμασιν γυμνασίων ὧδ' ἔχει· πλεῖον
375K μὲν χρὴ πονεῖν διὰ τὴν ὑγρότητα τῶν σωμάτων, σύν-
τονα δ' οὐ χρὴ διὰ τὴν θερμότητα. προσέχειν δ' ἀκριβῶς, ὅταν ἐξ ἀργοτέρας διαίτης εἰς τὰ γυμνάσια
μεταβάλλωσι· παραχρῆμα γὰρ ἁλίσκονται ῥευματικοῖς νοσήμασιν, εἰ μὴ προκενωθέντες ἅπτονται τῶν
γυμνασίων. ὅσον γὰρ ἂν ἐν τῷ σώματι συνεστὸς ᾖ
καὶ παχὺ κατὰ τοὺς χυμοὺς ἢ καὶ μετρίως ψυχρόν,
αὐτίκα τοῦτο πνευματοῦταί τε καὶ χεῖται. ταῦτ' ἄρα
καὶ κατὰ τὸ ἔαρ αἱ τοιαῦται φύσεις μάλιστα τοῖς ὑπὸ
πλήθους γινομένοις ἁλίσκονται νοσήμασι, συνάγ-
χαις κυνάγχαις κατάρροις αἱμορροῖσιν αἱμορραγίαις[59] αἵματος πτύσεσι ποδάγραις ἀρθρίτισιν ὀφθαλμίαις περιπνευμονίαις πλευρίτισι τοῖς τ' ἄλλοις
ἅπασιν, ὧν τὸ γένος ἐστὶ φλεγμονή. διὸ καὶ φθάνειν
χρὴ κατὰ τὴν ἀρχὴν τοῦ ἦρος αἵματος ἀφαιρεῖν
αὐτῶν ἢ φλέβα τέμνοντα ἢ ἀποσχάζοντα τὰ σφυρά.

[59] post αἱμορραγίαις: αἵματος πτύσεσι ποδάγραις add. Ko

the cold and dry.[30] The cold and the dry is similar to that of old people. For this reason too, such bodies quickly age; I have said enough about these previously. The moist and hot [nature] is readily affected by the rheumatic affections. The common principle in the case of all the *dyskratic* natures is better to apply even in this one. This is to correct it through the opposites with much rest but to preserve it through similars, whenever, due to too much activity, someone is prevented from resting himself. On the matter of exercises in such bodies, the following holds: where there must be greater exertion due to the moisture   375K of the bodies, [the exertions] must not be intense due to the heat. Pay close attention whenever people change from a more leisurely regimen to exercises, for they are seized at once with rheumatic diseases, if they engage in the exercises without prior evacuation. For however much a thickness of the humors and a moderate cold coexist in the body, this will immediately be pneumatised and liquefied. Due to these things, then, in the spring such natures particularly are seized by the diseases occurring due to excess: synanche (sore throat), kynanche (laryngitis), catarrhs, hemorrhoids, hemorrhages, hemoptysis, gout, arthritides, ophthalmias, peripneumonias, pleuritides[31] and all the others of which the class is inflammatory. On this account also, it is necessary beforehand, at the beginning of spring, to withdraw blood from them, either cutting veins or scar-

[30] There is some uncertainty about this final clause.

[31] The list is slightly different in Kühn: synanche (sore throat), kynanche (laryngitis), catarrhs, hemorrhoids, hemorrhages, arthritides, ophthalmias, peripneumonias, and pleuritides.

μὴ βουλομένων δ᾽ οὕτως κενοῦσθαι, καθαρτέον ἐστὶ
φαρμάκῳ ποικίλῳ, δυναμένῳ καὶ ξανθὴν χολὴν ἕλκειν καὶ φλέγμα[60] καὶ τῶν ὀρρωδῶν τι περιττωμάτων.

ἃ δ᾽ ἐπὶ τῶν γυμνασίων εἴρηται, ταῦτα κἀπὶ τῶν
376K βαλανείων εἰρῆσθαι χρὴ δοκεῖν. καὶ γὰρ καὶ ταῦτα
σφαλερά, πρὶν κενωθῆναι, κενωθεῖσι δὲ ὠφέλιμα, καὶ
μάλισθ᾽ ὅσων λουτρῶν οὐκ ἔστι πότιμον ὕδωρ, ἀλλ᾽
ἔχει τινὰ διαφορητικὴν δύναμιν. ὅ γε μὴν οἶνος ὅτι
μὲν ὠφελιμώτατός ἐστι[61] ταῖς ξηραῖς καὶ ψυχραῖς φύσεσιν, εἴρηται πρόσθεν. εἰ δὲ ταῖς θερμαῖς ἁπάσαις
οὐκ οἰκεῖος, ἀλλ᾽ ἄμεινον ἐπ᾽ αὐτῶν ἐστιν "ὕδωρ πο-
τόν," ὡς ἐν ταῖς Ἐπιδημίαις γέγραπται, σκεπτέον
ἐφεξῆς. ἴσως γάρ τινι δόξει παντάπασιν ἄτοπον εἶναι, νεανίσκον ἤτοι στρατιωτικὸν ἢ ἀθλητικὸν ἤ τινα
τῶν σκαπτόντων ἢ θεριζόντων ἢ ἀροτριώντων ἢ ὅλως
ὁτιοῦν ἔργον ἰσχυρὸν ἐργαζομένων ἐφ᾽ ὕδατος μόνου
διαιτᾶσθαι, καὶ ταύτῃ σφάλλεσθαι τὸν Ἱπποκράτην
περὶ τῶν θερμῶν κράσεων ἁπλῶς ἀποφηνάμενον, ὡς
ὑδροποσίας δεομένων. ἐμοὶ δ᾽ οὐχ ἁπλῶς Ἱπποκράτης
εἰρηκέναι δοκεῖ τοῦτο, ἀλλὰ περὶ τῆς ἄκρως θερμῆς
φύσεως κατὰ δυσκρασίαν δηλονότι τοιαύτης οὔσης,
οὐ τῷ πλεῖστον ἔχειν τὸ ἔμφυτον θερμόν, ὅπερ[62] αὐ-
ξάνεσθαί φησιν ἐν τῷ τῶν ἀθλητῶν ἐπιτηδεύματι.

377K ἀλλ᾽ ὅ γε δύσκρατος θερμὸς οὔτ᾽ ἀθλητὴς οὔτε
στρατιώτης οὔτε τῶν κατὰ γεωργίαν ἢ πολιτείαν ἔργων ἀγαθὸς ἄν ποτε γένοιτο ἐργάτης· ἰσχυρῶν γάρ

ifying the ankles. If someone doesn't wish to be evacuated in this way, you must purge with a complex medication capable of drawing yellow bile, phlegm and some of the serous superfluities.

It seems to me necessary to say also about baths what I said about exercises, for these are dangerous before 376K evacuation, but beneficial once there has been evacuation; and this applies particularly to those baths that are not of potable water but which have a certain diaphoretic (discutient) potency. I said before that wine is very beneficial for dry and cold natures. One must consider next whether it is not suitable for all hot natures but that a drink of water is better in these cases, as has been written in the *Epidemics*.[32] Perhaps it will seem to someone altogether absurd that a young man should use water alone in his diet, if he is a soldier, or an athlete, or a ditch digger, reaper or plowman, or in general does any sort of strenuous work whatsoever. And Hippocrates was wrong on this when he declared simply about the hot *krasias*, that they need to drink water. To me it seems Hippocrates did not say this generally, but about the very hot nature, which is obviously so in relation to a *dyskrasia* and not because it has too much innate heat, which he himself says is increased in the practice of athletes.

But the *dyskratic* heat is not of an athlete or soldier, 377K nor of the activities of farming or civil matters, where it is good sometimes to work hard. These kinds of pursuits are

32 Hippocrates, *Epidemics* 6.4(13).

60 καὶ φλέγμα Ko; τι, καὶ φλέγματος Ku
61 ἐστι add. Ko   62 post ὅπερ: αὐτὸς Ku

ἐστιν ἀνδρῶν τὰ τοιαῦτα ἐπιτηδεύματα· τοιοῦτοι δ᾽
οὐκ ἂν εἶεν ἄνευ τοῦ συμμέτρως κεκρᾶσθαι· συμμέτρως δ᾽ αὐτῶν κεκραμένων, πλεῖστον ἂν εἴη τὸ ἔμφυτον θερμόν. τῇ μὲν οὖν τοιαύτῃ φύσει σύμμετρον
δηλονότι δοτέον ἐστὶ τὸν οἶνον, ὥσπερ καὶ αὐτὴ σύμ-
μετρός ἐστι, καὶ μέχρι τοσούτου γε σύμμετρον, ἄχρι-
περ ἂν ᾖ σύμμετρος. ἀνάγκη γάρ ἐστι καὶ ταύτῃ, οὐκ
ἐν γήρᾳ μόνον, ἀλλὰ καὶ τῷ τῆς παρακμῆς χρόνῳ τῆς
συμμέτρου γενέσθαι ψυχροτέρα. ἥτις δ᾽ ἄν, ὡς ἐν ὑγιεινῇ δυσκρασίᾳ, θερμοτάτη κρᾶσις ᾖ, ταύτῃ συμφέρει μηδ᾽ ὅλως οἶνον διδόναι.

τριῶν δ᾽ οὐσῶν κατὰ γένος διαφορῶν τῆς θερμῆς
κράσεως, μιᾶς μέν, καθ᾽ ἣν ἡ ἑτέρα τῶν ἀντιθέσεων,
ἡ κατὰ τὸ ξηρὸν καὶ τὸ ὑγρόν, εὔκρατός ἐστι, δευτέρας δέ, ᾖ σύνεστιν ἀμετροτέρα τοῦ συμμέτρου ξηρότης, καὶ τρίτης, καθ᾽ ἣν ἅμα τῷ θερμῷ πλεονάζει
τὸ ὑγρόν, ἐν ᾖ μὲν ἡ ἑτέρα τῶν ἀντιθέσεων εὔκρατός
378K ἐστιν, οὐκ ἂν γένοιτό ποτε ἄκρως ἄμετρον, ἐπειδὴ τῷ
τοιούτῳ ξηρότης ἐπιγίνεται διὰ ταχέων, αὕτη δὲ οὐχ
ὑπόκειται ξηρά· ἐν ᾖ δὲ ἐστι καὶ ξηρότης ἅμα τῇ
θερμότητι, δυνατὸν ἐν αὐτῇ γενέσθαι τὴν παρὰ φύσιν
θερμότητα πλείστην, ὡς ἐν ὑγιεινῇ δυσκρασίᾳ, μέχρι
πλείονος χρόνου, πλείστην δὲ ἐν ὀλίγῳ χρόνῳ, καὶ
τὴν μεθ᾽ ὑγρότητος ἄκρως ἄμετρον θερμασίαν, ὡς ἐν
ὑγιεινῇ δυσκρασίᾳ, δυνατὸν γενέσθαι.

τῇ μὲν οὖν πρώτως ῥηθείσῃ τοσοῦτον ἐπιτρέψομεν
ὑδατώδους οἴνου προσφέρειν, ὅσον ἀποκεχώρηκε τῆς
ἄκρας δυσκρασίας· τῶν δ᾽ ἄλλων οὐδετέρᾳ συγχωρή-

for strong men. Such people would not be without a balanced mixture (*krasis*). When people are mixed in a balanced fashion, the innate heat is greatest. For such a nature, what one must obviously give is wine that is moderate, just as the nature itself is moderate, and to the degree that it is moderate, so should the wine be moderate. For it is inevitable also that this nature, not only in old age, but also during the post-prime period will become colder than moderate. If there is, as in a healthy *dyskrasia,* a very hot *krasis,* it will be of benefit for this not to give wine at all.

However, there are three *differentiae* of the hot *krasis* according to class. The first is in relation to the other of the antitheses—that pertaining to dry and moist—is *eukratic.* The second is what is linked with a dryness that is more excessive than normal. And the third is in relation to that in which the moisture is increased together with the heat, in which the other of the antitheses is *eukratic* and would not on occasion become extremely excessive, since 378K in such a person, dryness supervenes quickly, but this does not underlie the dryness. In that in which there is dryness together with the heat, it is possible for the greatest heat contrary to nature to occur, as in a healthy *dyskrasia,* and for the greatest heat to occur in a short time, and the very excessive heat along with moisture, as in a healthy *dyskrasia.*

In the one mentioned first, we shall rely on administering watery wine to the extent that it departs from the complete *dyskrasia.* Of the others, we shall assent to nei-

σομεν, ἄκρᾳ γε οὔσῃ, καθότι δέδεικται. ταῖς γὰρ μὴ
τοιαύταις δυσκρασίαις οἶνον δώσομεν ὀλίγον καὶ
ὑδατώδη· τοιοῦτος δ' ἐστὶν ὁ λευκὸς μὲν τὴν χροιάν,
λεπτὸς δὲ τὴν σύστασιν· καθάπερ γε καὶ ταῖς ψυχραῖς δυσκρασίαις τοὺς θερμοτέρους τῶν οἴνων προ-
σοίσομεν. ὁ μὲν οὖν ἄκρως θερμὸς οἶνος ταῖς ἄκρως
ψυχραῖς δυσκρασίαις σύμφορος, ἐπὶ δὲ τῶν ἄλλων ὁ
ἀνάλογος. οὐ γὰρ μόνον ἁπλῶς χρὴ μεμνῆσθαι τοῦ
379K   "τὰ ἐναντία τῶν ἐναντίων ἰάματα" ὑπάρχειν, ἀλλὰ καὶ
τοῦ καθ' ἑκάστην ἐναντίωσιν ποσοῦ. καθάπερ οὖν ἐπὶ
τῶν φαρμάκων ἐδείξαμεν, ὡς οὐ μόνον χρὴ σκοπεῖν,
εἰ θερμὸν ἢ ψυχρόν ἐστιν, ἀλλὰ καὶ τίνος ἐξ αὐτῶν
τάξεως, οὕτω κἀπὶ τῶν οἴνων ποιητέον, οὐ τὸν ὑδα-
τώδη μόνον ἢ τὸν θερμὸν αἱρουμένους, ἀλλὰ καὶ τὸν
τὴν ἀναλογίαν[63] ἔχοντα οἰκείαν τῆς ἑαυτοῦ θερμότη-
τος ἢ ψυχρότητος ὡς πρὸς τὸ τῆς δυσκρασίας εἶδος.

ταῦτα δὲ περὶ τῶν τὴν μέσην ἐχόντων ἡλικίαν τοῦ
γήρως εἴρηται, γινωσκόντων ἡμῶν τὸ μὲν πρῶτον
αὐτοῦ μέρος, ὃ τὸ[64] τῶν ὠμογερόντων ὀνομάζουσι, δυναμένων ἔτι τὰ πολιτικὰ πράττειν, τὸ δὲ δεύτερον, ἐφ'
οὗ σύμφορον τὸ ὄνομα φέρουσιν, αὐτὸ τοῦτο εἶναι,
καθ' οὗ λέγουσιν —"ἐπὴν λούσαιτο φάγοι τε, εὑδέμε-
ναι μαλακῶς—" οὐ μὴν ἐπί γε τῆς τρίτης, ἐν ᾗ τὸν
γραμματικὸν ἔφην Τήλεφον ὄντα ἐν τῷ μηνὶ δὶς ἢ
τρὶς λούεσθαι· διὰ γὰρ τὴν ἀρρωστίαν τῆς δυνάμεως
οὐ φέρουσιν οὗτοι τὰ συνεχῆ λουτρά. πρόσεστι δ'

[63] ἀναλογίαν Ko; ἀντιλογίαν Ku   [64] τὸ add. Ko

ther, being the peak, as has been shown. For we shall not give a small amount of watery wine in such *dyskrasias*— such a wine is white in color and thin in consistency. Also, in the cold *dyskrasias,* we shall give the warmer wines. Thus, the very hot wine is good for the very cold *dyskrasias* and analogously in the case of the others. For it is not only necessary to simply be mindful of that opposites are cures of opposites,[33] but also of the amount in relation to each opposition. Thus, just as I showed in the case of medications, it is not only necessary to consider whether it is hot or cold, but also what the order (rank) of these is. And one must do the same in the case of the wines, not only choosing the watery or hot, but also the one having proper correspondence of its own heat or cold as regards the kind of *dyskrasia.* 379K

These things were said about those in the middle period of old age, if we recognize a first part of it, which people call active old age, when they are still able to carry out civic duties. The second part, however, for which they apply the name "old age" appropriately, is itself that of which they say: "when he has bathed and eaten let him sleep softly."[34] But I did not in fact speak about the third stage in which I said the grammarian, Telephus, is, bathing two or three times a month. Due to the weakness of the capacity, these people do not tolerate frequent bathing. In

[33] Hippocrates, *Aphorisms* 2.22, *Hippocrates* IV, LCL 150, 112–13; and *Breaths* 1, *Hippocrates* II, LCL 148, 228–29.

[34] This is the quote from Homer previously given—see note 6 above.

111

380K αὐτοῖς τὸ μηδ᾽ ἀθροίζειν δακνώδη περιττώματα διὰ
τὴν ψύξιν τῆς ἕξεως. ὀνομάζουσι δὲ τὸν κατὰ τὴν
ἡλικίαν ταύτην πέμπελον, ὡς οἱ ταῖς ἐτυμολογίαις
χαίροντές φασι, παρὰ τὸ ἐκπέμπεσθαι τὴν εἰς Ἅιδου[65] πομπήν.

65 *post* Ἅιδου: δηλονότι *add.* Ku

addition, in them the biting superfluities do not collect due to the coldness of the state. In this age, they call him aged (an old man), taking pleasure in etymological matters as he obviously goes forth along the road to Hades.   380K

Z

1. Ἑτέρας ὑποθέσεως ὑγιεινῶν διαιτημάτων[1] ἀρχὴν ἐν τῷδε ποιούμενος ἀναμνῆσαι δέομαι τὰ κεφάλαια τῶν ἔμπροσθεν εἰρημένων, ἀναγκαῖα τοῖς λεχθησομένοις ὄντα. πρῶτον μὲν οὖν ἐρρέθη, τί ποτ᾽ ἐστὶν ὑγεία· δεύτερον δὲ ἐπ᾽ αὐτῷ, καθόλου τίς ὑπογραφή· τρίτον δέ, πῶς ἂν φυλάττοιτο· καὶ πρὸς τούτοις, ⟨εἰ⟩, διότι μεταβάλλει διαπαντὸς τὰ τῶν ζῴων σώματα, κατὰ τοῦτο ἀναγκαῖον εἴη καὶ τὴν ὑγίαν αὐτῶν κινδυνεύουσαν φθαρῆναι τῆς ἐξ ἡμῶν ἐπικουρίας δεῖσθαι πρὸ

τοῦ μεγάλην γενέσθαι μεταβολήν, ὡς νοσεῖν ἤδη σαφῶς. ἐπικουρία δ᾽ ἐστὶν ἐξ ἐδεσμάτων τε καὶ πομάτων ἀναπληρούντων ὅσον ἀπερρύη τῆς τοῦ σώματος οὐσίας. ἑτέρα δ᾽ ἀβοήθητος ἐδείχθη μεταβολὴ κατὰ τὸν τῶν ἡλικιῶν λόγον γινομένη, ξηραινομένου τοῦ παντὸς ζῴου μετὰ τὴν πρώτην γένεσιν ἄχρι τῆς τελευτῆς ἐν τῷ μεταξὺ χρόνῳ παντί. τοὺς δ᾽ ἀγνοοῦντας τὴν ὑγιεινὴν δίαιταν εἰκός ἐστι θᾶττον ἢ κατὰ τὸν τῆς φύσεως λόγον ἀποθνήσκειν. ἐπεὶ δ᾽, ὡς ἔφην, ἀναγκαῖον μέν ἐστι τρέφεσθαι ζῷον ἅπαν γεννητόν, ἡ δ᾽ οὐσία τῶν ἐδεσμάτων οὔκ ἐστιν ὅλη τρόφιμος, καὶ διὰ τοῦτο τὸ περιττὸν αὐτῆς ὑπολείπεταί τι μοχθη-

BOOK VI

1. Since I am making, in this book, a beginning of another subject pertaining to hygienic regimes, I need to call to mind the chief points of those things I said previously, that are essential for those things that will be said. The first thing stated was what, at any time, health is; second, following this, its outline in general terms; and third, how it may be preserved. And in addition to these things, because the bodies of animals are continuously changing, and inevitably in relation to this, their health is in danger of being destroyed, we need to provide our help before so great a change occurs that there is already clearly disease. Help is through foods and drinks which replenish as much of the substance of the body as flows away. However, another irremediable change was shown to occur by reason of the stages of life, the whole animal becoming dry the whole time between birth and death. And it is likely that those who are ignorant of a healthy regimen will die sooner than they would in the natural course of things. As I said, it is necessary for every animal engendered to be nourished, while the substance of the foods is not all nutritious. Because of this, what is redundant of the nutriment remains as something bad, which people call specifically

---

[1] διαιτημάτων Ko; θεωρημάτων Ku

ρόν, ὃ καλοῦσιν ἰδίως "περίττωμα," παρεσκευάσθη τῇ
φύσει μόρια τοῦ σώματος εἰς τὴν διάκρισίν τε καὶ
κένωσιν αὐτοῦ. πολλῆς δὲ οὔσης ἐν τῇ τῶν σωμάτων
φύσει διαφορᾶς, εὔλογόν ἐστι καὶ τὴν ἑκάστου πρόνοιαν ὑγιεινὴν ἰδίαν ὑπάρχειν ἑκάστῳ.

πρῶτον οὖν ὑποθέμενοι τῷ λόγῳ τὸν ἄριστα κατεσκευασμένον ἄνθρωπον, ὅπως ἄν τις τοῦτον ὑγιαί-
383K νοντα διαφυλάττοι, διελθεῖν προὐθέμεθα. πολλῶν δ'
αἰτίων κατὰ πολλοὺς τρόπους διακοπτόντων τὴν
ὑγιεινὴν ἀγωγὴν ὑπεθέμεθα τὸν ἄριστα κατεσκευασμένον ἑαυτῷ σχολάζειν[2] ἐλεύθερον ὄντα πολιτικῆς
ἀσχολίας ἁπάσης. καὶ τοῦτον ὅπως ἄν τις ἀπὸ πρώτης ἡλικίας ἄχρι γήρως ἐσχάτου φυλάξειεν ὑγιαίνοντα διελθόντες ἐν τοῖς πρώτοις ὑπομνήμασι πέντε.
μεταβησόμεθα νῦν ἐπί τε τοὺς κατὰ περίστασίν τινα
πραγμάτων ἀδυνατοῦντας ἐν τοῖς προσήκουσι και-
ροῖς ἐσθίειν τε καὶ πίνειν καὶ γυμνάζεσθαι καὶ τοὺς
εὐθὺς ἐξ ἀρχῆς νοσώδη κατασκευὴν σώματος ἔχοντας. συντομώτερος δὲ τοῦ πρόσθεν ὁ περὶ τούτων ἐστὶ
λόγος, καίτοι γ', ὅσον ἐφ' ἑαυτῷ, μακρότερος ὢν διὰ
τὸ προειρῆσθαι τῶν πλείστων ὑλῶν τὰς δυνάμεις, αἷς
εἰς τὴν ὑγιεινὴν πρόνοιαν χρώμεθα. διά τε γὰρ τρίψεων καὶ λουτρῶν καὶ γυμνασίων ἐδεσμάτων τε καὶ
πομάτων θάλψεών τε καὶ ψύξεων ἀφροδισίων τε χρήσεως καὶ ἀποχῆς, εἴ τί τε τοιοῦτόν ἐστιν ἄλλο, τὰς
μεταβολὰς τοῦ σώματος ὁρῶμεν γινομένας, ὧν, ὡς
ἔφην, εἴρηνται πρόσθεν αἱ δυνάμεις.

384K 2. Αἱ μοχθηραὶ δὲ τῶν σωμάτων κατασκευαὶ διτταὶ

"superfluity," and parts of the body are prepared by Nature for the separation and evacuation of this. As there is a great difference in the natures of bodies, it is reasonable for there to be specific precautionary hygiene of each in each case.

First, then, in the discussion, assuming the person having the best constitution, what I proposed to go over was how we might preserve such a person in good health. Since there are many causes acting in many ways which cut across the healthy regimen, we assumed the person having the best constitution would have time for himself and be free of all civic duties. And we went over how this [hypothetical] person would maintain himself in a healthy state from the first stage of life up to the extreme of age in the first five books. Now we shall move on to those who, in relation to some particular state of affairs, are unable to eat, drink and exercise at the appropriate times, and those who, right from the start, had a diseased constitution of the body. The discussion about these [two cases] is briefer than the previous one, although in itself it is longer on account of what was previously said about the potencies of the very many materials we use in taking care of health. For we see the changes occurring in the body due to massages, baths, exercises, foods, drinks, heating and cooling agents, indulging in or abstaining from sexual activity, and whatever other such thing there may be. As I said, the powers of these things were discussed previously.

2. The bad (abnormal) constitutions of bodies are two- 384K

---

2 σχολάζειν Ko; φυλάττειν Ku

κατὰ γένος εἰσίν· ἔνιαι μὲν γὰρ αὐτῶν ὁμαλῶς, ἔνιαι
δὲ ἀνωμάλως ἔχουσι κεκραμένα τὰ στοιχειώδη τε καὶ
πρῶτα τοῦ σώματος μόρια, καλούμενα δ' ὑπ' Ἀριστοτέλους "ὁμοιομερῆ." λέγω δ' ὁμαλῶς μέν, ὅταν ἐπί
τινα δυσκρασίαν ἐκτραπῇ πάντα ὁμοίως τὰ τοῦ σώματος μόρια, ψυχρότερα τοῦ προσήκοντος ἢ θερμότερα γινόμενα ἢ ξηρότερα ἢ ὑγρότερα καὶ κατὰ
συζυγίαν[3] δὲ ἢ ψυχρότερα καὶ ὑγρότερα ἢ θερμότερα
καὶ ξηρότερα ἢ θερμότερα καὶ ὑγρότερα ἢ ψυχρότερα
καὶ ξηρότερα· ἀνωμάλως δέ, ὅταν μὴ πάντως συνεκτραπῇ, ἀλλὰ τινὰ μὲν ᾖ θερμότερα, τινὰ δὲ ψυχρότερα ἢ τὰ μὲν ξηρότερα, τὰ δὲ ὑγρότερα καὶ μὲν
δὴ καὶ κατὰ συζυγίαν τινὰ μὲν ὑγρότερα καὶ ψυχρότερα, τινὰ δὲ ξηρότερα καὶ θερμότερα γινόμενα
καὶ τινὰ μὲν ὑγρότερα καὶ θερμότερα, τινὰ δὲ ψυχρότερα καὶ ξηρότερα. καὶ μέντοι καὶ κατὰ τὴν τῶν
ὀργανικῶν μορίων σύνθεσιν ἔνια μὲν ὁμαλῶς, ἔνια δὲ
ἀνωμάλως σύγκειται.

πρῶτον μὲν οὖν ἐρῶ καὶ νῦν, ὁποῖαι κατασκευαὶ
σωμάτων εἰσὶν αἱ νοσωδέσταται, καθάπερ ἔμπροσθεν
ἐδήλωσα τὴν ὑγιεινήν. ἀλλ' αὕτη μὲν μία· τὸ γὰρ

---

[3] post κατὰ συζυγίαν, pre καὶ μέντοι: δὲ ἢ ψυχρότερα καὶ
ὑγρότερα ἢ θερμότερα καὶ ξηρότερα ἢ θερμότερα καὶ
ὑγρότερα ἢ ψυχρότερα καὶ ξηρότερα· ἀνωμάλως δέ, ὅταν μὴ
πάντως συνεκτραπῇ, ἀλλὰ τινὰ μὲν ᾖ θερμότερα, τινὰ δὲ
ψυχρότερα ἢ τὰ μὲν ξηρότερα, τὰ δὲ ὑγρότερα καὶ μὲν δὴ
καὶ κατὰ συζυγίαν τινὰ μὲν ὑγρότερα καὶ ψυχρότερα, τινὰ
δὲ ξηρότερα καὶ θερμότερα γινόμενα καὶ τινὰ μὲν ὑγρότερα

fold in class. Some have the elemental and primary parts of the body, which Aristotle calls *homoiomeres*,[1] mixed regularly (uniformly) while some have them mixed irregularly (nonuniformly). I say "regularly" when all the parts of the body are similarly deviated to some *dyskrasia*, being colder, hotter, drier or moister than is appropriate, or there is a conjunction of colder and moister, or hotter and drier, or hotter and moister or colder and drier. I say "irregularly" when they are not all deviated together, but some are hotter, some colder, some drier or some moister; and further, when there is a conjunction, with some being moister and colder, some drier and hotter, some moister and hotter and some colder and drier. Furthermore, in respect of the composition of the organic parts, some are compounded regularly and some irregularly.

What I shall now state first is what kinds of constitutions of bodies are the most morbid, just as I showed before with regard to the healthy constitution.[2] But the

---

[1] Aristotle used this term in relation to both inanimate (*Meteorologica* 388a10–90b20) and animate (*Parts of Animals* 648a6–55b7) things. For a summary of the use of the term up to Galen, see Johnston, *Galen: On Diseases and Symptoms*, 45, and also the General Introduction to the present work, vol. 1 (LCL 535), xliv–xlvi. [2] See Galen's *Opt. Const.*, IV.737–49K (English trans., R. J. Penella and T. S. Hall, *Bulletin of the History of Medicine* [1973], 47; and Singer, *Galen: Selected Works*).

καὶ θερμότερα, τινὰ δὲ ψυχρότερα καὶ ξηρότερα. Κο; τινὰ μὲν ἅμα θερμότερα τε καὶ ξηρότερα, τινὰ δὲ ὑγρότερα καὶ ψυχρότερα γιγνόμενα, καὶ τινὰ μὲν ὑγρότερα καὶ θερμότερα, τινὰ δὲ ψυχρότερα καὶ ξηρότερα. Κυ

ἄριστον ἐν παντὶ γένει πράγματος ἕν ἐστι, τὰ μο-
χθηρὰ δὲ δηλονότι πάμπολλα. διττὴ δ' ἐν αὐτοῖς
ἐστιν ἡ γενικὴ διαφορά, καθάπερ ἀρτίως εἶπον, ἐνίων
385K μὲν ἐν ὁμοίᾳ δυσκρασίᾳ πάντ' ἐχόντων τὰ μόρια, τινῶν δ' ἐν διαφερούσῃ. τῶν μὲν οὖν ὁμοίαν ἐχόντων
δυσκρασίαν εὔδηλον ὡς χείριστά ἐστι τὰ σφοδρὰς
ἔχοντα ταύτας καὶ μάλιστα τὰς ψυχρὰς ἅμα καὶ ξη-
ράς· τῶν δ' ἀνωμάλους ἐχόντων κατασκευὰς σωμάτων
οὐδ' ἀριθμῆσαι ῥᾴδιόν ἐστι τὰς ἐπαλλάξεις. ἀλλὰ
κἀκεῖνα δ' ὑπαλλάξειεν ἄν τις διττοῖς εἴδεσιν ἢ γένεσιν ἢ ὅπως ἄν τις ὀνομάζειν ἐθέλοι. νοσωδέστατα μὲν
γὰρ αὐτῶν ἐστιν, ὅσα ταῖς ἐναντίαις κράσεσι τὰ κυριώτατα τῶν μορίων ἔχει πλεονεκτούμενα, μετριώτερα
δ', ἐν οἷς οὕτω διάκειται τὰ μὴ κύρια. τεθέαμαι γάρ
τινας ἤδη ἔχοντας κοιλίαν μὲν ψυχράν, κεφαλὴν δὲ
θερμήν, ὥσπερ ἐνίους ἔμπαλιν ἐπὶ κεφαλῇ ψυχρᾷ
θερμὴν κοιλίαν. ἐθεασάμην δὲ καὶ κατὰ συμβεβηκὸς
οὐ πρώτως οὐδὲ κατὰ τὸν ἴδιον λόγον τῆς κράσεως
ἐκχολουμένην γαστέρα συνεχῶς οὐκ οὖσαν φύσει
θερμήν, ὥσπερ γε καὶ ψυχομένην ἑτέραν καίτοι οὐκ
οὖσαν[4] ψυχράν· οὕτω δὲ καὶ κεφαλὴν ἧπάρ τε καὶ
σπλῆνα καὶ ἄλλο μόριον ἀπολαῦον ἐνίοτε τῆς ἑτέρας
386K δυσκρασίας ἤτοι μηθὲν αὐτὸ φύσει βεβλαμμένον εἰς
τὴν προσήκουσαν κρᾶσιν ἢ διακείμενον ἐναντίως
αὐτῇ.

περὶ πρώτων οὖν ὁ λόγος ἔσται μοι τῶν ὁμαλὴν
τὴν δυσκρασίαν ἐχόντων, ἀπὸ τῶν θερμοτέρων ἀρ-
ξαμένῳ. λέλεκται δ' ἐν τοῖς Περὶ κράσεων ἀδύνατον

healthy constitution is one single thing, for the best in
every class of thing is one, whereas the bad are obviously
very many. The generic difference in them is twofold, as
I said just now, some having all the parts in a similar *dys-* 385K
krasia, and some in a different *dyskrasia*. Of those having
the parts in a similar *dyskrasia,* it is clear that the worst
are those having severe *dyskrasias,* and particularly the
simultaneously cold and dry. It is, however, not easy to
enumerate the varieties of the irregular constitutions of
bodies, although one might resolve those varieties into
two kinds or classes, or whatever one might wish to call
them. Thus, the most morbid of these are the ones in
which the most important parts are involved in opposing
krasias, whereas the more moderate are those in which
the nonimportant parts are so disposed. I have already
seen some with a cold abdomen but a hot head, just as I
have seen others in turn who have a hot abdomen in addition to a cold head. And I saw a stomach which was not
hot in nature incidentally and not primarily, nor by the
specific reason of the *krasis,* continuously charged with
bile, just as I saw another cooled, although not cold by
nature. In the same way too, I have sometimes seen a
head, liver and spleen, and some other part, having the
"benefit" of a different *dyskrasia,* although not itself being 386K
harmed in nature in respect of the prevailing *dyskrasia,* or
being in a state opposed to this.

My discussion will first be about those who have a
regular *dyskrasia,* beginning with those that are too hot. I
have said in the treatises *On Mixtures* that it is impossible

---

⁴ *post* οὖσαν: *add.* φύσει Ku

εἶναι διαμένειν ἐπὶ πολὺ δυσκρασίαν τινὰ μίαν. αὐτὴ
γὰρ ἑαυτῇ προσκτᾶταί τινα ἐξ ἀνάγκης ἑτέραν. διὸ
καὶ τοῖς πλείστοις τῶν ἰατρῶν τέτταρες ἔδοξαν εἶναι
μόναι δυσκρασίαι σύνθετοι·[5] τήν τε γὰρ θερμὴν ἐξ-
ικμάζουσαν ἀεὶ τὰς ὑγρότητας ἐπικτᾶσθαι ξηρότητα,
τήν τε ψυχρὰν ἅτε οὐκ ἐκδαπανῶσαν ὑποτρέφειν
ὑγρότητα· κατὰ ταὐτὰ δὲ καὶ τὴν ξηρὰν ἐν μὲν τῷ
χρόνῳ τῶν αὐξητικῶν ἡλικιῶν ἑαυτοῦ θερμότερον
ἀποφαίνειν τὸ ζῷον, ἐν δὲ τῷ τῶν παρακμαστικῶν[6]
ἀποξηραίνειν μὲν τὰ στερεὰ μόρια τοῦ σώματος,
ἀθροίζειν δὲ περιττωμάτων πλῆθος· ὥσπερ γε καὶ τὴν
ὑγράν, ὅταν ἅμα θερμότητι συμμέτρῳ συστῇ, γίνε-
σθαί ποτε κατ' ἀμφοτέρας τὰς ἀντιθέσεις εὔκρατον.
ἀμφοτέρας δ' ἀντιθέσεις λέγω δηλονότι τὴν κατὰ τὸ
387K θερμὸν καὶ τὸ ψυχρὸν καὶ τὴν κατὰ τὸ ξηρὸν καὶ τὸ
ὑγρόν. ἔσται δ' εὔκρατος ἡ τοιαύτη κατὰ τὸν τῆς
ἀκμῆς καιρόν, οὐχ ἥκιστα δὲ κἂν τῷ κατὰ τὴν παρ-
ακμήν, ὡς ἐν ἐκείνῳ πρέπει.

τὴν μὲν γὰρ ἀρίστην κρᾶσιν ἴσχει τὸ σῶμα κατὰ
τὴν τῶν μειρακίων ἡλικίαν, αἱ δ' ἄλλαι πᾶσαι χείρους
ταύτης εἰσίν, ὡς ἐδείχθη. καὶ μέντοι καὶ τοῦτ' ἀνα-
μεμνήσθω, τῶν εἰρημένων ἕν τι ὂν καὶ αὐτό· διττῶν
γὰρ οὐσῶν ἐνεργειῶν κατὰ τὰ ζῷα, τὰς μὲν φυσικὰς[7]
ἀρίστας οἱ παῖδες ἔχουσι, τὰς δὲ ψυχικὰς ἡ μετὰ τοὺς
παῖδας ἡλικία μέχρι τῆς παρακμαστικῆς. οὐκ ἔστι δὲ
ἐτῶν ἀριθμῷ περιορίσαι ταύτας, καθάπερ ἔνιοι πε-

for a *dyskrasia* to remain single for long; it inevitably adds another to itself.[3] And this is why it seemed to the majority of doctors that there are only the four compound *dyskrasias* [and not those that are simple],[4] for the heat always causes a deprivation of moisture and the acquisition besides of dryness, and the cold *dyskrasia,* insofar as it doesn't exhaust it, fosters the moisture. In this way too, dryness in the time of the growing years, makes the animal seem hotter than normal, whereas in the years after its prime, the solid parts of the body dry out, collecting an abundance of superfluities, Likewise also, when the moist *dyskrasia* coexists with moderate heat, sometimes *eukrasia* arises in relation to both the antitheses. When I say "both antitheses," I am clearly referring to hot and cold, 387K and dry and moist. There will be such a state of *eukrasia* in the time of the prime of life, and no less also in the period of decline after the prime, as is fitting for that.

The body maintains the best *krasis* during the period of adolescence—all the other stages of life are worse than this, as was shown. And indeed, of what has been said, let this one thing be remembered—of the two groups of functions existing in the organism, children have the best physical functions, while the ages following childhood up to the postprime period have the best psychical functions. But it is not possible to define these by the number of

---

3 *Mixt.*, 1.8 (I.558K).

4 The translation here follows the Kühn text.

---

5 *post* σύνθετοι: τὰς δὲ ἁπλᾶς οὐκ εἶναι (Ku) *om.*

6 *post* τῶν παρακμαστικῶν: ἀποξηραίνειν Ko; ἀποψύχειν· ξηραίνειν Ku   7 φυσικὰς Ko; σωματικὰς Ku

ποιήκασι, πλὴν ἢ κατὰ τὸ πλάτος. ἡβάσκειν οὖν ἄρ-
χονταί τινες ἅμα τῷ πληρῶσαι τὸ τεσσαρεσκαιδέκα-
τον ἔτος, ἔνιοι δὲ μετ᾽ ἐνιαυτὸν ἢ καὶ πλείονα χρόνον,
ἀρχήν τε τῆς παρακμῆς ἴσχουσιν ἔνιοι μὲν εὐθέως
μετὰ τὸ τριακοστὸν ἔτος, ἔνιοι δὲ μετὰ τὸ πέμπτον
καὶ τριακοστόν. τὴν μὲν οὖν ἰσχὺν ἅπαντες ἄνθρωποι
καθαιροῦνται μετὰ τὴν ἀκμαστικὴν ἡλικίαν, οὐ μὴν
τήν γε ὑγείαν ἀπολλύουσιν, ἀλλ᾽ ἧττον μὲν ἐπαινετὴν
388K   ἔχουσιν ἢ πρόσθεν, ἔχουσι δ᾽ οὐ μόνον ἄχρι γήρως
ἀρχῆς, ἀλλὰ καὶ κατ᾽ αὐτὸ τὸ γῆρας ὅλον, ὃ δὴ δοκεῖ
τισιν εἶναι νόσημα φυσικόν. ὁπόταν γὰρ μήτ᾽ ὀδυνῶνταί τι μήτε τινὰ τῶν ἐνεργειῶν, αἷς εἰς τὰς κατὰ τὸν
βίον πράξεις χρώμεθα, τελέως ἀπολέσωσιν ἢ μὴ
παντάπασιν ἄρρωστον ἔχωσιν, ὑγιαίνουσιν ὑγείαν
οἰκείαν γήρᾳ. μεμνῆσθαι γὰρ δεῖ που καὶ τῶν περὶ
τῆς ὑγείας ἀποδεδειγμένων, ὡς πάμπολυ τὸ πλάτος
αὐτῆς ἐστιν.

ἔστι δὲ καὶ τρίτη τις διάθεσις σώματος, ἣν οἱ περὶ
τὸν Ἡρόφιλον "οὐδετέραν" ὀνομάζουσι, τοῖς τε ἐκ πυ-
ρετῶν χαλεπῶν διασωθεῖσιν ὑπάρχουσα κατ᾽ αὐτὸν
τὸν τῆς ἀναλήψεως χρόνον καὶ τῇ τοῦ γήρως ἡλικίᾳ.
<καὶ> τὸ μὲν ἔξω νόσου τοῖς γέρουσιν εἶναι πάντως
ὑπάρξει, τὸ δὲ τὰς ἐνεργείας ἰσχυρὰς ἔχειν ὁμοίως
τοῖς ἀκμάζουσιν οὐχ ὑπάρχει. ἀλλ᾽ ἐάν, εἰς ὅσον γέροντι χρεία βλέπειν τε καὶ ἀκούειν καὶ βαδίζειν καὶ
τἆλλα πράττειν δύνασθαι, μηδὲν ἐλλίπῃ τὸ σῶμα, καὶ
τοῦτ᾽ ἄν τις εἰκότως ὀνομάζοι γέροντος ὑγείαν, ὅλον
τοῦτο λέγων ὑγείαν γέροντος, οὐχ ἁπλῶς ὑγείαν.

years, as some have done, other than in broad terms. Thus, some begin to mature on completion of the fourteenth year, some a year later, and some after a longer time. Some begin their decline immediately after the thirtieth year and some after the thirty-fifth year. Now all men suffer a reduction of strength after the prime of life, but not all lose health. They do have less laudable health than before, 388K but they do not have health only up to the beginning of old age; they may also have health throughout old age itself, which seems to some to be a physical disease, when they neither feel pain nor lose completely any of the functions we use in the affairs of life, nor are altogether weak, but are healthy in a way that is proper for old age. For it is necessary to be mindful to some degree at least of how very wide the range of those things demonstrated as pertaining to health actually is.

There is also a third condition of the body which the Herophileans call "neither" ("neutral").[5] This exists in the recovery phase of those recuperating from severe fevers, and at the time of old age, when those who are aging are altogether free of disease but do not have functions that are as strong as they were during the prime of life. But if an old person is able to see, hear and walk, and do the other things to the degree necessary for an old person, and has no bodily deficiency, one might reasonably call this health of an old person, meaning by this wholly the health

[5] See Galen's *Ars M.* 1–4 (I.305–18K), and Von Staden, *The Art of Medicine in Early Alexandria*, 89–108.

389K ἐκείνη μὲν γὰρ ἄμεμπτός ἐστι διὰ τὴν τῶν ἐνεργειῶν
ἀρετήν, μεμπτὴ δ' ἡ τῶν γερόντων· ἔχουσα γὰρ ἀπά-
σας τὰς ἐνεργείας οὐδεμίαν ἐρρωμένην ἔχει. καὶ δὴ
καὶ τὸ γηροκομικὸν ὀνομαζόμενον μέρος τῆς ἰατρικῆς
σκοπὸν ἔχει τὴν γεροντικὴν ὑγείαν εἰς ὅσον οἷόν τε
διαφυλάττειν. ὅσα δὲ σώματα νοσώδη κατασκευὴν
ἔσχηκεν εὐθὺς ἐν τῇ πρώτῃ γενέσει, ταῦτα τὴν ἀρχὴν
μὲν οὐδ' ἀφικνεῖταί ποτε εἰς γῆρας, εἰ δ' ἀφίκοιτο,
πάντως ἕν γέ τι νόσημα χρόνιον ἴσχει. προὐθέμεθα
δ' ἐν τῷ νῦν λόγῳ περὶ τῶν μοχθηρὰν ἐχόντων κατασκευὴν διελθεῖν σκοποῦντες, ὅπως ἄν τις αὐτὰ φυλάτ-
τοι κατὰ τὸ πλεῖστον ὑγιαίνοντα.

3. Τὴν ἀρχὴν οὖν ἀπὸ τῶν ὁμαλὴν ἐχόντων τὴν
δυσκρασίαν ἐν ἅπασι τοῖς μέρεσι τοῦ σώματος[8] ποι-
ησάμενοι λέγωμεν ἤδη περὶ τῆς θερμοτέρας κράσεως,
οὐ μὴν ἐν τῇ καθ' ὑγρότητα καὶ ξηρότητα συμμετρίᾳ
μεμπτῆς. εὐθὺς μὲν οὖν ἀπ' ἀρχῆς ἡ τοιαύτη φύσις
τοῦ σώματος ὑγιεινοτέρα φαίνεται τῆς κατ' ἀμφοτέρας τὰς ἀντιθέσεις δυσκράτου· λέγω δ' ἀμφοτέρας
390K τῆς τε κατὰ θερμότητα καὶ ψυχρότητα καὶ τῆς καθ'
ὑγρότητα καὶ ξηρότητα. καὶ τούς τε ὀδόντας φύσει
θᾶττον καὶ φθέγξεται διηρθρωμένην φωνὴν καὶ βαδι-
εῖται θᾶττον αὐξηθήσεταί τε κατὰ τὴν ἀναλογίαν ἑκάστοτε τῶν ἐτῶν. ἐπειδὰν δὲ τὴν τῶν μειρακίων ἡλικίαν
διεξέλθῃ, τὸ μεταξὺ πᾶν ἄχρι τῆς παρακμῆς σαφοῦς
πλέον ἤδη φανεῖται θερμόν, ὡς τοῖς ἀπὸ ξανθῆς χολῆς νοσήμασί τε καὶ συμπτώμασιν εὐάλωτον ὑπάρχειν. ἡ γάρ τοι πολλὴ θερμότης ἐκδαπανῶσα τὴν

of old age and not simply health. For the latter is faultless   389K
due to the excellence of the functions, whereas it is faulty
in old people, who may have all the functions but none
of them are strong. Furthermore, the so-called geriatric
component of medicine has as its objective the preservation of geriatric health as far as is possible. Those bodies
that have a diseased constitution right from the time of
birth, generally never reach old age in the first place, but
if they do, one chronic disease lasts throughout. In the
present book, I now intend to go over those who have
a bad constitution, considering how one might maintain
these in the greatest possible degree of health.

3. Therefore, making the start from those who have a
regular *dyskrasia* in all parts of the body, let me speak now
about the hotter *krasis* which is without fault in terms of
the balance involving moist and dry. Immediately, then,
right from the start, such a nature of the body seems
healthier than the *dyskrasia* involving both antitheses—I
refer to both hot and cold, and moist and dry. In this (hot-   390K
ter) nature, the teeth appear quicker, as does articulate
speech, [children] walk sooner, and there will be growth
in proportion each year. When the period of adolescence
has been gone through, in the whole intervening period
up to the beginning of decline (postprime stage), there
will clearly seem to be more heat, so there will be a ready
susceptibility to diseases and symptoms caused by yellow
bile. For certainly, the great heat, consuming the mois-

---

8 *post τοῦ σώματος*: ἐχόντων (Ku) *om.*

ὑγρότητα ξηροτέραν ἐργάζεται τὴν κρᾶσιν. οὔσης δὲ
τῆς τῶν ἀκμαζόντων ἡλικίας θερμῆς ἡ συζυγία τῶν
κράσεων αὐτῶν ἔσται θερμή τε ἅμα καὶ ξηρά· χολὴ
δὲ ἐν ταῖς τοιαύταις κράσεσιν ἡ ὠχρά τε καὶ ξανθὴ
πλεονάζει. τοὺς οὖν οὕτω πεφυκότας ἄχρι μὲν τῆς τῶν
μειρακίων ἡλικίας ὁμοίως χρὴ τοῖς ἄριστα πεφυκόσι
διαιτᾶν, ὑπὲρ ὧν εἴρηται κατὰ τὸν ἔμπροσθεν λόγον·
ἐπειδὰν δὲ τέλειον αὐτῶν ᾖ τὸ σῶμα, διασκέψασθαι
προσήκει, πότερον ὑπέρχεται τὸ περιττὸν τῆς χολῆς
αὐτοῖς ἅμα τοῖς διαχωρήμασιν ἢ πρὸς τὴν ἄνω κοιλίαν ὁρμᾷ. κάτω μὲν ὑπιόντος αὐτοῦ πρόδηλον, ὡς
391K οὐδὲν χρὴ περιεργάζεσθαι, πρὸς δὲ τὴν ἄνω γαστέρα
φερομένου, δι' ἐμέτων ἐκκενοῦν, μακρὰ χαίρειν εἰπόν-
τας ἐκείνοις τῶν φιλοσόφων, ὅσοι κωλύουσι μετὰ τὰ
γυμνάσια πρὸ τροφῆς ἀφ' ὕδατος χλιαροῦ ἐμεῖν.

οἴνῳ γὰρ οὐδ' ἐγὼ συμβουλεύω τηνικαῦτα χρῆσθαι, πλὴν εἰ δυσχερῶς ἐμοῖεν ἀφ' ὕδατος. εἰσὶ γὰρ
καὶ τοιαῦται φύσεις σωμάτων, αἷς συγχωρητέον οἴνου γλυκέος προσφέρεσθαι προπίνοντας αὐτοῦ τὸ
ὕδωρ. ἔτι δὲ καὶ μᾶλλον, ὅταν ἡ κρᾶσις ἐξ ἀρχῆς ᾖ
θερμοτέρα τε καὶ ξηροτέρα, πρὸς τοὺς ἐμέτους ἔρχε-
σθαι· καὶ γὰρ καὶ μᾶλλόν εἰσιν οὗτοι χολώδεις ἐπὶ
τῆς ἀκμαστικῆς ἡλικίας. καὶ δὴ καὶ γυμνάζεσθαι
βέλτιον αὐτούς, οὐκ ὀξὺ καὶ σύντονον γυμνάσιον,
ἀλλὰ σχολαιότερόν τε καὶ μαλακώτερον· εἰσὶ γὰρ δὴ
καὶ ἰσχνότεροι πάντως οἱ τοιοῦτοι. συμπεφώνηται δὲ
καὶ τοῖς γυμνασταῖς ἅπασι, λεπτύνειν μὲν τὰ ὀξέα
γυμνάσια, σαρκοῦν δὲ τὰ βραδέα. τινὲς δὲ τῶν σφό-

ture, makes the *krasis* drier. And since the period of life of those in their prime is hot, the conjunction of these same *krasias* will be hot and dry. In such *krasias,* bile, whether pale yellow or yellow, is in excess. Therefore, until the time of adolescence, such natures must follow a regimen similar to those natures that are the best—I spoke about these in the previous discussion. However, when their bodies are fully developed, it is appropriate to consider whether the excess of bile is eliminated in them along with the feces or is urged upward to the belly. If it clearly passes downward, it is not necessary to do anything, but if it is carried upward to the stomach, it must be evacuated by vomiting, largely dismissing from one's mind the statements of those philosophers who forbid vomiting induced by lukewarm water after exercises and before food.

391K

Under the circumstances, I do not recommend the use of wine unless it proves difficult to induce vomiting with water. There are also natures of bodies such that one must agree to give them sweet wine, if they drink water before it. Even more, when the *krasis* is hotter and drier from the beginning, have recourse to vomiting, for these [natures] are more bilious in the prime of life. Furthermore, it is better for them to exercise—not rapid and strenuous exercise but slower and more gentle exercise, for such people are altogether thinner. And it is accepted by all gymnastic trainers that rapid exercises thin, while slow exercises enflesh. Some of those who are very hot in terms

δρα θερμῶν τὴν κρᾶσιν οὐδ᾽ ὅλως χρῄζουσι γυμνασίων, ἀλλ᾽ ἀρκεῖ περίπατός τε καὶ λουτρὸν αὐτοῖς,
392K ἀνατριψαμένοις ἐλαίῳ μαλακαῖς τρίψεσι. δριμὺ γὰρ
καὶ δακνῶδες καὶ θερμὸν τὸ διαπνεόμενον ἀπ᾽ αὐτῶν
ἐστιν, οὐκ ἀτμῶδες οὐδ᾽ ἡδὺ καὶ ἄδηκτον. οὗτοι δὲ καὶ
τοῖς μετὰ τροφὴν λουτροῖς χαίρουσι. καί τινες αὐτῶν
ἄπιστον ὅπως ἐπαχύνθησαν ἐπὶ τῇ τοιαύτῃ διαίτῃ.
παλαιὸς δὲ οἶνος ἐναντίος αὐτοῖς ἐστι, οἰκεῖος δὲ ὁ
λευκὸς καὶ λεπτός. ἐπιβλέπειν δὲ ἐπὶ πάντων τῶν
μετὰ τροφὴν λουομένων, μή πως κατὰ τὸ δεξιὸν ὑποχόνδριον, ἔνθα κεῖται τὸ ἧπαρ, αἰσθάνονταί τινος ἀλ-
γήματος ἢ βάρους ἢ τάσεως. αἱ γὰρ τοιαῦται κατασκευαὶ τῶν σωμάτων ἡπατικαῖς ἁλίσκονται νόσοις,
ἐὰν ἐδηδοκότες λούωνται. κἂν αἴσθωνται δέ ποτε τοιούτου τινὸς συμπτώματος, αὐτίκα διδόναι τι τῶν ἐκφραττόντων τὸ ἧπαρ ἀπέχειν τε καὶ τῶν παχυχύμων
ἐδεσμάτων καὶ μάλισθ᾽ ὅσα γλίσχρα. λέλεκται δὲ
περὶ αὐτῶν ἐπὶ πλέον ἔν τε τοῖς τρισὶν ὑπομνήμασιν,
ἃ Περὶ τῶν ἐν ταῖς τροφαῖς δυνάμεων ἔγραψα, καὶ τῷ[9]
Περὶ τῆς εὐχύμου τε καὶ κακοχύμου διαίτης ἔτι τε κἂν
τῷ Περὶ τῆς λεπτυνούσης.

393K   καὶ γὰρ οὖν κεχρῆσθαι ἐν τοῖς ἀλγήμασι τοῖς καθ᾽
ἧπαρ ἀναγκαῖόν ἐστι λεπτυνούσῃ διαίτῃ, μέχριπερ
ἂν αὐτοῖς ἀνώδυνόν τε καὶ κοῦφον γένηται πᾶν τὸ
δεξιὸν ὑποχόνδριον. ἀγαθὸν δὲ καὶ τῆς κόμης τοῦ

---

[9] post καὶ τῷ: Περὶ τῆς εὐχύμου τε καὶ κακοχύμου διαίτης
Ko; περὶ τῆς εὐχυμίας τε καὶ κακοχυμίας Ku

of *krasis* do not need exercises at all; walking about and bathing is enough for them, after rubbing down and soft 392K massage with oil. For the exhalation from them is acrid, biting and hot, and not vaporous, sweet and nonbiting. These people also take pleasure in baths after food, and it is incredible how fat some of them have become with such a regimen. Old wine is contraindicated for them whereas thin, white wine is suitable. One must keep an eye on all those who bathe after food in case they somehow feel pain, heaviness or tension in the right hypochondrium, for the liver lies there. Such bodily constitutions are seized by hepatic diseases, if they bathe after eating. And if they sense at any time such a symptom, immediately give one of those agents that unblock the liver and make them abstain from foods with thick juices—especially those that are viscous. I have spoken about these at greater length in three treatises I wrote: *The Powers of Foods, On the Good and Bad Humors of Nutriments,* and *On the Thinning Diet.*[6]

It is also necessary to use the thinning diet in pains 393K related to the liver, until the whole right hypochondrium becomes pain free and light in these patients. It is also good to administer the infusion of leaf of absinth and that

[6] The three works referred to are *Alim. Fac.*, VI.453–748K; *Bon. Mal. Suc.*, VI.749–815K, CMG V.4.2 (Italian translation); *Vict. Att.*, CMG V.4.2. All three are referred to several times in the *Hygiene,* usually as a group, and are included in the same CMG volume as the *Hygiene.* There are English translations of the first by Grant, *Galen on Food and Diet,* and by Powell, *On the Properties of Foodstuffs,* and of the third by Singer, *Galen: Selected Works.*

ἀψινθίου τὸ ἀπόβρεγμα προσφέρεσθαι καὶ τὸ διὰ
ταύτης τε καὶ ἀνίσου καὶ πικρῶν ἀμυγδάλων συν-
τιθέμενον, ὃ δι᾿ ὀξυμέλιτος ἄμεινον πίνειν ἐν τῷ μέσῳ
χρόνῳ τῆς τε ἐκ τῶν ὕπνων ἀναστάσεως καὶ τοῦ λουτροῦ. προκατειργασμένων τε γὰρ τῶν ἀναδεδομένων
εἰς ἧπαρ ἐκ τῆς γαστρὸς ἄμεινον προσφέρεσθαι τὰ
τοιαῦτα, τῇ τε ἐξ αὐτῶν ἐνεργείᾳ χρόνον τινὰ δίδο-
σθαι πρὸ τῆς τῶν σιτίων προσφορᾶς. ἀγαθὸν δὲ καὶ
τὸ διὰ τῆς καλαμίνθου φάρμακον, οὗ τὴν σύνθεσιν[10]
ἐν τῷ τετάρτῳ τῶνδε τῶν ὑπομνημάτων ἔγραψα. ἀλλ᾿
ἐπί γε τῶν χολωδῶν κράσεων προσήκει φυλάττεσθαι
τὴν συνεχῆ τοῦδε χρῆσιν. καὶ μέντοι καὶ ὁπότε τις
αὐτῷ χρῆται τῆς καθ᾿ ἧπαρ ἐκφράξεως ἕνεκεν, ἄμεινόν ἐστι δι᾿ ὀξυμέλιτος πίνειν. τὸ δὲ τῶν τροφῶν εἶδος, εἰ μὲν ὑπαλλάττειν ἐπὶ τὸ βέλτιον ἐθέλοις τὴν
κρᾶσιν τῶν οὕτως ἐχόντων, ἐναντίον ἔστω τῇ δυσκρασίᾳ. κατὰ βραχὺ δὲ γίνεται τοῦτο χωρὶς βλάβης
ἐπιστατοῦντος μὲν ἰατροῦ, σχολὴν δ᾿ ἄγοντος αὐτῷ
τοῦ βοηθουμένου τοσαύτην, ὡς ἅπαντα ποιεῖν ἐν τῷ
προσήκοντι καιρῷ.

394K

πράττοντι δὲ ἀνθρώπῳ τὰ πολιτικὰ καὶ πολλαῖς
ἀσχολίαις δουλεύοντι κάλλιόν ἐστι μηδ᾿ ἐπιχειρεῖν
ὑπαλλάττειν τὴν κρᾶσιν, ἀλλὰ τὰς οἰκείας αὐτῇ τροφὰς δοτέον. οἰκεῖαι δ᾿ εἰσὶν αἱ μὲν ὑγραὶ ταῖς ὑγραῖς,
αἱ δὲ ξηραὶ ταῖς ξηραῖς. ἡ γάρ τοι θρέψις ὁμοιουμέ-
νου γίνεται τοῦ θρέψοντος ἐδέσματος τῷ τρεφομένῳ
σώματι· θᾶττον δὲ ὁμοιοῦται τὰ μὲν ξηρὰ[11] τοῖς ξηροῖς, τὰ δ᾿ ὑγρὰ τοῖς ὑγροῖς. ἐπὶ δὲ τῶν ὁμαλὴν τὴν

compounded with this, anise and bitter almonds. It is better to drink *oxymel* during the time between rising from sleep and bathing. When prepared beforehand for distribution to the liver from the stomach, it is better, in administering such things, to allow some time for their function before the provision of food. And the medication made from catmint is good—I wrote about the composition of this in the fourth book of these treatises.[7] But in the bilious *krasias*, it is appropriate to guard against the continuous use of this. And indeed, if someone does use this for a blockage of the liver, it is better to drink it with *oxymel*. The kind of nutriments, if you wish to change for the better the *krasis* of those so disposed, should be opposite to the *dyskrasia*. This occurs gradually and without 394K harm if the doctor in charge and the person being treated by him have the time available to do everything at the appropriate juncture.

For a man involved in civic affairs and subject to many pressing matters, it is better not to attempt to change the *krasis,* but it is necessary to give this *krasis* the proper nutriments. Moist foods are right for the moist *dyskrasias* and dry foods for the dry. For nourishment certainly occurs when the nourishing food is assimilated into the body being nourished. Dry foods are assimilated quicker by those who are dry and moist foods by those who are moist.

---

[7] See VI.281–82K.

---

[10] *post* τὴν σύνθεσιν: ἐν τῷ τετάρτῳ τῶνδε Ko; ἐν τῷ δ' Ku
[11] *post* τὰ μὲν ξηρὰ: τοῖς ξηροῖς, τὰ δ' ὑγρὰ τοῖς ὑγροῖς. Ko; ταῖς ξηραῖς, τὰ δ' ὑγρὰ ταῖς ὑγραῖς. Ku

κρᾶσιν ἐχόντων, ὅσωπερ ἂν ἥδιον ᾖ τὸ βρῶμα, τοσ-
ούτω τροφιμώτερον γίνεται· τῶν δ' ἀνώμαλον τὴν
κρᾶσιν ἐχόντων σώματος, ὡς ἄλλην μὲν εἶναι τὴν τοῦ
ἥπατος, ἄλλην δὲ τὴν τῆς γαστρὸς ἤ τινος τῶν καθ'
ἧπαρ, ἕτερον μέν ἐστι τὸ κατὰ τὴν προσφορὰν ἥδιον,
ἕτερον δὲ τὸ καθ' ἕκαστον μόριον οἰκεῖον. ἐπεὶ δὲ τὸ
καθ' ἕκαστον οἰκεῖον ἐδείχθη διττόν, ἕτερον μὲν τὸ
395K   κατὰ τὰς ἁπλᾶς ποιότητας, ἕτερον δὲ τὸ κατὰ τὴν
ὅλην οὐσίαν, ἡ μὲν οὖν κατὰ τὰς ἁπλᾶς ποιότητας
οἰκειότης¹² ὁποία τίς ἐστιν ὀλίγῳ πρόσθεν εἴρηται
μεταβάλλειν τε βουλομένοις αὐτὴν ἐπὶ σχολῆς ὑπ-
ηρετεῖν τε ἀναγκαζομένοις ἐν ἀσχολίᾳ, ἡ δὲ κατὰ τὴν
ὅλην οὐσίαν τῇ πείρᾳ μόνῃ γινώσκεται. μεγίστην οὖν
δύναμιν εἰς τὴν τῶν ἐσθιομένων τε καὶ πινομένων
κατεργασίαν οὐ κατὰ τὴν γαστέρα μόνον, ἀλλὰ καὶ
κατὰ σύμπαν τὸ σῶμα¹³ καὶ τὸ ἧπαρ ἡ κατὰ τὴν ὅλην
οὐσίαν οἰκειότης ἔχει, δι' ἣν αἱ τροφαὶ τοῖς ζῴοις
ὑπηλλαγμέναι τε καὶ πολὺ διαφέρουσαι ταῖς ἰδέαις
εἰσὶν οὐδὲν ὅμοιον ἐχούσης τῆς ἐκ τῶν ἀχύρων τε καὶ
πόας τῇ ἐξ ὀστῶν τε καὶ σαρκῶν ἢ τῆς ἐξ ἄρτων τῇ
ἐκ κωνείου τε καὶ ἐλλεβόρου· καὶ γὰρ ταῦτα ζῴοις
τισίν εἰσι τροφαί. γίνονται δὲ καὶ κατὰ τὰς ποιότητας
οἰκειότητες, εἰ μὲν φυλάττειν ἐθέλοις τὰς κράσεις,
ταῖς μὲν ὑγραῖς τῶν ὑγραινόντων ἐδεσμάτων ἁρμοτ-
τόντων, ταῖς δὲ ξηραῖς τῶν ξηραινόντων, εἰ δὲ ὑπαλλάττειν, τῶν ἐναντίων.

396K   ἐπὶ δὲ τῆς κατὰ τὸ θερμόν τε καὶ ψυχρὸν εἰς δυσκρασίαν ἐκτροπῆς ἡ τῶν ἐναντίων προσφορὰ δια-

In those who have a regular (uniform) *krasis,* the more pleasant the food is, the more nourishing it is. In those who have an irregular (nonuniform) *krasis* of the body, so the *krasis* of the liver is different to that of the stomach, or another of the structures in relation to the liver, the pleasantness of what is offered differs in suitability in relation to each part. And since the suitability in relation to each part was shown to be twofold, in one case in relation to the simple qualities and in the other in relation to the 395K whole substance, what kind of thing is suitable in respect of the simple qualities for those wishing to change the *krasis* in a time of leisure and for those compelled to serve in an occupation was stated a little earlier. However, what kind of thing is suitable in respect of the whole substance is learned by experience alone. What is suitable in respect of the whole substance has, then, the greatest power for the working up of foods and drinks—and not only in the stomach alone, but also in the whole body and the liver. Due to this, the nutriments for animals are changed and differ greatly in their forms; that from brans and grasses is nothing like that from bones and flesh, while that from breads is nothing like that from hemlock and hellebore, and yet these are foods for some animals. Suitabilities are also related to qualities, if you wish to preserve the *krasias,* since moistening foods are suitable for moist *krasias* and drying foods for dry *krasias.* If, however, you wish to change the *krasias,* opposites are suitable.

In the case of the deviation of hot and cold to a *dyskra-* 396K *sia,* the exhibition of opposites is always appropriate. Such

---

12 οἰκειότης Ko; οἰκειοτάτη Ku
13 *post* τὸ σῶμα: καὶ τὸ ἧπαρ *add.* Ko

135

παντὸς ἁρμόττει. δραστικώταται γάρ εἰσι καὶ ὡς ἂν
εἴποι τις δυναστικώτεραι τῶν καθ᾽ ὑγρότητα καὶ ξηρότητα δυσκρασιῶν αἱ τοιαῦται· ῥᾳδίως γὰρ ὑπὸ τῶν
ὁμοίων τροφῶν εἰς νοσώδη διάθεσιν ἀφικνοῦνται. ξηρότερον δ᾽ ἐργασάμενος ὁτιοῦν τῶν μορίων οὐδέν τι
βλάψεις σαφές, ὥσπερ οὐδ᾽ εἰ φυλάττοις ὑγρότερον.
ἐναργέστερον δέ σοι τεκμήριον αἱ ἡλικίαι γινέσθωσαν, αἱ μὲν ἀπὸ τῆς πρώτης γενέσεως ἄχρι τῆς τῶν
μειρακίων ὑγροτάτην ἔχουσαι τὴν σάρκα, ξηρὰν δὲ
ἱκανῶς αἱ πρεσβύτεραι. μεμνῆσθαι δ᾽ ἐν τῷ τοιούτῳ
λόγῳ χρὴ τῶν εἰρημένων ἐν τοῖς Περὶ κράσεων, ὅπως
μή τις ὑπολάβῃ τοὺς ὑπὸ περιττῶν ὑγρῶν βαρυνομένους[14] ὑγρὰν ἔχειν τὴν κρᾶσιν, ὃ καὶ τοὺς οἰηθέντας
ὑγρὸν καὶ ψυχρὸν εἶναι τὸ γῆρας ἐξηπάτησεν. οὐ γὰρ
αὐτὰ τὰ μόρια τοῖς γέρουσιν ὑγρότερα ταῖς κράσεσίν
ἐστιν, ἀλλ᾽ αἱ μεταξὺ τῶν σωμάτων χῶραι τῆς περιτ-
τῆς ὑγρότητος ἐμπίπλανται γερόντων τε καὶ νοσούντων νόσους ὑγράς.

397K   ὁποῖον δέ τι τὸ ξηρὸν νόσημά ἐστιν, ἐπιστημονι-
κὴν γνῶσιν ἕξεις τὸ γεγραμμένον μοι Περὶ μαρασμοῦ
βιβλίον ἐπιμελῶς ἀναγνούς, οὐχ ἅπαξ ἢ δὶς οὐδὲ
παρατρέχων, ἀλλ᾽ ἐφιστάμενός τε καὶ προσέχων τὸν
νοῦν ἑκάστῳ τῶν λεγομένων. προγεγυμνάσθαι δὲ χρὴ
τὸν μέλλοντα καλῶς ἀναλέξασθαι ταῦτα τῷ δευτέρῳ
Περὶ κράσεων ὑπομνήματι, καθ᾽ ὃ δέδεικται τοῦ γέροντος ἡ κρᾶσις εἶναι ξηρά. τὸ δὲ προκείμενον ἐν τῷ

----

[14] post βαρυνομένους: ὑγρὰν ἔχειν Ko; ὑγροὺς γίνεσθαι Ku

dyskrasias are very active and, one might say, more potent than those relating to moistness and dryness, and come more easily to a disease condition through similar nutriments. If you make any of the parts whatsoever drier, you will clearly do no harm, just as you will not if you keep a part more moist. The stages of life should be quite clear evidence for you; those from birth to adolescence have flesh that is very moist, whereas the older stages are very dry. And we must bear in mind what was said in the discussion such as this in the work *On Mixtures,* so someone does not assume that those who are burdened by moist superfluities have a moist *krasis*—something which also deceived those who thought old age is moist and cold. For it is not the parts themselves that are more moist in the *krasias* in those who are old, but the spaces between bodies are filled with excessive moisture in those who are old and those sick with moist diseases.

You will have a scientific knowledge as to which kind 397K of disease is a dry disease, if you read carefully the book I wrote, *On Marasmus,*[8] and not once or twice, or in a cursory fashion, but applying and directing your attention to each of the things said. However, it is necessary for someone who intends to pick these things up well to be practiced beforehand in the second book of my work, *On Mixtures,*[9] in which it was shown that the *krasis* of old age is dry. But the body before us in the present discussion is

---

[8] *Marc.,* VII.666–704K (English trans., T. C. Theoharides, "Galen on Marasmus," *Journal of the History of Medicine and Allied Sciences* [1971]: 26).

[9] *Mixt.,* 509–684K, Book 2, 582K.

νῦν λόγῳ σῶμα θερμότερον τοῦ δέοντος, ἐὰν μὲν ἐν
τῇ πρώτῃ συστάσει μέσον ὑγρότητός τε καὶ ξηρότητος ᾖ, πάντως τοῦτο κατὰ τὴν ἀκμαστικὴν ἡλικίαν
γίνεται ξηρόν, ἔτι δὲ μᾶλλόν τε καὶ θᾶττον, εἰ καὶ
φύσει ξηρότερον εἴη. καὶ δὴ καὶ γηράσει τοῦτο θᾶττον, ὅσῳ καὶ θᾶττον εἰς ἀκμὴν ἀφίκοιτο. πάντων γὰρ
ἐν τῇ παρακμῇ ξηραινομένων εὔλογόν ἐστι τὸ φύσει
ξηρότερον εἰς τὴν τῆς γεροντικῆς ἀμετρίας ξηρότητα
ἀφικνεῖσθαι θᾶττον. δέονται τοίνυν οὗτοι μάλιστα
κατὰ τὴν ἀκμαστικὴν ἡλικίαν ὑγρᾶς διαίτης, ἧς τὸν
398K τύπον ἀρτίως ὑπέγραψα, διά τε τῶν ὑγραινόντων ἐδεσμάτων καὶ λουτρῶν ἀποχῆς τε γυμνασίων συντόνων
καὶ πολλῶν γινομένης, ὥστε θέρους ὥρᾳ καὶ θᾶττον
λούεσθαι καὶ μετὰ τροφὴν αὖθις.[15] ὀνίνησι δὲ τούτους
καὶ ἡ τοῦ ψυχροῦ πόσις. ἐναντιώτατα δὲ ταῖς ξηραῖς
κράσεσίν εἰσιν ἀφροδίσια. φείδεσθαι δὲ χρὴ μάλιστα αὐτοὺς ἐγκαύσεών τε καὶ κόπων καὶ φροντίδων
καὶ ἀγρυπνιῶν καὶ κινήσεων ἁπασῶν ὀξειῶν. καὶ οἱ
θυμοὶ δὲ μάλιστα τὰς χολώδεις φύσεις ἐκπυροῦντες
ὀξεῖς γεννῶσι πυρετούς.

ἅπερ οὖν ἐπήνηται διαιτήματα τοῖς θερμοῖς τὴν
κρᾶσιν ἐπὶ τῆς ἀκμῆς, κἂν σύμμετρον ἐξ ἀρχῆς ἔχωσιν τὴν ὑγρότητα, ταῦτα μᾶλλον ἁρμόττει τοῖς φύσει
θερμοῖς τε καὶ ξηροῖς. εὔδηλον δ', ὅτι τὴν ποσότητα
τῆς ὑπεροχῆς τοῦ κρατοῦντος στοιχείου σκεπτέον ἐν
τοῖς μάλιστα. τοσοῦτον γὰρ ἐπιτείνειν τε καὶ ἐκλύειν
ἢ αὐξάνειν χρὴ τὸ τῆς διαίτης εἶδος, ὅσον ὑπὲρ τὸ
κατὰ φύσιν ηὔξηταί τε καὶ μεμείωται τὰ τῆς κράσεως

hotter than it ought to be, and if in the first formation it is midway between moist and dry, it becomes altogether dry in the prime of life, and more so and quicker, if it is drier in nature. And in particular, this will age quicker to the extent that it comes more quickly to its prime. Since all bodies become dry in the period after the prime, it is to be expected that a body which is drier in nature comes to the dryness of extreme old age more quickly. Accordingly, these people particularly need a moist regimen during the prime of life, (I wrote the outline of this just now), through moistening foods and baths, and abstaining from exercises   398K that are strenuous and prolonged. As a consequence, in summer they should bathe quicker and again after food. A cold drink benefits such people. Sexual activities are particularly inimical to the dry *krasias*. It is especially necessary for them to avoid heatstrokes, fatigues, anxieties, lack of sleep and all rapid movements. Also rages, particularly when they inflame the bilious natures, generate acute fevers.

Therefore, the regimens recommended for those who are hot in *krasis* in the prime of life and have a moderate moistness from the beginning are more suitable for those who are hot and dry in nature. It is clear, however, that particular consideration must be given to the quantity of excess of the prevailing element in them. For it is necessary to intensify, relax or augment the kind of regimen to the extent that the elements of the *krasis* are increased or

---

15 τό δεύτερον *add.* Ku

στοιχεῖα. διὸ καὶ τοὺς ὑγροτέρους τε καὶ φύσει θερμοτέρους, εἰ μὲν ὑπαλλάττειν αὐτῶν ἐθέλοις τὴν κρᾶσιν ἐπὶ τὸ ψυχρότερόν τε καὶ ξηρότερον εἶδος, χρὴ διὰ τῆς ἐναντίας διαίτης ἄγειν, εἰ δὲ φυλάττειν, διὰ τῶν ὁμοίων.

αἱ δὲ τοιαῦται φύσεις ἐν τῇ τῶν παίδων ἡλικίᾳ ῥευματικοῖς τε καὶ πληθωρικοῖς ἁλίσκονται νοσήμασι, καὶ πρὸς τούτοις γε σηπεδονώδεσι· δέονται τοίνυν γυμνασίων πλειόνων ἀκριβοῦς τε τῆς ἐν τῇ γαστρὶ πέψεως· ὅσα γὰρ ἐν ταύτῃ διαφθείρεται, νοσημάτων σηπεδονωδῶν ὑπόθεσις γίνεται παντὶ τῷ σώματι. διὸ καὶ πρὸ τροφῆς δὶς καὶ τρὶς οὗτοι λουσάμενοι καὶ τοῖς αὐτοφυέσι θερμοῖς ὕδασι χρώμενοι μάλιστα ὠφελοῦνται. σκοπὸς γὰρ ἐπ' αὐτῶν ἐστιν αὐτὰ μὲν τὰ μόρια τοῦ ζῴου φυλάττειν ὑγρά, κοινὸν ὂν τοῦτο πάσης κράσεως, εἴ γε καὶ τοὐναντίον αὐτῷ θᾶττον ἐπὶ τὸ γῆρας ἄγει τὰ σώματα τῶν ζῴων ἁπάντων αὐτοῦ τε τοῦ θνητὸν εἶναι τὸ γεννητὸν σῶμα τὴν αἰτίαν ἔχει· ὡς, εἴγε δυνατὸν ἦν ἀεὶ διαφυλάττειν ὑγρὰν τὴν κρᾶσιν τοῦ σώματος, ὁ τοῦ σοφιστοῦ λόγος, ὃν ἐν ἀρχῇ διῆλθον, ἀθάνατον ἐπαγγελλομένου ποιήσειν τὸν αὐτῷ πειθόμενον, ἀληθὴς ἦν. ἀλλ' ἐπεὶ τὴν φυσικὴν ὁδὸν τὴν ἐπὶ τὸ ξηραίνεσθαι τὸ σῶμα φυγεῖν οὐκ ἔστιν, ὡς ἐδείχθη, διὰ τοῦτο γηρᾶν ἀναγκαῖον ἡμῖν ἐστι καὶ φθείρεσθαι, πολυχρονιώτατος δ' ἂν ὁ ἥκιστα ξηραινόμενος γένοιτο.

τῆς δ' ὑγρᾶς διαίτης περιττώματά τε καὶ πλῆθος χυμῶν γεννώσης χαλεπὸν γίνεται κρατῆσαι τῆς συμ-

diminished beyond an accord with nature. On this account too, if you wish to change the *krasis* of those who are more moist and hotter in nature to the colder and drier kind, 399K you must do this through the opposing regimen, whereas if you wish to preserve it, you must do so through similars.

Such natures, during childhood, are seized by rheumatic and plethoric diseases, and in addition to these, by putrefactive diseases. Accordingly, they need much exercise and completion of concoction in the stomach, for those things that are corrupted in the stomach become a basis for putrefactive diseases in the whole body. On this account also, when these people bathe two or three times before food and use natural warm waters, they are particularly benefitted. An objective in these cases is to keep the actual parts of the organism moist, which is a common aim for every *krasis,* if in fact the opposite to this brings the bodies of all animals to old age quicker, and is the reason why the engendered body is mortal. So if it were really possible to always keep the *krasis* of the body moist, the argument of the Sophist, which I went over at the start, promising that he would make anyone persuaded by him immortal, would be true.[10] But since it is impossible for the body to escape the natural road to being dried, as was 400K shown, and because of this, it is inevitable that we grow old and die, the one who is dried the least lives longest.

However, when a moist regimen generates superfluities and an abundance of humors, it becomes difficult to

[10] This is the second reference to the unnamed philosopher— see Book 1, section 12, 63K, note 32.

μετρίας, ὡς μήτε νόσοις ἁλίσκεσθαι μήτε ταχέως
γηρᾶν. ὅσον δ' οὖν ἐπὶ τῇ κράσει, πολυχρονιώτατοι
πάντων τ' εἰσὶν οἱ ὑγρότατοι καὶ μέντοι καὶ ὑγιαίνουσιν, ἐπειδὰν κρατυνθῇ τὸ σῶμα, τῶν ἄλλων μᾶλλον,
ὥσπερ καὶ ἰσχυρότεροι μέχρι γήρως ἐσχάτου τῶν
τὴν αὐτὴν ἡλικίαν ἐχόντων εἰσί. καὶ κατὰ τοῦτο
ἐπῄνηται σχεδὸν ὑπὸ πάντων ἰατρῶν τε καὶ φιλοσόφων, ὅσοι τὰ στοιχεῖα τοῦ σώματος ἀκριβῶς ἔγνω-
σαν, ἡ κρᾶσις αὕτη· καὶ μέντοι καὶ διὰ τοῦτο μόνη
τισὶν ἔδοξε κατὰ φύσιν εἶναι. κρείττων γὰρ ἐν τῷ
χρόνῳ τῶν ἄλλων γίνεται τό γε κατ' ἀρχὰς οὖσα χείρων. ὥστε καὶ τὸν ἐπιστατοῦντα τῷ τοιούτῳ σώματι
τῶν ἀπορροιῶν αὐτοῦ προνοεῖσθαι χρὴ διά τε γυμνα-
401K σίων, ὡς ἔφην, καὶ λουτρῶν πρὸ τροφῆς πλεόνων ἐκ-
κρίσεώς τε τῶν δι' οὔρων καὶ γαστρὸς γινομένων.
οὐδὲν δὲ κωλύει καὶ ἀποφλεγματισμοῖς ποτε χρῆσθαι
καὶ καθάρσεσιν εὐχύμοις τε πρὸ πάντων ἐδέσμασιν
οἴνων τε πόσει τῶν οὔρησιν κινούντων.

4. Περὶ μὲν οὖν τῶν θερμῶν κράσεων αὐτάρκως
εἴρηται· περὶ δὲ τῶν ψυχρῶν ἐφεξῆς ἂν εἴη λεκτέον.
εἰσὶ δὲ καὶ τούτων αἱ μέγισται διαφοραὶ τρεῖς· ἤτοι
γὰρ εὔκρατοι κατὰ τὴν ἑτέραν ἀντίθεσίν εἰσιν, ὡς μηδὲν μᾶλλον ὑγροὺς ἢ ξηροὺς ὑπάρχειν, ἢ τὸ ὑγρὸν ἢ
τὸ ξηρὸν ἐν αὐτοῖς ἐπικρατεῖ. χειρίστη δ' εἰκότως
ἐστὶν ἡ ξηρὰ κρᾶσις· ὃ γὰρ ἐν τῷ χρόνῳ τοῖς γηρῶσι
γίνεται, τοῦτ' εὐθὺς ἐξ ἀρχῆς ὑπάρχει τούτοις. ὑγραί-

---

11 The old term is retained here in part because there is no

attain moderation, so that the person is neither seized by diseases nor ages quickly. Thus, to the extent that is due to the *krasis,* the very moist are the longest lived of all, and are healthy as well, since the body is made strong more than the others, as they are stronger than others of the same time of life right up to the extreme of age. And on these grounds, it has been praised by almost all doctors and philosophers who know accurately the elements of the body. Indeed, because of this alone, this *krasis* has seemed to some to be in accord with nature. Although in the beginning it is worse, in time it becomes stronger than the others. As a result, it behooves someone responsible for the care of such a body to give forethought to the outflows from it, through exercises, as I said, and baths before food, and the excretion of many things, which occurs through the urine and stomach. Also, nothing prevents the use of apophlegmatics[11] and cathartics at times, and above all, the use of *euchymous* foods and the drinking of wines that promote urination. 401K

4. Enough has been said about the hot *krasias.* Next I must speak about the cold *krasias.* There are three major *differentiae* of these: (1) when people are *eukratic* in the other antithesis—that is to say, neither more moist nor more dry; (2) when moisture prevails in them; (3) when dryness prevails in them. In all likelihood, the dry *krasis* is the worst. It occurs in time to those who are growing old, but in some it is present right from the start. It is

adequate modern equivalent. The essential meaning is the action of purging phlegm from the head. The term in several forms is listed in the 1933 OED, where reference is made to Swift's *Gulliver's Travels* and Samuel Johnson's dictionary.

νειν τε οὖν αὐτοὺς χρὴ καὶ θερμαίνειν. ἔσται δὲ τοῦτο
γυμνασίοις συμμέτροις καὶ τροφαῖς ὑγραῖς τε καὶ
θερμαῖς οἴνου τε πόσει τῶν θερμοτέρων ὕπνοις τε
πλείοσι προνοουμένων ἡμῶν, ὅπως τὰ καθ' ἑκάστην
ἡμέραν ἐν τῷ σώματι γεννώμενα περιττώματα τῆς τε
τροφῆς καὶ τοῦ πόματος ἐκκενῶται πάντα.

402K περὶ δὲ τῶν ἀφροδισίων εἴρηται μὲν καὶ πρόσθεν,
ὡς ἅπαντας μὲν τοὺς ξηροτέρους τῇ κράσει βλάπτει,
νυνὶ δ', ὅτι τούτων αὐτῶν μάλιστα τοὺς πρὸς τῇ ξη-
ρότητι καὶ ψυχρούς. ἀβλαβῆ γὰρ ἀφροδίσια μόνοις
τοῖς ὑγροῖς καὶ θερμοῖς ἐστι καὶ ὅσοι φύσει πολύ-
σπερμοι, περὶ ὧν ἐν ταῖς ἀνωμάλοις κράσεσιν αὐτίκα
λεχθήσεται. μοχθηραὶ δὲ καὶ αἱ μεθ' ὑγρότητος ψυ-
χραὶ κράσεις εἰσὶ καὶ μάλιστα αὗται τοῖς ῥευματικοῖς
ἁλίσκονται νοσήμασιν, ὀνίνησι δ' αὐτὰς ἀλουσία τε
καὶ γυμνάσια καὶ δίαιτα λεπτοτέρα καὶ χρίσματα μετρίως θερμαίνοντα, περὶ ὧν τῆς ἰδέας ἔμπροσθεν εἴρηται κατὰ τὸν περὶ τῶν κόπων λόγον. ὅσοι δὲ ψυχρότεροι μέν εἰσι φύσει, συμμέτρως δ' ἔχουσι τῆς
κατὰ τὸ ξηρόν τε καὶ ὑγρὸν κράσεως, ἧττον οὗτοι τῶν
κατὰ τοῦτο δυσκράτων εἰσὶ μοχθηροὶ πρὸς ὑγείαν τε
καὶ ῥώμην σώματος. ἐπεγείρειν τε οὖν αὐτῶν χρὴ καὶ
ῥωννύναι τὴν θερμασίαν, ἐν δὲ τῇ καθ' ὑγρότητα καὶ
ξηρότητα τῆς ὅλης διαίτης ἰδέᾳ τὸ μέσον ἐκλέγεσθαι.

403K τοιαῦτά σοι σύμπαντα περὶ τῶν ὁμαλὴν τὴν δυσκρα-

---

12 For Galen's definition of the term, see his *Ars M.*, I.339K,
Johnston, *On the Constitution*, 207.

necessary to moisten and heat these cases. This will be by moderate exercises, by moist and hot foods, by drinking one of the hotter wines, by having plenty of sleep, and by giving forethought as to how all the superfluities generated in the body each day from food and drink are emptied out.

On the matter of sexual relations, I also said before that 402K they harm all those who are drier in *krasias;* now I say they are particularly harmful to those who are cold in addition to being dry. Sexual relations are without harm only in those who are moist and hot, and are by nature polyspermous.[12] We shall speak about these forthwith in connection with the irregular *dyskrasias.*[13] Also bad are the cold *krasias* with moistness; these are particularly prone to the rheumatic diseases. Avoidance of bathing benefits these, as do exercises, a thinner diet, and moderately heating unguents. I have spoken previously about the specific kinds of these in the discussion on the fatigues.[14] Those who are colder in nature but have a balanced *krasis* in terms of dryness and moisture are less bad in the matters of health and strength of the body than those who are *dyskratic* in this respect. It is, then, necessary to stir up and strengthen their heat, choosing a middle course in the specific nature of the whole regimen in terms of moistness and dryness. These are all the things I have said to you about those having a regular *dyskrasia*—that is to say, 403K

[13] Galen has a short work on the anomalous (uneven, irregular, nonuniform) *dyskrasias*—*Inaequal. Intemp.*, VII.733–52K (English trans., Novo, *On the Anomalous Dyskrasia*).

[14] See the present work, Book 3, chaps. 5–9.

σίαν ἐχόντων εἴρηται, τουτέστιν ὁμοίως ἅπαντα μόρια πρὸς τὸ θερμότερον ἢ ψυχρότερον ἢ ὑγρότερον ἢ
ξηρότερον ἐκτετραμμένα.

5. Περὶ δὲ τῶν ἀνώμαλον ἐχόντων τὴν κατασκευὴν
τοῦ σώματος οὐκ ἐγχωρεῖ διὰ βραχέων εἰπεῖν, ἐπειδὴ
πολλαὶ τῶν τοιούτων ὑπαλλάξεις εἰσὶν ἄλλων ἄλλο
μόριον ἐχόντων δύσκρατον. ἀλλὰ κἂν δύο μόρια
ἔχωσί τινες δύσκρατα, ὑπαλλάξεις καὶ τούτων γίνονται πολλαί. καὶ διὰ τοῦτο πρότερον εἰπόντες, ὅσα τῷ
τε τὴν ἄμεμπτον ἔχοντι κατασκευήν, βιοῦντι δὲ βίον
δουλικὸν ἢ περισπαστικὸν ἢ ὅπως ἄν τις ὀνομάζειν
ἐθέλοι ποιητέον ἐστίν, καὶ μετὰ ταῦτα, ὅσα τοῖς ὁμαλὴν τὴν δυσκρασίαν ἐν ὁμοίῳ βίῳ, μεταβησόμεθα
τηνικαῦτα πρὸς τοὺς ἀνώμαλον ἔχοντας ἐν τοῖς τοῦ
σώματος μέρεσι τὴν κατασκευήν. ἄμεμπτον δὲ κατασκευὴν σώματος, ὡς πολλάκις εἴρηται, κατὰ πλάτος
νοητέον ἐστίν, ὥσπερ καὶ αὐτὴν τὴν ὑγείαν. οὔτε γὰρ
ὑγιαίνειν τις ἡμῶν ἂν δόξειε τήν γ᾽ ἀκριβῶς ἄμεμ-

404K πτον ὑγείαν οὐκ ἔχων οὔτ᾽ ἀμέμπτως κατεσκευάσθαι
τὸ σῶμα.

λέγουσι δ᾽ ὑγιαίνειν, ὅσοι μήτ᾽ ὀδυνῶνταί τι μέρος τοῦ σώματος εἴς τε τὰς κατὰ τὸν βίον ἐνεργείας
ἀπαρεμπόδιστοι τυγχάνουσιν ὄντες, οὕτω δὲ καὶ
κατασκευὴν ἔχειν σώματος ἄμεμπτον, ὅταν μήθ᾽ ὑπὸ
τῶν ἔξωθεν αἰτίων ἑτοίμως εἰς νόσον ἄγηται μήθ᾽ ὑπὸ
τῶν ἐξ αὐτοῦ, διωρισμένου δηλονότι τοῦ νοσεῖν ἐνίους
συνεχῶς οὐ διὰ τὴν οἰκείαν κατασκευὴν τοῦ σώματος,
ἀλλὰ καὶ διὰ μοχθηρὰν δίαιταν ἤτοι γ᾽ ἀργὸν βίον

those in whom all parts are similarly deviated toward being hotter, colder, moister or drier.

5. It is not possible to speak briefly about those who have an irregular (nonuniform, anomalous) constitution of the body since there are many varieties of such constitutions, some having one part *dyskratic* and others other parts. But even if two parts were to have certain *dyskrasias,* the varieties of these would also be many. And because of this, after first stating what must be done for someone having a faultless constitution, whether living the life of a slave or a man engaged in affairs, or however someone might wish to term it, the next thing to state is what must be done for those with a regular (uniform) *dyskrasia* living a similar life, I shall then pass on to those having an irregular constitution in the parts of the body. We must think of a faultless constitution of the body in broad terms, as was often said, just like health itself, otherwise none of us would seem healthy if we did not have perfectly faultless health, nor would the body seem to faultlessly constituted. 404K

People call healthy those who never feel pain in any part of the body and happen to be unhindered in the functions of life, and in this way have a faultless constitution of the body, when they are not readily affected either by diseases through external causes, or by causes within themselves. Obviously, we distinguish some who are continuously diseased, not due to the intrinsic constitution of the body, but through a bad regimen, or living an idle life,

βιοῦντας ἢ ὑπερπονοῦντας ἢ περὶ τὰς τῶν σιτίων
ποιότητας ἢ ποσότητας ἢ καιροὺς ἁμαρτάνοντας ἢ
τινα τῶν ἐπιτηδευμάτων βλαβερῶς ἐπιτηδεύοντας ἢ
περὶ τὴν τῶν ὕπνων ποσότητα σφαλλομένους ἢ χρῆσιν ἀφροδισίων ἄμετρον ἢ καὶ συντήκοντας ἑαυτοὺς
ἐν λύπαις καὶ φροντίσιν οὐκ ἀναγκαίαις. οἶδα γὰρ
παμπόλλους διὰ τοιαύτην αἰτίαν νοσοῦντας καθ᾽ ἕκαστον ἐνιαυτόν· ἀλλ᾽ οὐ φήσομεν αὐτοὺς κακῶς κατεσκευάσθαι τὸ σῶμα, καθάπερ ἐκείνους, ὅσοι μηδὲν
ἁμαρτάνοντες ὧν κατέλεξα συνεχῶς νοσοῦσιν.

   ὑποκείσθω τοίνυν ἐν τῷ λόγῳ πρῶτος ὁ τὴν κατὰ
405K πλάτος ἄμεμπτον ἔχων κατασκευὴν σώματος, ἐν βίῳ
δουλικῷ δι᾽ ὅλης ἡμέρας ὑπηρετῶν ἤτοι τῶν μέγιστα
δυναμένων τισὶν ἢ μοναρχῶν, χωριζόμενος δὲ[16] περὶ
τὰ πέρατα τῆς ἡμέρας. ὁρίσαι δὲ πάλιν ἐπὶ τούτου
χρή, τίνα λέγω πέρατα· παρακοὴν γὰρ ὁ λόγος ἐργάσεται τοῖς ἀναλεγομένοις αὐτόν, εἰ μὴ τύχοι διορισμοῦ προσήκοντος. ἐὰν γοῦν εἴπω χωρίζεσθαι τηνικαῦτα πρῶτον εἰς τὴν ἐπιμέλειαν τοῦ σώματος, ἡνίκα
ὁ ἥλιος δύνῃ, μὴ προσθείς, ὁποίας ἡμέρας λέγω,
πότερον τῆς περὶ τὰς θερινὰς τροπὰς ἢ χειμερινὰς ἢ
κατά τινα τῶν ἰσημερινῶν ἢ χρόνον, ἑκάτερον ἐν τῷ
μεταξὺ τῶν εἰρημένων καιρῶν, ἀδύνατον ἔσται συμ-
φερούσας ποιήσασθαι ὑποθήκας. κατὰ γοῦν τὴν
Ῥωμαίων πόλιν αἱ μέγισται μὲν ἡμέραι καὶ νύκτες
βραχὺ μείζους ὡρῶν ἰσημερινῶν πεντεκαίδεκα γίνονται, καθάπερ γε πάλιν αἱ ἐλάχισται μικρὸν ἀπο-
δέουσι τῶν ἐννέα, κατὰ δὲ τὴν μεγάλην Ἀλεξάνδρειαν

148

or working too hard, or being in error regarding the qualities, quantities or times of foods, or practicing some activity that is harmful, or erring in regard to the amount of sleep, or excessive indulgence in sex, or needlessly tormenting themselves with grief and anxiety. Every year I see very many who are sick through such a cause, but we shall not say they have a bad bodily constitution, like those who are continuously ill, despite committing none of the mistakes I listed.

Therefore, let us assume first in the discussion someone who has a faultless constitution of the body in a broad sense, living the life of a slave, serving throughout the day some of the greatest magistrates or monarchs, but parting from his master around the end of the day. Again in this case, I must define what I call "end," for the discussion will create a misunderstanding in those reading it, if it fails to provide a suitable definition. Anyway, if I say he leaves, primarily for the care of his body, at the time when the sun sets, without adding what kind of day I am speaking of—whether of the kind around the summer or winter solstice, or at one of the equinoxes, or at each of the times between those stated—it will not be possible to make beneficial suggestions. At any rate, in the city of the Romans, the longest days and nights are a little more than fifteen equinoctial hours, just as, on the other hand, the shortest are a little less than nine. However, in the great city of Alex-

405K

---

16 *post* χωριζόμενος δὲ: τοῦ δεσπότου (Ku)

τεσσάρων καὶ δέκα μὲν ὡρῶν αἱ μέγισται, δέκα δὲ
αἱ σμικρόταται. ὁ μὲν οὖν ἐν ταῖς σμικροτάταις μὲν
ἡμέραις μεγίσταις δὲ νυξὶν ἀφιστάμενος τῆς ὑπηρεσίας ἡλίου δυομένου καὶ τρίψασθαι κατὰ σχολὴν καὶ
406K λούσασθαι δύναται καὶ κοιμηθῆναι συμμέτρως,[17] ὁ δ᾽
ἐν ταῖς μεγίσταις οὐδ᾽ ἓν τούτων οἷός τ᾽ ἐστὶ πρᾶξαι
μετρίως· οὐ μὴν οὐδ᾽ ἔγνων τινὰ τοιαύτῃ δυστυχίᾳ
βίου χρησάμενον.

ὁ γοῦν ὧν ἴσμεν αὐτοκρατόρων ἑτοιμότατα πρὸς
τὴν τοῦ σώματος ἐπιμέλειαν ἀφικόμενος Ἀντωνῖνος
ἐν μὲν ταῖς μικραῖς ἡμέραις ἡλίου δύνοντος εἰς τὴν
παλαίστραν παρεγίνετο, κατὰ δὲ τὰς μεγίστας ἐνάτης
ὥρας ἢ τὸ πλεῖστον δεκάτης, ὡς ἐξεῖναι τοῖς παραμέ-
νουσιν αὐτῷ κατὰ τὰς ἡμερησίας πράξεις ἀπαλλα-
γεῖσι προνοήσασθαι τοῦ σώματος ἐν τῷ λοιπῷ μέρει
τῆς ἡμέρας, ὡς ἅμα τῷ δῦναι τὸν ἥλιον εἰς ὕπνον
τρέπεσθαι. τῆς γάρ τοι σμικροτάτης νυκτὸς ἴσης οὔσης ἐννέα ταῖς ἰσημεριναῖς ὥραις αὐτάρκης ὁ τοσ-
οῦτος χρόνος αὐτοῖς ὕπνου τυχεῖν. ἐπισκεπτέον οὖν,
εἴτε γυμνάζεσθαι κατὰ τὸν ἔμπροσθεν βίον ὁ τοιοῦτος
ὑπηρέτης ἔθος εἶχεν εἴτ᾽ ἀγύμναστος λούεσθαι. τινὲς
γὰρ οὐδὲ ἄχρι τοῦ τρίψασθαι προνοοῦνται σφῶν αὐτῶν, ἀλλ᾽ ἄντικρυς εἰς τὸ βαλανεῖον εἰσίασιν περι-
χεόμενοι τὸ ἔλαιον ἢ καὶ στλεγγίδα μόνον λαβόντες,
407K ὡς κατ᾽ αὐτὸ τὸ βαλανεῖον ἀποστλεγγίζεσθαι τὸν
ἱδρῶτα. καὶ μέντοι καὶ μετρίως φέρουσι τὸ τοιοῦτον
ἔθος ἔνιοι. ὥστε τινὰς αὐτῶν μηδὲ νοσεῖν συνεχῶς,
ὅταν ὦσιν εὐδιάπνευστοι. καὶ καλοῦσί γε αὐτοὺς

andria, the longest days are fourteen hours and the shortest, ten. Therefore, the one who, in the shortest days and longest nights, gets away from service when the sun sets, can be massaged at leisure, bathe, and sleep in moderation. But there is no one who can do any of these things moderately in the longest days, nor have I known anyone employed in such unfortunate life circumstances [do so].

406K

Anyway, Antoninus,[15] who, of the autocrats I have known, attended most promptly to the care of his body, came to the wrestling school at sunset on the short days, and on the longest days at the ninth or, at the latest, the tenth hour. Consequently, it was possible for those who attended him during his daily affairs, once they were dismissed, to take care of the body in the remaining part of the day, so they could retire to sleep at once when the sun sets. For surely, since the shortest night is equal to nine equinoctial hours, this amount of time is enough for them to sleep. Therefore, we must consider whether such a servant was accustomed to exercise in his previous life, or to bathe without exercising. For some do not take care of themselves to the point of being massaged, but come straight to the bath house after anointing themselves with oil, using only a strigil to scrape off the sweat in the bath house itself. Nevertheless, there are some who tolerate such a custom moderately well. As a result, some of them are not continually sick whenever they perspire freely. In

407K

---

15 This is presumably Antoninus Pius, who died in AD 161.

---

17 *post* συμμέτρως,: ὁ δ᾿ ἐν ταῖς μεγίσταις οὐδ᾿ ἐν τούτων Κο; οὐδείς δ᾿ ἐστιν ὅς οὐδ᾿ ἐν ταῖς μεγίσταις, οὐδ᾿ ἐν τούτοις Κu

"ἀραιοσυγκρίτους" ἔνιοι τῶν ἰατρῶν τε καὶ γυμναστῶν. εἴρηται δὲ καὶ ὑφ᾽ Ἱπποκράτους ὑγιεινοτέρους
εἶναι τοὺς τοιούτους· "ἀραιότης γάρ" φησί "σώματος
ἐς διαπνοὴν οἷς πλεῖον φέρεται ὑγιεινότεροι, οἷς δ᾽
ἐλάσσω νοσερωδέστεροι." τὴν μὲν δὴ τοιαύτην φύσιν
τοῦ σώματος οὐ χρὴ μετάγειν ἐφ᾽ ἕτερον ἔθος, οὐδ᾽
ὅλως οὐδεμίαν, ἀλλ᾽ ὅσαι διὰ πολλοῦ χρόνου νοσοῦσιν.

6. Εἰ δὲ συνεχῶς τις νοσοίη, σκεπτέον τὴν αἰτίαν
ἐπ᾽ αὐτοῦ. εὑρεθήσεται δὲ τὴν ἀρχὴν τῆς ζητήσεως
ἡμῶν ἀπὸ τοῦ τῶν νόσων εἴδους ποιησαμένων. ὅσα
γὰρ[18] ἐκ τῆς ἐν αὐτῷ κακίας νοσεῖ τὸ σῶμα, διττὴν
ἔχει τὴν πρόφασιν, ἤτοι πλῆθος ἢ κακοχυμίαν. εἰ μὲν
δὴ πληθωρικὰς νόσους νοσεῖν φαίνοιτο, σκοπὸς ἔστω
σοι καθ᾽ ὅλην τὴν ὑγιεινὴν δίαιταν αὐτοῦ, ὅπως συμ-
408K μέτρους ἔχῃ τοὺς χυμούς· εἰ δὲ διὰ κακοχυμίαν, ὅπως
ἀρίστους. ἡ δὲ κατὰ μέρος πρόνοια τῆς μὲν ἐν ποσότητι συμμετρίας εἰρήσεται πρότερον, τῆς δ᾽ ἐν ποιότητι δεύτερον. λέγωμεν οὖν ἤδη περὶ τῆς προτέρας
ἐπ᾽ ἀρχὴν ἀναγαγόντες τὸν λόγον. ὅταν ἐλάττω τὰ
διαπνεόμενα τοῦ σώματος ᾖ τῶν λαμβανομένων, αἱ
πληθωρικαὶ νόσοι γίνονται. φυλακτέον οὖν ἐστι τὴν
συμμετρίαν τῶν ἐσθιομένων τε καὶ πινομένων πρὸς
τὰ κενούμενα· σύμμετρον δὲ ἔσται σκεψαμένων ἡμῶν
ἐν ἑκατέροις τὰς ποσότητας. τοὺς μὲν γὰρ ἄντικρυς

[18] post ὅσα γάρ: ἐκ τῆς ἐν αὐτῷ Ko; ἐκ τῆς ἐντὸς Ku

fact, some doctors and gymnastic trainers call them "loose-textured."[16] It was stated by Hippocrates that such people are healthier. "Loose texture of the body," he says, "allows greater transpiration, making them healthier, whereas those in whom there is less transpiration are more sickly."[17] It is not necessary to change a body of such a nature to a different custom, nor to change any nature at all, other than those which are diseased over a long period.

6. If someone is continually sick, one must consider the cause in his case. This will be discovered, if we make the start of our investigation from the kind of diseases. For those things that make the body diseased from a badness within it have a twofold cause: either excess (*plēthos*) or *kakochymia.* If the person seems to be sick with plethoric diseases, your objective should be a regimen relating to his health as a whole—how he might have a balance of humors. If, however, it is due to a *kakochymia,* [the regimen is determined by] how to have the best humors. The care of balance in quantity will be spoken of separately first, and that in quantity second. Let me now bring forward the discussion of the former, starting from the beginning. Whenever those things dispersed in vapor from the body are less than those things taken in, the plethoric diseases arise. What must be preserved, then, is the balance between foods and drinks on the one hand and those things evacuated on the other. There will be balance when we give consideration to the quantities in each. In the case

408K

[16] LSJ defines the term ἀραιοσύγκριτος as "with loose tissues" citing this passage.

[17] Hippocrates, *On Nutrition* 28. *Hippocrates* III, LCL 147, 352–53.

εἰς τὸ βαλανεῖον εἰσιόντας[19] ἐπιτάξομεν τρίψει τε κεχρῆσθαι καί τι καὶ βραχὺ κινεῖσθαι πρότερον, ὅσοι
δ' ἔφθανον ταῦτα ποιεῖν, ἐπ' ὀλίγον μὲν αὐξῆσαι καὶ
αὐτὰ ταῦτα συμβουλεύσομεν. ἀφελεῖν δὲ χρὴ καὶ τῶν
τροφῶν ἤτοι τῆς ποιότητος ἢ τῆς ποσότητος ἢ καὶ
συναμφοτέρων· τοὺς μὲν γὰρ ταχέως ἀθροίζοντας
ἀξιόλογον πλῆθος ἑκατέρων· ὅσοι δ' οὐ ταχέως ἢ οὐκ
ἀξιόλογον, ἀρκέσει θάτερον, ὃ ἂν αὐτὸς ὁ ἄνθρωπος
409K   αἱρῆται. τὸ μὲν οὖν τῆς ποσότητος ἀφελεῖν γνώριμον,
τὸ δὲ τῆς ποιότητος ἐν τῇ τῶν ὀλιγοτρόφων ἐδεσμάτων προσφορᾷ γίνεται. πολλοὶ γὰρ ἐπὶ χοιρείοις
κρέασι διαιτώμενοι πλῆθος ἀθροίζουσι τάχιστα, δεο-
μένης τῆς τοιαύτης τροφῆς τρίψεών τε καὶ γυμνασίων
ἰσχυρῶν. ἐπὶ λάχανα τοίνυν αὐτοὺς ἀκτέον ἐστὶ καὶ
χέδροπα τὰ μὴ πολύτροφα καὶ ἰχθύας ὁμοίως καὶ
ὄρνιθας, ὅσοι μὴ πολύτροφοι.

τῶν δὲ κακοχυμίαν ἀθροιζόντων, οὐχ ὡσαύτως τοῖς
τὸ πλῆθος ἀθροίζουσιν, οὐχ εἷς ἐστιν ὁ σκοπός, ὅτι
μηδὲ τῆς κακοχυμίας ἰδέα μία. τινὲς μὲν γὰρ ψυχρο-
τέραν τε καὶ φλεγματικωτέραν ἀθροίζουσι κακοχυμίαν, τινὲς δὲ θερμοτέραν τε καὶ χολωδεστέραν, ἔνιοι
δὲ ὑδατωδεστέραν, ὥσπερ ἄλλοι μελαγχολικωτέραν.
ἀφεκτέον οὖν ἐστιν ἑκάστου τῶν σιτίων τε καὶ ποτῶν,
ὅσα πέφυκε γεννᾶν ἑτοίμως τὸν ἀθροιζόμενον αὐτοῖς
χυμόν. εἴρηται δὲ περὶ αὐτῶν ἱκανῶς ἐν τρισὶ μὲν
ὑπομνήμασιν, ἃ Περὶ τῶν ἐν ταῖς τροφαῖς δυνάμεων
ἐπιγέγραπται, καθ' ἕτερον δέ, ἐν ᾧ Περὶ εὐχυμίας καὶ
κακοχυμίας ἡ διαίρεσις γίνεται. καὶ τούτων δ' ἔξωθέν

of those who go directly to the bath house, we should direct them to have used massage and a little movement first. We shall advise those who were already doing these things beforehand, to increase these same things a little. It is also necessary to reduce their nutriments, either in quality or quantity, or in both together. For those who quickly collect a significant excess, reduce both; for those who collect neither quickly nor a large amount, reduction of either will suffice—whichever the person himself chooses. The reduction of quantity is well-known; the reduction of quality lies in the provision of less nutritious foods. Many who eat pork collect an excess very quickly; such nutriment creates a need for massages and vigorous exercises. Moreover, one must direct such people to vegetables and legumes which are not very nutritious, and similarly to fish and birds which are not very nutritious. 409K

However, those who build up *kakochymia* are not like those who build up an excess. There is not a single objective because there is not a single kind of *kakochymia*. Some build up a *kakochymia* that is colder and more phlegmatic; some, one that is hotter and more bilious; some, one that is more watery; and others, one that is more melancholic. There must, then, be abstention from each of the foods and drinks that are of a nature to readily generate the collected humor in these people. Enough was said about these in the three books I wrote entitled, *The Powers of Foods*. In another respect, there is the classification in the work, *On the Good and Bad Humors of*

---

[19] *post* εἰσιόντας: ἐπιτάξομεν τρίψει τε κεχρῆσθαι Κο; ἔπειτα τρίψει τε χρῆσθαι Κu

410K ἔστι τὸ Περὶ τῆς λεπτυνούσης διαίτης, ἥτις ἐπιτή-
δειος τοῖς τὸν καλούμενον ὠμὸν χυμὸν ἀθροίζουσίν
ἐστι, παχὺν μὲν ὑπάρχοντα πάντως, οὐκ ἀεὶ δὲ καὶ
γλίσχρον. ἐπὶ πάντων δὲ τούτων κοινὸν βοήθημα γαστρὸς ὑπαγωγὴ καὶ μάλισθ᾽, ὅσοις ἐστὶ φύσει σκληροτέρα, κοινὸν δὲ καὶ τὸ τῶν ἀφροδισίων σύμμετρον.

7. Ἐπὶ μὲν γὰρ τῶν ἀμέμπτων κατασκευῶν τοῦ
σώματος οὐ χρὴ παντάπασιν ἀφροδισίων ἀπέχεσθαι,
καθάπερ ἐπὶ τῶν ξηρῶν ἔμπροσθεν εἴρηται. σκεπτέον
δ᾽ ἐν τοῖς μάλιστα, πότερον ἅπαξ ἢ δὶς ἐσθίειν τοῖς
ἐν τῷ τοιούτῳ βίῳ συμφέρει. τῆς σκέψεως δ᾽ ἐστὶ κεφάλαιον ἥ τε τοῦ σώματος αὐτῶν φύσις ἥ τ᾽ ἐπὶ
ταύτῃ συνήθεια τῆς διαίτης καὶ τρίτον ἐπ᾽ αὐτοῖς, ἐάν
σοι φανῇ μετακινητέον εἶναι τὸ ἔθος, ὅπως αὐτοῦ
φέρει τὴν ὑπαλλαγὴν ὁ ἄνθρωπος. ἀπὸ μὲν οὖν τῆς
φύσεως ἡ τοῦ συμφέροντος ἔνδειξις γίνεται τῶν μὲν
ἐκχολουμένης τῆς γαστρός, τῶν δὲ οὔ. ὅπως δὲ χρὴ
προνοεῖσθαι γαστρὸς ὑπαγωγήν, εἴρηται μὲν ἤδη καὶ
διὰ τοῦ τετάρτου γράμματος, εἰρήσεται δὲ καὶ νῦν·
ἔστω μὲν ἀεὶ τὰ πρῶτα τῶν ἐσθιομένων τε καὶ πινομένων, ὅσα λαπάττει τὴν γαστέρα, τῶν μὲν οἴνων οἱ
411K γλυκεῖς καὶ ὑπακτικοί (οὐ γὰρ ἅπαντές εἰσι τοιοῦτοι),
ἐδεσμάτων δὲ τὰ λάχανα δι᾽ ἐλαίου καὶ γάρου. φυλακτέον δὲ τῶν οἴνων τοὺς αὐστηρούς, ὥσπερ γε καὶ τῶν
ἐδεσμάτων τὰ στύφοντα, πλὴν εἰ μετὰ τὴν ἐδωδὴν
ἅπασαν ἕνεκα τοῦ τονῶσαι τὸ στόμα τῆς γαστρὸς ἡ

Nutriments. Apart from these, there is the work, *On the Thinning Diet*,[18] which is suitable for those who build up 410K the so-called crude humor, which is always thick but not always also viscous. The common remedy for all these is downward purging of the stomach, and particularly in those who are harder in nature. Common also is moderation in sexual activity.

7. It is not necessary for those with faultless constitutions of the body to abstain altogether from sexual intercourse, as was previously said for those with dry constitutions. What must be particularly considered in them is whether it is beneficial for those who lead such a life to eat once or twice [a day]. The chief point to consider is the nature of their bodies. In addition to this, there is their customary regimen, and third, with these people, whether it seems to you that the ethos must be changed, and how the person might tolerate this change. The indication of what is beneficial arises from the nature in those who have the stomach charged with bile and those who do not. How you must give forethought to downward purging of the stomach was already stated in the fourth book and will also be stated now. The first foods and drinks should always be those that empty the stomach—among wines, those that are sweet and aperient (for not all wines are such), and 411K among foods, vegetables with oil and fish sauce. However, one must avoid wines that are harsh, just as one must also avoid foods that are astringent, unless the presentation occurs after all food for the sake of strengthening the

18 For these three works, referred to as group or individually, and included in the same CMG volume along with the *Hygiene,* see note 6 above.

προσφορὰ γίνοιτο. κριτέον δὲ καὶ τοῦτο τῇ πείρᾳ·
τισὶ γὰρ οὐ μόνον ἄλυπός ἐστιν ἡ τοιαύτη χρῆσις
τῶν στυφόντων, ἀλλὰ καὶ συντελεῖ τι πρὸς ὑπαγωγὴν
γαστρὸς ἐκ τοῦ τονωθῆναι κατὰ τὸ στόμα τῆς κοιλίας. ἐναργέστατον δὲ ἐπὶ τῶν ἀτονώτερον ἐχόντων
αὐτὸ φαίνεται γινόμενον. ὅσοι δὲ τοῖς κατὰ παλαίστραν γυμνασίοις ἐχρῶντο πρὸ τοῦ λουτροῦ, πρὶν
ἐμπεσεῖν ἀσχόλῳ βίῳ, τούτοις ἀποστῆναι μὲν αὐτῶν
βλαβερώτατον, ὁμοίως δὲ χρῆσθαι πρὸς τῷ βλαβερῷ
καὶ ἀδύνατον. ὅπερ οὖν ἔφην ὀνομάζεσθαι γυμνασίων
εἶδος ἀποθεραπευτικόν, τούτῳ χρηστέον αὐτοῖς ἅμα
τῷ καὶ τοῦ πλήθους τῶν χοιρείων κρεῶν ἀφαιρεῖν.
ἐπεὶ δέ, ὡς ἔφην, ἐνίοις ἐσθίειν ἄμεινόν ἐστι πρὸ τοῦ
λουτροῦ, περί τε τῆς ὥρας, ἐν ᾗ τοῦτο πρακτέον αὐ-
412K τοῖς ἐστι, καὶ περὶ τῆς ποσότητός τε καὶ ποιότητος,
ὧν χρὴ προσφέρεσθαι, λεκτέον ἐστίν.

ὅπερ οὖν εἴωθα ποιεῖν αὐτὸς ἐγὼ καθ᾽ ἣν ἂν ἡμέραν
ὀψιαίτερον ἡγῶμαι λούσασθαι δι᾽ ἀρρώστων ἐπισκέ-
ψεις ἤ τινα πολιτικὴν πρᾶξιν, εἰπεῖν οὐκ ὀκνήσω.
ὑποκείσθω γοῦν ἡμέρα, καθ᾽ ἣν τοῦτο γίνεται, τριῶν
καὶ δέκα τῶν ἰσημερινῶν ὡρῶν, ἐλπιζέσθω δὲ περὶ
δεκάτην ὥραν ἡ τοῦ σώματος ἐπιμέλεια γενήσεσθαι.
κατὰ ταύτην τὴν ὑπόθεσιν ἔδοξέ μοι περὶ τετάρτην
ὥραν προσφέρεσθαι τροφὴν ἁπλουστάτην, ἥτις ἐστὶν
ἄρτος μόνος. ἐγὼ μὲν οὖν οὕτως αὐτὸς ἔπραξα, τινὲς
δὲ οὐχ ὑπομένουσιν ἄρτον ἐσθίειν μόνον, ὄψου χωρίς,
ἀλλ᾽ ἤτοι μετὰ βαλάνων φοινίκων ἢ ἐλαιῶν ἢ μέλιτος
ἢ ἁλῶν ἐσθίουσιν, εἶτα καὶ πίνουσιν ἔνιοί τινες αὐτῶν,

opening of the stomach (cardiac orifice). This must be judged by experience. For some, such a use of astringents is not only painless, but also contributes something to the downward purging of the stomach through the opening of the stomach being strengthened. This is most clearly apparent in those who are obviously weaker in this respect. If those who were in the habit of using the exercises at the wrestling school prior to bathing, before they succumbed to a busy life, give these up, it is very harmful for them, while to use them similarly, in addition to being harmful, is also impossible. Therefore, what I said is termed the restorative (apotherapeutic) kind of exercises is what must be used for them, along with setting aside most forms of pork. Since however, as I said, it is better for some to eat before bathing, what I must speak about is the hour in which this is to be done for them, and about the quantity   412K
and quality of those things that must be administered.

Therefore, I shall not hesitate to say what I myself am accustomed to do on a day when I decide to bathe later due to visiting patients or being engaged in some civil matter. Anyway, let us suppose the day on which this happens is one of thirteen equinoctial hours, and let us hope the care of the body will take place around the tenth hour. On the basis of this same assumption, I expected to take the simplest nourishment around the fourth hour, this being bread alone. This, then, is what I did, although some cannot bear to eat bread alone without something to give it flavor, but eat it either with the fruit of the date palm, olive oil, honey or salt. Then some of them also drink. I

ἐγὼ δὲ οὐκ ἔπιόν ποτε ἐπὶ τοιαύτῃ τροφῇ καὶ μόνον ἔφαγον τὸν ἄρτον. ἔστω δ' αὐτῶν πλῆθος ἑκάστου τοσοῦτον, ὅσον ἄχρι τῆς δεκάτης ὥρας ἐν τῇ γαστρὶ πεφθῆναι δύναται· καὶ γὰρ εἰ γυμνάζεσθαι βού- λοιντο, μάλιστα ἂν οὕτως ἀβλαβῶς γυμνάσαιντο. βλάβη γὰρ ἐνίοις οὐ σμικρὰ γίνεται γυμναζομένοις

413K ἐν πλήθει σιτίων, ἐνίοις δὲ καὶ ἡ κεφαλὴ πληροῦται καὶ κατὰ τὸ ἧπαρ αἴσθησις γίνεται τάσεως ἢ βάρους ἢ ἀμφοῖν. ὅταν οὖν τι τοιοῦτον συμβαίνῃ, θεραπευ- τέον αὐτὸ παραχρῆμα, τὰς μὲν κατὰ τὸ ἧπαρ ἐμφρά- ξεις διὰ τῶν ἐκφραττόντων, τὰς δὲ τῆς κεφαλῆς πλη- ρώσεις διὰ περιπάτων, μάλιστα μὲν πρὸ τροφῆς, οὐδὲν μὴν κωλύει καὶ μετὰ τροφήν. ἀλλὰ τούτους μὲν ἐσχάτως βραδεῖς ποιεῖν προσήκει, τοὺς δὲ πρὸ τῆς τροφῆς συντονωτέρους μὲν τούτων, οὐ μὴν τοιούτους, ὁποίους ἐπειγόμενοι πρός τινα πρᾶξιν ἐνεργοῦμεν.

ὅσα δ' ἐκφράττει τὸ ἧπαρ, ἁρμόζει ταῦτα ταῖς βραδυπεψίαις, ἄριστα δ' αὐτῶν ἐστιν ὀξύμελί τε καὶ τὸ διὰ τριῶν πεπέρεων, ἐν ᾧ μηδὲν μέμικται τῶν ἀήθων φαρμάκων, καὶ ἡ λεπτύνουσα δίαιτα· ταῦτα γὰρ βραδυπεψίας ἐπανορθοῦται καὶ τὰς κατὰ τὸ ἧπαρ ἐμφράξεις ἰᾶται. διαφθορᾶς δὲ τῶν ἐν τῇ γα- στρὶ σιτίων γινομένης, οἷς μὲν ὑπέρχεται τὰ διεφθαρ- μένα, μέγιστον ἐφόδιον εἰς ὑγείαν ἔχουσιν, οἷς δὲ οὐχ ὑπέρχεται, διὰ τῶν ἀλύπως ὑπαγόντων ἐρεθιστέον.

---

19 The transliterated Greek terms, *bradypepsia* and *apepsia*, are used to signify abnormal slowness and complete failure of

didn't ever at any time drink after such food and ate only
the bread. The amount of each of these things should be
such that it can be concocted in the stomach by the tenth
hour. Also, if people wish to exercise, they will particularly
do so in this way without harm. In some, considerable
harm occurs, if they exercise when replete with food. In
some also, the head is filled and a sensation of tension or 413K
heaviness, or both arises in the liver. Whenever such a
thing happens, one must treat it immediately—the obstructions in the liver by agents that remove obstructions
and the fullnesses of the head by walking around, particularly before food, and if nothing prevents it, also after
food. But it is appropriate to do these things very slowly;
the former may be done more vigorously than the latter,
but not in the way we do these things when we are con-
ducting some business.

Those things that clear obstructions in the liver are also
suitable for the *bradypepsias*.[19] The best of these are oxymel, the medication made from the three peppers in
which none of the unusual medications have been mixed,
and the thinning diet. These things correct the *bradypepsias* and cure the obstructions in the liver. When corruption of foods occurs in the stomach, those who pass the
products of corruption have the best road to health. In
those who do not pass these, one must stimulate their passage through agents that effect downward purging pain-

concoction, respectively. The reason is that "concoction" and "digestion" have different meanings, and these terms are in fact
technical terms specific to the pathophysiological ideas espoused
by Galen and others of the time.

414K ἔστι δὲ ταῦτα τό τε Διοσπολιτικὸν ὀνομαζόμενον
φάρμακον, ὅταν ἴσον τοῖς ἄλλοις λάβῃ τὸ νίτρον, τό
τε διὰ ἰσχάδων καὶ κνίκου, ὅσα τ᾽ ἄλλα διὰ κνίκου ἢ
δι᾽ ἐπιθύμου συντίθεται. τοῖς δ᾽ οὕτω διακειμένοις ἀνθρώποις συμφέρουσι καὶ οἱ πρὸ τῆς τροφῆς ἔμετοι δι᾽
οἴνου πόσεως γλυκέος γινόμενοι. συμβουλεύειν δὲ
αὐτοῖς χρὴ μηδὲν κνισσῶδες ἢ βρομῶδες ἢ ὅλως εὔ-
φθαρτον ἔδεσμα προσφέρεσθαι, τὰ καλούμενα δ᾽ ὑπὸ
τῶν ἰατρῶν εὔχυμα, ταῦτ᾽ ἐκλέγεσθαι. γέγραπται δὲ
περὶ αὐτῶν ἑτέρωθι κατὰ τὸ Περὶ τῆς εὐχυμίας τε καὶ
κακοχυμίας ὑπόμνημα· γέγραπται δέ, ὡς ἔφην ἤδη,
κἂν τοῖς τρισίν, ἐν οἷς Περὶ τῆς ἐν ταῖς τροφαῖς δυνάμεως ὁ λόγος ἐστίν.

ἐπιτήδειον δὲ τοῖς οὕτως ἔχουσιν ἀνθρώποις ἐκ
διαλειμμάτων χρόνου συμμέτρου τῆς γαστρὸς ὑπαγωγὴ διὰ τῶν μετρίως καθαιρόντων, ὁποῖόν ἐστι τὸ
δι᾽ ἀλόης πικρόν· οὕτω γὰρ αὐτὸ καλοῦσι πολλοὶ τῶν
ἰατρῶν, ὥσπερ ἔνιοι πικρὰν θηλυκῶς. ἐὰν δ᾽ ἐπιτρέ-
ψῃς πολλῷ χρόνῳ τὴν κακοχυμίαν ἀθροίζεσθαι, νό-
415K σημά τι χαλεπὸν αὐτοῖς συμβήσεται. καὶ δύνανταί
γε τὴν τοιαύτην ἑαυτῶν πρόνοιαν ποιεῖσθαι κατὰ τὰς
ἡμέρας ἐκείνας, ἐν αἷς ἑορτή τίς ἐστι δημοτελὴς ἐλευ-
θεροῦσα τῆς δουλικῆς ὑπηρεσίας αὐτούς. ἀλλὰ δι᾽
ἀκρασίαν οὐ μόνον οὐδὲν ποιοῦσιν εἰς ἐπανόρθωσιν
τῶν κατὰ τὸ σῶμα μοχθηρῶς ἠθροισμένων, ἀλλὰ καὶ

---

[20] Two meanings of the term ἀκρασία are listed in LSJ: (1) as
the opposite of εὐκρασία (referring to Hippocrates, *Ancient*

lessly. These are the so-called Diospoliticum medication   414K
when niter is mixed equally with the other ingredients,
and the medication made with dried figs and safflower
(*Carthamus tinctorius*), and others that are compounded
with safflower or epithyme (*Cuscuta epithymum*). Vomiting before food also benefits people in this state when it
occurs due to a drink of sweet wine. It is necessary to
advise them not to take foods that are smoky (fuliginous),
malodorous or, to speak generally, easily corruptible, but
to choose those called *euchymous* by doctors. I have written about these elsewhere, in the treatise, *On the Good
and Bad Humors of Nutriments*. I have also written about
them, as I already said, in the three books of my work, *The
Powers of Foods*.

It is advantageous for people in such a state to evacuate
the stomach downward after a moderate time interval
through those things that are moderately purging, like bitter aloes, as the majority of doctors call it, just as others
call it "female bitterness." If you allow the *kakochymia* to
build up over a long time, a disease of some severity will
come about in these people. And they are able to make   415K
such provision for themselves during those days on which
there is some public festival, when they free themselves
from the services of a slave. But due to lack of control
(*akrasia*)[20] they not only do nothing to correct those things
collecting deleteriously in the body, but they also fill them-

Medicine 7) and (2) as a form of ἀκράτεια meaning lack of power
or control, or debility (referring to Hippocrates, *Aphorisms* 5.16).
Context would seem to demand the second usage here. Galen
does not use the term in relation to *eukrasia / dyskrasia*.

163

προσεπιπληροῦσιν αὐτὰ κακῶς διαιτώμενοι κατὰ τὰς
ἑορτάς. ἁλίσκονται τοιγαροῦν νοσήμασιν ἔνιοι μὲν
αὐτῶν οὕτως χρονίοις, ὡς ἐν ἁπάσῃ τῇ ζωῇ παραμέ-
νειν, ὁποῖόν ἐστι κακὸν ἥ τε ποδάγρα καὶ ἡ ἀρθρῖτις
καὶ ἡ νεφρῖτις, ἔνιοι δὲ ὀξέσι μέν, ἀλλ᾽ ἐνοχλοῦσιν
αὐτοῖς ἤτοι καθ᾽ ἕκαστον ἔτος ἢ πάντως γε διὰ δυοῖν,
τινὲς δὲ καὶ δὶς οὕτω νοσοῦσιν καθ᾽ ἕκαστον ἔτος.

8. Ἡ δ᾽ ὑγιεινὴ τέχνη τοὺς πειθομένους αὐτῇ διαφυλάττειν ὑγιαίνοντας ἐπαγγέλλεται, τοῖς δ᾽ ἀπει-
θοῦσιν ἐν ἴσῳ καθέστηκεν, ὡς εἰ καὶ μηδ᾽ ὅλως ἦν.
ἀπειθοῦσι δ᾽ ἔνιοι μὲν ὑπὸ τῆς ἐν τῷ παραχρῆμα
νικηθέντες ἡδονῆς, οὓς ἀκρατεῖς τε καὶ ἀκολάστους
ὀνομάζομεν, ἔνιοι δὲ ὑπὸ φιλοτιμίας, ἣν ὀνομάζουσιν
416K   οἱ νῦν Ἕλληνες κενοδοξίαν, ὁποῖος ἦν καὶ ὁ πάντα
μᾶλλον ὑπομείνας παθεῖν ἢ πιττοῦσθαι συνεχῶς ὅλον
τὸ σῶμα συμβουλευόντων αὐτῷ τῶν ἰατρῶν τὸ βοήθημα τοῦτο διὰ τὴν ἰσχνότητα. γίνεται δ᾽ αὕτη τισὶ
μὲν ἐπὶ τῇ καθ᾽ ὅλον τὸ σῶμα δυσκρασίᾳ κατὰ τὴν
ἐπὶ τὸ ξηρόν τε καὶ ψυχρὸν ὑπερβολήν, ἐνίοις δὲ διὰ
τὴν ἀναδοτικὴν δύναμιν ἢ θρεπτικὴν ἢ ἀμφοτέρας
ἀρρωστοτέρας οὔσας φύσει. πάντας δὲ τοὺς οὕτως
ἔχοντας ὀνίνησιν ὁ καλούμενος ὑπὸ τῶν νῦν Ἑλλήνων "δρῶπαξ"· καὶ γὰρ ἀναδόσει συντελεῖ καὶ θρέ-
ψει καὶ πολλούς γε τῶν ἰσχνῶν ἔμπροσθεν ἔστιν ἰδεῖν
παχυνθέντας ἐπὶ τῷ βοηθήματι καὶ μάλισθ᾽ ὅσοις ἡ

---

21 For Galen's ideas on the "faculties" or "capacities" responsible for the intake, processing, and distribution of nutriments,

selves full of these things by eating badly at the festivals. For that very reason, they are seized by diseases, some of which are so chronic that they remain throughout their lives—bad diseases like gout, arthritis and nephritis. In others, the diseases are acute but afflict them either every year, or at all events every two years, whereas some become ill in this way twice each year.

8. The art of hygiene professes to keep healthy those who put their trust in it. For those who do not put their trust in it, it is as if it did not exist at all. Some do not put their trust in it because they are overcome by the pleasure of the moment—we call such people weak-willed and ill-disciplined. In others, it is because of their love of honor, which the Greeks now call vanity, like the man who endured all manner of suffering rather than have his whole body continuously covered in pitch, which was the remedy the doctors recommended for him on account of his thinness. In some, this occurs from a *dyskrasia* involving the whole body, related to an excess of dryness and coldness. In some, however, it is due to the distributive capacity or the nutritive,[21] or both being quite weak in nature. The *drōpax* (pitch plaster),[22] as it is called by the Greeks of the present-day, benefits all those affected in this way; for this also contributes to both distribution and nutrition. Many of those who were previously thin are seen to be fattened up by the remedy. This is particularly so in those in whom

416K

see particularly Book 1 of his *Nat. Fac.,* II.1–73K, and Brock's introductory discussion of the subject.

[22] On *drōpax*, see Hippocrates, *Epidemics* 19; Galen, *Hipp. Off. Med.,* XVIIIB.894K; and Oribasius, frag. 75.

μὲν ἀναδοτικὴ δύναμις ἄρρωστός ἐστιν, ἡ θρεπτικὴ
δὲ οὐκ ἄρρωστος μέν, ἀλλὰ δι᾽ ἀπορίαν ὕλης ἐπιτη-
δείου τρέφειν αὐτάρκως οὐκ ἐδύνατο, τούτου συμβαί-
νοντος αὐτῇ δι᾽ ἔνδειαν τῆς ἀναδιδομένης τροφῆς.
ἀλλ᾽ ἔνιοι, ὡς ἔφην, ὑπὸ φιλοτιμίας, ἵνα μὴ δόξωσιν
ὁμοίως τοῖς τρυφῶσιν ἢ καλλωπιζομένοις πιττοῦσθαι, φεύγουσι τὴν ἐξ αὐτοῦ βοήθειαν ἄλλο τι κελεύοντες[20] εὑρεῖν αὐτοῖς βοήθημα τῆς ἰσχνότητος.

417K ἔστι δὲ οὐδὲν τῷδε δύναμιν ἔχον ἴσην· ὠφελεῖν μέντοι
δύναται καὶ τὰ τοιαῦτα· πρὸ τοῦ λουτροῦ χερσὶ μὴ
πάνυ μαλακαῖς, ὥσπερ αὖ μηδὲ τραχείαις, ἀνατρίβειν
τὸ σῶμα, μέχριπερ ἂν ἔρευθος σχῇ, κᾆπειτα τρίψει
σκληρᾷ μὴ πολλῇ πιλοῦντα τὸ δέρμα πυκνὸν καὶ
σκληρὸν ἐργάζεσθαι καὶ μετὰ ταῦτα γυμνασίοις
συμμέτροις χρησάμενον, εἶτα λουσάμενον ἄνευ τοῦ
χρονίσαι κατὰ τὸ βαλανεῖον, ἐκμάξασθαι παραπλησίως τῇ προειρημένῃ ξηροτριβίᾳ καὶ μετὰ ταῦτα ἐπα-
λειψάμενον ἐλαίῳ βραχεῖ προσφέρεσθαι τὰ σιτία.

σκοπὸς γὰρ ἐπὶ τῶν τοιούτων ἐστὶν εἰς τὸ σαρκῶ-
δες γένος ἑλκύσαντας αἷμα χρηστὸν ἐπιρρῶσαι τὴν
ἐν αὐτοῖς θρεπτικὴν δύναμιν, ὅπως μὴ διαφορηθῇ τὸ
ἑλχθέν. ῥώννυται μὲν οὖν ἡ θρεπτικὴ δύναμις ἐκ τοῦ
θερμανθῆναι τὰς σάρκας, οὐ διαφορεῖται δὲ τὸ παρ-
αγόμενον εἰς αὐτὰς αἷμα χρισαμένων ἐλαίῳ δύναμιν
ἐμπλαστικὴν ἔχοντι φαρμάκου. εἰ δὲ τὰ τῆς ἡλικίας
ἐπιτρέπει, καὶ ψυχρολουσίαις ὁ αὐτὸς ἄνθρωπος χρώ-

[20] post κελεύοντες: add. ἡμᾶς (Ku)

the distributive capacity is weak, while the nutritive capacity is not weak, but unable to nourish adequately due to lack of suitable material. This happens to it due to a deficiency of distributed nutriment. But there are some, as I said, who due to vanity, do not accept being pitched—people such as those who are delicate and beautify themselves. Rather, they eschew the remedial effect of pitching, directing us to discover some other remedy for their thinness. There is, however, nothing with a potency equal to pitching. Nonetheless, such things [as the following] are able to bring benefit: before the bath massaging the body with the hands, neither very gently nor again very roughly, until it becomes red; and then by firm massage (but not much) compressing the skin to make it dense and hard; and after these things using moderate exercises, then bathing without spending a long time in the bath, and wiping off like the previously mentioned dry massage; and after these things, anointing with a little oil and giving food.

417K

The aim of such measures is, by drawing useful blood to the fleshy class [of structures], to strengthen the nutritive capacity in them in such a way that what is drawn is not dispersed. Thus, the nutritive capacity is strengthened by the fleshes being heated, while the blood diverted into them is not dispersed by rubbing with oil which has the potency of an emplastic medication.[23] If, however, the factors pertaining to age permit, and the person himself uses

23 For the term ἐμπλαστικός meaning "to cause to adhere," see Dioscorides, 1.102.

μενος ἅμα τοῖς εἰρημένοις ὀνήσεται μεγάλως. ἔμπα-
418K λιν δὲ ἐπὶ τῶν ἀμέτρως παχυνθέντων ἐκλύειν μὲν χρὴ
τὴν ἀνάδοσιν, αὐξάνειν δὲ τὰς ἀπορροίας τοῦ σώματος. ἡ μὲν οὖν ἀνάδοσις ἐκλύεται συνεχέσιν ὑπαγω-
γαῖς γαστρὸς ἐθισάντων ἡμῶν τὰ τῆς τροφῆς ὄργανα
κάτω ταχέως ὠθεῖν, ὅσα περιέχεται κατὰ ταῦτα. τὰς
δὲ ἀπορροίας τοῦ σώματος αὐξήσομεν ὀξέσι τε γυμνασίοις, ὧν ἐστι καὶ δρόμος, ἀλείμμασί τε διαφορητικοῖς τρίβοντες ἐπὶ πλεῖστον. εἶναι δὲ χρὴ δηλονότι
τὴν τρῖψιν μαλακήν τε καὶ ἀραιωτικήν.

ἐγὼ γοῦν παχύν τινα ἱκανῶς ὀλίγῳ χρόνῳ συμμέτρως εὔσαρκον εἰργασάμην ὀξεῖ δρόμῳ χρῆσθαι
καταναγκάζων, εἶτ’ ἀπομάττων μὲν τὸν ἱδρῶτα σινδό-
σιν ἤτοι λίαν μαλακαῖς ἢ λίαν τραχείαις, ἐφεξῆς δὲ
τρίβων ἐπὶ πλεῖστον ἀλείμμασι διαφορητικοῖς, ἃ καλοῦσιν ἤδη συνήθως οἱ νεώτεροι τῶν ἰατρῶν "ἄκοπα."
καὶ μετὰ τὴν τοιαύτην τρῖψιν ἐπὶ τὸ λουτρὸν ἦγον, ἐφ’
ᾧ λουτρῷ τροφὴν μὲν εὐθέως οὐ προσέφερον, ἡσυχάζειν δὲ μεταξὺ κελεύων ἢ καί τι τῶν συνήθων αὐτῷ
πράττειν αὖθις ἐπὶ τὸ δεύτερον ἦγον λουτρόν, εἶτ’
419K ὄγκους ἐδεσμάτων ὀλιγοτρόφους ἐδίδουν, ὡς ἐμπίπλασθαι μέν, ὀλίγον δὲ ἐξ αὐτῶν εἰς ὅλον ἀναδίδο-
σθαι τὸ σῶμα.

καὶ τὰς ἄλλας γε δυσκρασίας ἐν ὅλῳ τῷ σώματι
γινομένας ὁμοίως ἐπανορθοῦσθαι χρή, πρῶτον μὲν

cold baths along with the things mentioned, he will be greatly benefitted. Contrariwise, in those who are excessively fat, it is necessary, on the one hand, to reduce the 418K distribution and, on the other, to increase the emanations of the body. Thus, the distribution is diminished by continuing downward purgings of the stomach, since our organs of nutrition are accustomed to quickly dispel downward those things contained in them. We shall increase the emanations of the body by rapid exercises, one of which is running, and much massage with diaphoretic unguents; obviously the massage must be gentle and rarefying.

At any rate, I have made someone who was very fat become moderately well-fleshed in a short time by compelling him to use fast running, and then, having wiped away the sweat with fine muslin cloths, either very soft or very rough, and next massaging still more with diaphoretic unguents which younger doctors are accustomed to call *akopa*.[24] After such massage, I am accustomed to lead the person to the bath. Following the bath, I did not give food immediately, but directed him to rest in between times or do one of the things he is accustomed to doing. Then, having led him to a second bath, I gave him large amounts of poorly nourishing foods so as to fill him up but have 419K little of the foods distributed from them to the whole body.

It is necessary to correct the other *dyskrasias* occurring in the whole body similarly, first discovering the objective

[24] Among the several uses listed in LSJ, that of an application that is refreshing and relieves weariness is probably most applicable here, although there may be an element of pain relief—see Hippocrates, *Aphorisms* 2.48. The term is defined and exemplified in detail in Galen's *Comp. Med. Gen.*, XIII.1005–9K.

ὧν προσήκει πράττειν εὑρίσκοντα τὸν σκοπόν, εἶτα
τὰς δυναμένας αὐτὰ δρᾶν ὕλας. διὸ καὶ σύντομος ὁ
νῦν λόγος ἅπασι γίνεται τοῖς μεμνημένοις τῶν τε τρίψεων καὶ τῶν γυμνασίων καὶ τῶν λουτρῶν τὰς διαγω-
γὰς ἔτι τε πρὸς τούτοις ἐδεσμάτων τε καὶ φαρμάκων.
περὶ μὲν οὖν τρίψεών τε καὶ γυμνασίων ἐν τῇδε τῇ
πραγματείᾳ πρόσθεν εἴρηται, περὶ δὲ τῶν τροφῶν
ἑτέρωθι, καθάπερ γε καὶ περὶ τῶν φαρμάκων. ἡ γάρ
τοι κατὰ μέθοδον διδασκαλία τὰ κοινὰ καὶ καθόλου
περιλαμβάνουσα εἰς πολλὰ τῶν κατὰ μέρος εὐμνημό-
νευτός τε καὶ σύντομος γίνεται. ἐγὼ δὲ οὐ μόνα τὰ
καθόλου λέγω, ἀλλὰ καὶ τὰ παραδείγματα τῶν κατὰ
μέρος αὐτοῖς προστιθεὶς τελειοτάτην ἡγοῦμαι ποιεῖσθαι τὴν διδασκαλίαν.

9. Καιρὸς οὖν ἤδη μεταβαίνειν ἐπὶ τὰς ἀνωμάλους
κατασκευὰς τῶν σωμάτων, αἳ δὴ καὶ νοσώδεις πάντως εἰσίν. ἀνώμαλοι δὲ κατασκευαὶ γίνονται τρισταί,
διότι καὶ ἡ τοῦ σώματος ἡμῶν σύνθεσίς ἐστι τριττή·
μία μὲν ἡ ἐκ τῶν πρώτων στοιχείων, ἐξ ὧν γέγονε τὰ
καλούμενα πρὸς Ἀριστοτέλους "ὁμοιομερῆ"· δευτέρα
δὲ ἡ ἐκ τούτων αὐτῶν τῶν ὁμοιομερῶν, ἃ δὴ καὶ αὐτὰ
στοιχεῖα πάλιν ἐστὶν αἰσθητὰ τῶν ἀνομοιομερῶν,
ὅθεν ἡ τῶν ὀργανικῶν γίνεται σύνθεσις· τρίτη δ' ἐπ'
αὐτοῖς ἡ τοῦ παντὸς σώματος, ἡ ἐκ τῶν ὀργανικῶν.
εὐκολωτέρα μὲν οὖν ἐστιν ἡ τρίτη λεγομένη καὶ δια-
γνωσθῆναι καὶ προνοίας τυχεῖν, δυσκολωτέρα δὲ ἡ
δευτέρα, χαλεπωτάτη δὲ ἡ πρώτη. βέλτιον οὖν ἀπὸ
τῆς τρίτης ἄρξασθαι ῥάονα τήν τε διάγνωσιν ἐχού-

of those things it is appropriate to do, then the materials that are able to do these things. On which account also, the present brief discussion is for all those calling to mind the instructions about massages, exercises and baths, and in addition to these, foods and medications. I spoke before about massages and exercises in this treatise, and about foods elsewhere, just as I did about medications. Certainly, the teaching by method, taking in what is common and general, is easy to remember and concise for the many things individually. I do not only refer to general principles but also add to these examples of individual matters, thinking to make the teaching absolutely complete.

9. Now it is time to pass on to the irregular (nonuniform, anomalous) constitutions of bodies—those which, in fact, are also altogether morbid. Irregular constitutions   420K
exist in three forms because the composition of our body is also threefold.[25] First, it is from the primary elements, from which what Aristotle called *homoiomeres* have arisen. Second, it is from these same *homoiomeres*, which are in turn the actual perceptible elements of the *an-homoiomeric* parts, from which the compounding of the organs arises. Third, in addition to these, there is the compounding of the whole body, which is from the organs. What I have termed the third is easier to understand, in terms of being both diagnosed and cared for. The second is more difficult and the first very difficult. It is better, then, to begin from the third, which is easier in respect of

[25] For an outline of Galen's tripartite division of bodily structure as a basis for the classification of diseases and symptoms, see Johnston, *Galen: On Diseases and Symptoms*, 69–72.

σης καὶ τὴν πρόνοιαν, οἷον εὐθέως ἐπὶ τῆς κεφαλῆς (οὐδὲν γὰρ χεῖρον ἐντεῦθεν ἄρξασθαι) δυσκράτου φύσει γινομένης, ὡς πολλὰ γεννᾶν περιττώματα, ἐξ ὧν αἱ βλάβαι γίνονται πᾶσι τοῖς ὑποκειμένοις ὀργάνοις, ἐφ' ὅ τι ἂν αὐτῶν τρέπηται τὸ περιττόν. ἑτοιμοτάτη μὲν οὖν ἡ εἰς τὸ στόμα καὶ τὰς ῥῖνας φορά, παραγίνεται δὲ καὶ εἰς ὀφθαλμούς, ἐνίοις καὶ εἰς ὦτα. τὴν δὲ

421K εἰς τὸ στόμα φορὰν τῶν περιττωμάτων ὑποδέχεται στόμαχός τε καὶ ἡ τραχεῖα ἀρτηρία (καλοῦσι δὲ αὐτὴν καὶ βρόγχον), ἧς τὸ συνάπτον τῷ στόματι πέρας ὑψηλὸν (ὀνομάζεται δὲ λάρυγξ) ὄργανον φωνῆς ὑπάρχει, ὡς ἐν τῷ Περὶ φωνῆς ὑπομνήματι δέδεικται. οὗτός τε οὖν αὐτὸς ὁ λάρυγξ ὑπὸ τῶν εἰς αὐτὸν ἐκ τῆς κεφαλῆς καταρρεόντων ὑγραινόμενος ἐν ἀρχῇ μὲν βραγχώδη τὴν φωνὴν ἐργάζεται, προϊόντος δὲ τοῦ χρόνου μικράν· ἢν δ' ἐπὶ πλέον ἥκῃ τὸ κακόν, ἀπόλλυται πᾶσα· συνδιαβρέχεται γὰρ τῷ λάρυγγι καὶ ἡ ἀρτηρία.

δριμέος δὲ τοῦ ῥεύματος ὄντος, οὐκ εἰς φωνὴν μόνον ἡ βλάβη τοῖς εἰρημένοις μορίοις, ἀλλὰ καὶ χαλεπωτάτη τις ἀνάβρωσις προσέρχεται τοιαύτη τὴν φύσιν, οἵας καὶ κατὰ τὸ δέρμα[21] θεώμεθα γινομένας πολλάκις ἄνευ τῆς ἔξωθεν αἰτίας. οὕτω δὲ καὶ πνεύμων αὐτοῖς ἑλκοῦται, καὶ φθόη καλεῖται τὸ πάθος. ἐὰν δὲ εἰς στόμαχόν τε καὶ γαστέρα τρέπηται τὸ ῥεῦμα,

---

[21] post κατὰ τὸ δέρμα: θεώμεθα Ko; βλέπομεν βλαπτόμενον Ku

diagnosis and care. For example, directly in the case of the head (it is no bad thing to start here), when it becomes *dyskratic* in nature, it generates many superfluities from which harms occur to whichever of all the organs lying below [the head], to which some of the superfluity is diverted. The easiest passage is to the mouth and nose, but there is also passage to the eyes, and in some cases to the ears. The esophagus and the rough artery (people also call   421K this "bronchus") receive below the passage of the superfluities to the mouth. The upper end of the rough artery (which is called the "larynx") that is joined to the mouth is the organ of voice, as I have shown in my work, *On the Voice*.[26] This then, the larynx itself, when it is moistened by the things flowing down to it from the head, at first makes the voice hoarse, and as time progresses, small. If the badness comes down to it even more, it is destroyed altogether because the rough artery (trachea and main bronchi) is wet through completely, along with the larynx.

If the flux is acrid, the injury is not to the voice alone among the parts mentioned, but also a very troublesome erosion develops, which is of the same nature as we often see occurring in the skin without an external cause. In this way too, the lungs are ulcerated in these cases, and the affection is called *phthoe* (*phthisis,* consumption). However, if the flux is diverted to the esophagus and stomach,

[26] The work *De voce* (Περὶ φωνῆς) was in four books and dedicated to Boethus. The original has been lost, although some fragments and an Arabic summary remain. See Boudon, *Galien,* 419n3.

ψυχρὸν μὲν ὑπάρχον εἰς δυσκρασίαν ψυχρὰν ἄγει τὰ
σώματα, θερμὸν δὲ εἰς θερμήν. ἑλκοῖ δὲ καὶ τοῦτο τῷ
χρόνῳ, κατ᾽ ἀρχὰς δὲ τάς τε ὀρέξεις βλάπτει καὶ τὰς

422K   πέψεις. ἐὰν μὲν οὖν ψυχρὸν ᾖ τὸ ῥεῦμα, βραδυπεψίας
τε καὶ ἀπεψίας καὶ ὀξυρεγμίας ἐργάζεται· ἐὰν δὲ καὶ
διεφθαρμένον, εἰς διαφθορὰν ἄγει τὴν τροφὴν κνισ-
σώδεις ἢ ὀξώδεις ἐρυγὰς ἀναπέμπουσαν ἤ τινος ἑτέρας ἀρρήτου τε καὶ ῥητῆς ποιότητος. ὑποκαταβαί-
νουσα δὲ ἡ βλάβη καὶ τὴν νῆστιν ἀδικεῖ καὶ τὸ
κῶλον. ἅπτεται δὲ καὶ τῶν κατὰ τὸ μεσεντέριον ἀγγείων, δι᾽ ὧν εἰς ἧπαρ ἡ ἀνάδοσις γίνεται· καὶ τισὶ
μὲν ἀνορεξίαι παρακολουθοῦσιν, ἐνίοις δὲ ὀρέξεις
παρὰ φύσιν, ἃς ὀνομάζουσι "κυνώδεις," ἢ καὶ μοχθη-
ρῶν ἐδεσμάτων ἐπιθυμία καθάπερ ταῖς κιττώσαις.
εὔδηλον δέ, ὅτι καὶ σταφυλὴ καὶ παρίσθμια καὶ
παρουλίδες ἀντιάδες τε καὶ βρώσεις ὀδόντων ἑλκώσεις τε καὶ σηπεδόνες ἐν τῷ στόματι τοῖς ἐκ τῆς κεφαλῆς εἰς αὐτὸ καταρρέουσιν ἰχῶρσιν ἕπονται. καὶ
τό γε πολὺ πλῆθος τῶν ἰατρῶν ἢ τὸν γαργαρεῶνα
τέμνουσιν ἢ φάρμακα διδόασιν, ὅπως ἀναπτύηται τὰ
διὰ τῆς τραχείας ἀρτηρίας εἰς τὸν πνεύμονα κατ-
ενεχθέντα. τῆς γαστρὸς δὲ ἄλλοι προνοοῦνται, καθάπερ ἄλλοι τῶν ὀδόντων τε καὶ τοῦ στόματος ἢ καὶ

423K   τῶν ἐν τῇ ῥινὶ συνισταμένων, ἵνα τοὺς ὀφθαλμοὺς ἢ
τὰ ὦτα παραλείπω, βλαπτομένων οὐκ ὀλίγων εἰς
αὐτά.

βέλτιον δ᾽ ἦν, οἶμαι, τὴν οἷον πηγὴν τῶν κακῶν
ἐκκόψαι ῥώσαντα τὴν κεφαλὴν ἤ, εἴπερ ἀδύνατον εἴη

174

and is cold, it leads the bodies to a cold *dyskrasia,* and if
it is hot, to a hot *dyskrasia.* This may ulcerate in time, but
at the start harms the appetites and concoctions. Thus, if   422K
the flux is cold, it produces *bradypepsias, apepsias* and
heartburns.[27] And if it is corrupted and leads the nutriment to corruption, it sends forth greasy and acidic eructations, or others of an unnamed or named quality. The
damage, descending by degrees, also harms the jejunum
and colon, and involves the vessels in the mesentery,
through which the distribution to the liver occurs. In
some, anorexias follow as well; in others there are unnatural appetites which people call "ravenous,"[28] or the desire
for bad foods, as in pregnant women. And it is clear that
inflammation of the uvula, tonsillitis, gumboils, throat in-
fections, dental caries, and ulcers and putrefaction in the
mouth follow the downward flow of ichors from the head
to the mouth. And in fact, the great majority of doctors
either excise the uvula or give medications to promote
expectoration of those things carried to the lungs via the
trachea. Others direct their care to the stomach, just as
others again do to the teeth and mouth, and also to those   423K
things existing in the nose. Let me leave aside the eyes and
ears; there are quite a number of harms that befall them.

It would, I think, be better to cut off the source, as it
were, of the bad things by strengthening the head, or if

---

[27] "Heartburn" is the term used to translate ὀξυρεγμία, defined in LSJ as "the sour fumes caused by indigestion, heartburn."
[28] Literally, "dog-like" (Aristotle, *Generation of Animals,*
746a33). For Galen's description see his *Sympt. Caus.,* VII.131K;
Johnston, *Galen: On Diseases and Symptoms,* 228.

175

τοῦτο διὰ τὴν σφοδρότητα τῆς φυσικῆς δυσκρασίας,
προνοεῖσθαι γοῦν αὐτῆς διὰ παντὸς τὴν ἔνδειξιν τῆς
προνοίας ἀπὸ τῆς κατ' αὐτὴν ἰδέας ποιούμενον, οὐχ
ὡς ἔνιοι τῶν ἰατρῶν ἁπάσαις ἀεὶ ταῖς κεφαλαῖς τὰ
διὰ θαψίας καὶ νάπυος προσφέρουσιν, οὕτω δὴ καὶ
αὐτοὺς πράττοντας. εἰ γὰρ ὑπὸ θερμῆς δυσκρασίας ἡ
κεφαλὴ κακῶς ἔχει, βλαβερὰ τὰ τοιαῦτα φάρμακα.
χρὴ τοίνυν αὐτοὺς λουτροῖς πολλοῖς ποτίμων ὑδάτων
παρηγορεῖν διαφοροῦντάς τε ἅμα τοὺς ἐν τῇ κεφαλῇ
γενομένους ἀτμοὺς θερμοὺς καὶ τὴν κρᾶσιν ἐργαζομένους ὅλην βελτίω. βλαβερὰ δὲ τούτοις ἡ τῶν αὐτοφυῶν ὑδάτων θερμῶν χρῆσις. ὅσα μὲν γὰρ αὐτῶν
θειώδη τέ ἐστι καὶ ἀσφαλτώδη, τῷ θερμαίνειν ἐναντιώτατα ταῖς φύσει θερμαῖς κεφαλαῖς, ὅσα δὲ στυ-
424K πτηριώδη, τῷ στεγνοῦν. μόνοις δ' ἂν, εἴπερ ἄρα, τοῖς
γλυκέσι τῶν αὐτοφυῶν ὑδάτων ἀβλαβῶς χρῶντο·
τοῦτο γὰρ ἀσφαλὲς εἰπεῖν, ὡς τό γέ τι καὶ ὠφελεῖσθαι παρ' αὐτῶν οὐχ ὁμοίως ἀσφαλές. οὐ γὰρ ἂν ἦν
ἴσως θερμὰ μὴ μετέχοντά τινος δυνάμεως φαρμακώδους θερμῆς.

ἄμεινον δὲ τῇ πείρᾳ κρίνειν τὰ τοιαῦτα τῶν ὑδάτων·
καὶ γὰρ καὶ σπανίως εὑρίσκεται παρ' ἡμῖν ἀπὸ στα-
δίων τῆς πόλεως πολὺ πλειόνων ἑκατόν, ἐν Προύσῃ
δὲ ἐλαττόνων ἢ δέκα. τὸ μὲν οὖν παρ' ἡμῖν ἐν Ἀλλι-
ανοῖς (οὕτω γὰρ ὀνομάζεται τὸ χωρίον) ἑνὸς εἴδους
ἐστὶν ὅλον, ἐκ μιᾶς πηγῆς ἀνίσχον· ἐν δὲ τῷ τῆς
Προύσης προαστείῳ[22] καὶ ἄλλη πηγὴ φαρμακώδους

this is impossible due to the great severity of the natural *dyskrasia,* to at least take care of this throughout, taking the indication of the care from the specific kind of this and not, as some doctors do, always giving [the medication made with] *thapsia* (deadly carrot) and mustard for all headaches—and this is in fact what they do. For if the head is badly affected by a hot *dyskrasia,* such medications are harmful. Accordingly, you must alleviate these cases with many baths of potable waters, which at one and the same time disperse the hot vapors arising in the head and make the *krasis* as a whole better. For them, the use of natural hot waters is harmful—those waters that are sulfurous and full of asphalt are, due to their heating effect, most inimical to heads which are hot in nature, while those that are alum containing are harmful due to their constricting effect. Use only the natural waters that are 424K sweet, which are without harm, if any are. It is safe to say this: the benefit derived from these waters is not equally safe. Perhaps, if they were not hot, they would not partake of any power of medicinal heat.

It is better to judge such waters by experience. Furthermore, they are rarely found—with us (i.e., in Italy), they are much more than a hundred *stadia* from the city; in Prusa, they are less than ten. With us, in Allia (for this is what they call the place), there is just one kind and it comes from one spring. In the area around Prusa, there is another spring of medicinal water, as there is with us in

---

22 προαστείῳ Ko; ἑτέρου Ku

ὕδατος, ὥσπερ παρ᾽ ἡμῖν ἐν Λυκέτοις. ὅσοι δὲ σφόδρα θερμὴν καὶ διακαῆ ἔχουσι κεφαλήν, ἄμεινον
αὐτοῖς ὥρᾳ θέρους ἀλείφεσθαι ῥοδίνῳ χωρὶς στυμ-
μάτων ἐκ μόνων ῥόδων γεγονότι· καὶ τούτου γε αὐτοῦ
βέλτιόν ἐστι τὸ καλούμενον ὀμφάκινόν τε καὶ ὠμοτριβὲς ἔλαιον ἐμβεβλημένων τῶν ῥόδων καὶ πολύ γε
βέλτιον, εἰ καὶ χωρὶς ἁλῶν ἐσκευασμένον εἴη. τινὲς
425K δὲ κεφαλαὶ διὰ τὴν τῶν ἀρτηριῶν κρᾶσιν ἐν συμ-
πτώμασι γίνονται σκοτωματικοῖς τε καὶ κεφαλαλγικοῖς, ἐφ᾽ ὧν καὶ τῇ ἀρτηριοτομίᾳ χρώμεθα τῆς ὑγιεινῆς πραγματείας ἐκπεπτωκυιῶν.

10. Τὰς μέντοι συνεχῶς ὀδυνωμένας διὰ τὴν τῶν
νεύρων εὐαισθησίαν, ὅσα τῆς κεφαλῆς ἐκφυόμενα τῷ
πλείστῳ μὲν ἑαυτῶν μέρει τὸ τῆς γαστρὸς στόμα δια-
πλέκει, φέρεται δέ τις αὐτῶν μοῖρα καὶ πρὸς τὴν λοι-
πὴν γαστέρα, τῆς ὑγιεινῆς τέχνης ἔργον ἰάσασθαι,
μᾶλλον δ᾽ ὅπως μηδ᾽ ὅλως γίνηται ταῦτα παρασκευ-
άζειν. ἀρίστη δὲ προφυλακὴ προνοουμένων ἡμῶν τὸν
χολώδη χυμὸν ἤτοι μηδ᾽ ὅλως συρρεῖν εἰς τὴν κοιλίαν ἢ ὅτι τάχιστα κενοῦσθαι. τοῦ μὲν οὖν μηδ᾽ ὅλως
συρρεῖν ἡ πρόνοια γίνεται διὰ τοῦ θᾶττον ἐσθίειν.
ὅταν μὲν γὰρ ἡ μὲν ὅλου τοῦ σώματος κρᾶσις ᾖ χο-

29 Prusa was a settlement at a site called Cius that came under
the control of the king of Bithynia in 202 BC. The city was re-
named Prusias and bequeathed to the Roman Empire by Nico-
medes IV, the last king of Bithynia, in 74 BC. Allia was a river that
joined the Tiber. The confluence was the site of a major battle ca.

Lyceti.[29] With those who have a very hot and burning head, it is better for them to be anointed in summer with oil of roses made from roses alone without astringents. And better than this itself is the so-called *omphakinos* or *omotribes* (crude olive oil)[30] when oil of roses is added; and this is much better if it is prepared without salt. Due to the *krasis* of the arteries, some heads develop the symptoms of dizziness and headache, in which cases we also use arteriotomy, although this falls outside the scope of hygiene.

425K

10. With respect to the pains suffered continuously due to the hypersensitivity of the nerves,[31] which take their origin from the head and for the most part interweave themselves around the opening of the stomach, although some parts of them are also carried on to the rest of the abdomen, it is the task of the art of hygiene to treat, or rather, to make preparation, so that these do not occur at all. The best prophylaxis is when we give prior care to the bilious humor, so that it either does not flow to the stomach at all, or is very quickly evacuated. Prior care, so that it does not flow at all, is achieved by eating quite quickly. For whenever the *krasis* of the whole body is too bilious,

390 BC, in which the Gauls defeated the Romans. It is recorded by several historians (Livy, Tacitus, Polybius).

[30] ὀμφάκινον ἔλαιον is an oil made from unripe olives; ὀμοτριβὲς ἔλαιον is also oil made from unripe olives—for both terms, see Dioscorides, 1.30.

[31] The reference is to the vagus, or tenth cranial nerve arising bilaterally from the medulla oblongata and passing down to supply the gastrointestinal tract as far as the splenic flexure of the colon, as well as to the heart and respiratory structures.

λωδεστέρα, τὰ δὲ κατὰ κοιλίαν ἀτονώτερα, στηρίζε-
σθαι δεῖται διὰ τροφῆς εὐστομάχου. μὴ γινομένου δὲ
τούτου καταρρέουσιν ἰχῶρες εἰς αὐτήν, οὓς ἀθροίζει
τὸ σῶμα περιττούς. ἐκ δὲ τῆς τούτων ἀναθυμιάσεως
οὐ μόνον αἱ κεφαλαλγίαι γίνονταί τισιν, ἀλλὰ καὶ τὰ
426K τῶν ὑποχεομένων συμπτώματα, καί τισιν αὐτῶν ἐπι-
ληπτικοὶ σπασμοί.

τούτοις οὖν ἅπασιν ἡ μὲν ὅλη δίαιτα πρὸς τὸ ψυχρότερον μᾶλλον καὶ ὑγρότερον ὑπαλλαχθήτω, κε-
νούσθω δὲ τὰ συρρέοντα δι' ἐμέτων τε καὶ γαστρὸς
ὑπαγωγῆς, ῥωννύσθω δ' ὁ στόμαχος ἐδέσμασι μὲν
ἑκάστης ἡμέρας πρὶν ἐκχολωθῆναι, φαρμάκοις δ' ἐκ
διαλειμμάτων μακροτέρων, ἀψινθίου τε πόσει²³ καὶ
τοῦ διὰ τῆς ἀλόης, ἣν καλοῦσι πικράν. ἐκκαθαίρει
γὰρ τοῦτο καὶ τὸ διὰ βάθους χολῶδες ἐν τοῖς χιτῶσι
τῆς γαστρός, οὐ πάνυ τι τοῦ ἀψινθίου δυναμένου
τοῦτο ποιεῖν· ῥυπτικαί τε γὰρ αὐτῷ καὶ στυπτικαὶ δυνάμεις εἰσὶ μόναι, ὥστε τὸ μὲν ὡς ἂν εἴποι τις ῥύπον
τῶν χιτώνων τῆς γαστρὸς ἀποσμᾷ, οὐ μήν, εἴ τι κατὰ
βάθος αὐτοῦ ἐστιν, ἕλκειν ἔξω τοῦτο καὶ κενοῦν,
ὥσπερ ἡ ἀλόη. τὰ δ' ἔξωθεν τοῦ σώματος χρίσματα
τῆς γαστρὸς ἐπὶ τούτων ἔστω μετρίως στύφοντα.
μήλινον δέ ἐστι καὶ μαστίχινον καὶ νάρδινον καὶ τὰ
τοιαῦτα· θέρους μὲν τὸ μήλινον μᾶλλον, ἐν χειμῶνι δὲ
ἡ νάρδος, ἦρος δὲ τὸ μαστίχινον· ἐν τῷ μέσῳ γὰρ
427K τοῦτό ἐστι μηλίνου τε καὶ ναρδίνου μύρου, θερμαίνον-

²³ πόσει Ko; κόμης Ku

180

while the stomach is too weak, there is a need to settle this
with wholesome nutriment. If this is not done, ichors,
which the body collects as superfluities, flow down to it.
From the exhalation of these, not only do headaches occur
in some people, but also the symptoms of cataracts,[32] and   426K
in some of these cases, epileptic seizures [occur].

For all these, then, the whole regimen should be
changed—especially to one that is colder and more
moist—and those things flowing in should be evacuated
through vomiting and downward evacuation of the stomach, while the esophagus should be strengthened each day
with foods, before it becomes charged with bile, and with
medications at longer intervals—a drink of absinth and of
the medication made with aloes, which they call bitter. For
this also purges the bile deep in the tunics of the stomach;
the absinth is certainly not able to do this. Its powers are
only detergent and astringent, so that one might say, they
only clean out some dirt from the tunics of the stomach,
but if the dirt is in the depths of the stomach, they do not
draw this out and evacuate it, like the aloes does. Let the
unguents applied externally to the body of the stomach in
these cases be moderately astringent—apple, mastic, nard
and such things. In summer, apple is better; in winter,
nard; and in spring, mastic. The last is midway between
apple and oil of spikenard, since spikenard is heating, and   427K

---

[32] This reading of ὑποχεομένων is given in LSJ, citing this
passage and Chrysippus, SVF, 2.52.

τος μὲν τοῦ ναρδίνου, καὶ μάλισθ᾽ ὅταν ἀμώμου πλέον ἔχῃ, ψύχοντος δὲ τοῦ μηλίνου.

ταῖς δὲ πλουσίαις γυναιξὶν ὑπάρχει νάρδου κρεῖττον χρίσμα, φουλιᾶτα τε καὶ τὰ καλούμενα σπικᾶτα, θερμαίνοντα καὶ ῥωννύντα τὴν γαστέρα. ἐὰν δὲ μὴ μόνον ἡ κεφαλὴ θερμοὺς ἰχῶρας ἐπιπέμπῃ τοῖς κατὰ γαστέρα χωρίοις, ἀλλὰ καὶ αὐτὰ τύχῃ θερμὴν δυσκρασίαν ἔχοντα, τοῖς ἐμψύχουσιν ἐδέσμασί τε καὶ πόμασιν ἀεὶ χρῆσθαι προσήκει τὸ μᾶλλόν τε καὶ ἧττον ἐν αὐτοῖς ὑπαλλάττοντας κατὰ τὰς ὥρας τοῦ ἔτους, οὐ πρὸς τοὐναντίον ἀφικνουμένους, ὥσπερ ἐπὶ τῶν εὐκράτων ἐργαζόμεθα, χειμῶνος μὲν θερμαίνοντες, θέρους δὲ ψύχοντες. ὡσαύτως δὲ κἂν ἀμφότερα τὰ μόρια τὴν ψυχρὰν ἔχῃ δυσκρασίαν, ἀεὶ μὲν χρῆσθαι τοῖς θερμοῖς ἐδέσμασί τε καὶ πόμασι καὶ χρίσμασιν, ὑπαλλάττειν δὲ αὐτὰ κατὰ τὰς ὥρας, ἐπιτείνοντας καὶ ἀνιέντας. χαλεπὴ δὲ γίνεται μίξις, ὅταν ἤτοι γε εἰς θερμὴν γαστέρα καταρρέωσιν ἐκ τῆς κεφαλῆς οἱ ψυχροὶ καὶ φλεγματώδεις ἰχῶρες, εἰς τὴν
428K ψυχρὰν δὲ οἱ θερμοί· καὶ τούτων γ᾽ ἀμφοτέρων ἐπειράθην ἀεὶ δυσμεταχειριστοτέρας οὔσης τῆς ἑτέρας, καθ᾽ ἣν εἰς θερμὴν κοιλίαν οἱ φλεγματώδεις τε καὶ ψυχροὶ καταφέρονται χυμοί. χειρίστη δέ, ὅταν οὕτως ἔχῃ κατασκευῆς τε καὶ φύσεως ὁ ἄνθρωπος, ὡς μήτ᾽ εὔλυτον ἔχειν γαστέρα μήτ᾽ ἐμεῖν ἑτοίμως. διαφθείρονται γὰρ οἱ φλεγματώδεις χυμοί, κατὰ τὴν θερμὴν γαστέρα χρονίζοντες, ὡς δάκνειν ταύτην ἀναπέμπειν τε πρὸς τὴν κεφαλὴν ἀτμοὺς[24] μοχθηρούς.

particularly when it has more amomum, whereas the apple is cooling.

For rich women there is an unguent stronger than nard—*foliatum* and the so-called *spikata* (embrocation of spikenard),[33] which heats and strengthens the stomach. If, however, the head not only sends hot ichors to the places in the stomach, but these also happen to have a hot *dyskrasia*, it is always appropriate to use cooling foods and drinks, adjusting the amount of these according to the seasons of the year, without approaching the opposite, as we do in the case of the *eukrasias*, heating in winter and cooling in summer. And similarly, if both parts have a cold *dyskrasia*, always use hot foods, drinks and unguents, changing these according to the seasons, increasing and reducing. A troublesome mixing occurs when either cold and phlegmatous ichors flow down from the head to a hot stomach, or hot ichors to a cold stomach. It has always   428K
been my experience that of both of these, the one in which phlegmatous and cold humors flow down to a hot stomach was more difficult to deal with than the other. However, the worst is when the person is of such a constitution and nature that the stomach is not easily relaxed and does not vomit readily. For the phlegmatous humors are corrupted when they stay a long time in a hot stomach, so as to bite it and send noxious vapors up to the head.

---

[33] This is an ointment or oil made from the leaves of spikenard—see Pliny, 13.1.2, no. 15. Galen mentions its use in *MM*, X.574K.

ὡς ἐάν γε φλεγματώδης χυμὸς ἐκ τῆς κεφαλῆς εἰς ψυχρὰν φύσει γαστέρα καταρρέῃ, ῥᾴστη τοῖς τοιούτοις βοήθεια γίνεται, λαμβάνουσιν ἔωθεν τοῦ διὰ τριῶν πεπέρεων ἁπλοῦ φαρμάκου. καὶ πέπερι δὲ μόνον ἀκριβῶς λεῖον ὕδατι μίξαντα πίνειν ἐγχωρεῖ· κάλλιον δ' ἐστὶ καὶ εὐστομαχώτερον εἰς αὐτὰ τὸ λευκόν. ἀψινθίου δὲ πόσις ἐναντιωτάτη τοῖς οὕτω διακειμένοις· ἐμπλάττει γὰρ αὐτῶν τῇ γαστρὶ τὸν φλεγματώδη χυμόν, ὡς ἂν οὐκ ἔχον ἀξιόλογον ἐν ἑαυτῷ δύναμιν ῥυπτικήν. οὐ μὴν οὐδ' ἀλόη τούτους ὀνίνησιν

429K ἑλκτικὴν δύναμιν ἔχουσα χολώδους χυμοῦ. καὶ διὰ τοῦτ' ἐπενοήθη καλῶς ἡ καλουμένη πικρὰ τῇ τῶν μιγνυμένων δριμύτητι καὶ θερμότητι τέμνουσα καὶ ῥύπτουσα τοὺς παχεῖς καὶ γλίσχρους χυμούς, ὁποῖόν ἐστι τὸ κινάμωμον. ἀλλ' οὐχ οἷόν τε τοιούτῳ φαρμάκῳ χρῆσθαι συνεχῶς, ὥσπερ τῷ διὰ τριῶν πεπέρεων ἢ τῷ διὰ τῆς καλαμίνθης. εἰ γὰρ καὶ καθ' ἑκάστην ἡμέραν χρῷτο τούτοις ὁ τὴν ψυχρὰν ἔχων γαστέρα, βλάβης οὐδεμιᾶς πειραθήσεται. τοὐπίπαν μὲν οὖν εἰς ἑκατὸν δραχμὰς τῆς ἀλόης ἐξ ἑκάστου τῶν ἄλλων μίγνυνται, ὄντων καὶ αὐτῶν ἕξ. καὶ αὕτη γ' ἐστίν, ἣν ἅπαντες ἐν Ῥώμῃ σκευάζουσι πικράν. ἐγὼ δὲ καὶ δύο ἄλλας συντίθημι κατὰ μὲν τὴν ἑτέραν αὐτῶν πλέονα μιγνὺς τὰ θερμαίνοντα, κατὰ δὲ τὴν ἑτέραν ἐλάττονα. πλέονα μὲν οὖν ἐμβληθήσεται τῆς ἀλόης μόνης ἀφελόντων ἡμῶν τοῦ πλήθους, ὡς εἰς ὀγδοήκοντα[25] δραχμὰς ἐμβάλλεσθαι τῶν ἄλλων ἑκά-

25 post ὀγδοήκοντα: ἀλόης Ku

So if a phlegmatous humor flows down from the head to a stomach that is cold in nature, the remedy for such cases is very easy—early in the morning, they take the simple medication made from the three peppers. And it is possible to drink pepper alone, very fine and mixed with water. The white pepper is better for these and also better for the stomach. However, a drink of absinth is absolutely contraindicated for those in this state because it causes the phlegmatous humor to adhere to their stomach, as it doesn't have any significant cleansing power in itself. Nor does aloes benefit such cases because it has a drawing power for bilious humor. And because of this, the so-called *hiera-piera* (*higry-pigry*)[34] is thought to be good since, through the acridity and heat of the things that are mixed, it cuts and cleanses the thick and viscid humors; cinnamon is like this. But it is not possible to use such a medication continuously, as it is with that made from the three peppers or catmint. For if someone with a cold stomach were to use these every day, no harm would be experienced. In total then, in a hundred drachms of aloes, mix six drachms of each of the others, which are also themselves six in number. This is what everyone prepares in Rome as *picra*. I also make two other [preparations of aloes], mixing in one of them more of the heating agents and in the other, less. In the former, I only reduce the amount of aloes into which [the other ingredients] will be put, so that, into eighty drachms of aloes, I put six drachms of each of the

429K

[34] *Hiera-piera,* also known as *higry-pigry, hickery-pickery,* or *picra,* is basically a purgative made from aloes and canella bark, although sometimes other ingredients such as honey are added. Galen's particular recipe is given here.

στου τὰς ἕξ· ἐλάττονα δὲ τῆς ἀλόης τὸ πλῆθος αὐξη-
σάντων εἴκοσι δραχμαῖς, ὡς εἰς ἑκατὸν καὶ εἴκοσι τῆς
ἀλόης ἑκάστου τῶν ἄλλων ἐμβάλλειν ἕξ. εὔδηλον δέ,

430K ὅτι κιναμώμου τις ἀπορῶν ἐμβάλλει κασίας τῆς ἀρίστης τὸ διπλάσιον· ἔνιαι γὰρ εἰς τοσοῦτον ἥκουσι[26]
τῆς οἰκείας ἀρετῆς, ὡς μηδὲν ἀποδεῖν ἀτόνου κινα-
μώμου· τοῦ γὰρ εὐτόνου καὶ ἡ τοιαύτη κασία πολλῷ
λείπεται.

ἀλλ᾽ ὥσπερ, ὅταν ἀπορῶμεν ἄρτου καλοῦ, τὸν φαυ-
λότατον ἐσθίομεν, οὕτω καὶ κασίᾳ τῇ καλλίστῃ χρη-
σόμεθα κατὰ τὴν ἀπορίαν τοῦ κιναμώμου. ἐφ᾽ ὧν μὲν
οὖν σωμάτων ἡ γαστήρ[27] ἐστιν ἤτοι γ᾽ εὔκρατος μέν,
ἀλλ᾽ ἐπὶ τὸ ψυχρότερον ῥέπουσα, ἐπιτήδειόν ἐστι τὸ
καλούμενον Διοσπολιτικὸν φάρμακον, ὃ καὶ αὐτὸ δι-
τῶς εἴωθα συντιθέναι, τοῖς μὲν ἐπεχομένοις τὴν γαστέρα νίτρου τοῦ Βερνικαρίου καλουμένου μιγνὺς
ἴσον τῷ τε κυμίνῳ καὶ τῷ πηγάνῳ καὶ τῷ πεπέρει,
τοῖς δ᾽ εὔλυτον ἔχουσιν ἥμισυ. συμπεφώνηται γὰρ
ὑπὸ τούτου τοῦ φαρμάκου λεπτύνεσθαί τε τὸ φλέγμα
καὶ τὰ φυσώδη πνεύματα κενοῦσθαι. χαλεπὴ δ᾽ ἐστὶ
καὶ δυσμεταχείριστος ἡ ἐπιπλοκὴ τῆς δυσκράτου
κατὰ θερμότητα γαστρὸς τῇ ψυχρᾷ κεφαλῇ. τὸ μὲν
γὰρ φλέγμα τὸ καταρρέον εἰς τὴν γαστέρα χρῄζει
τοῦ διὰ τριῶν πεπέρεών τε καὶ καλαμίνθης, ὥσπερ

431K καὶ τοῦ προειρημένου Διοσπολιτικοῦ, τὸ δὲ στόμα
τῆς γαστρὸς ὑπὸ τῶν ἐκπυρούντων βλάπτεται. τέ-

[26] post ἥκουσι: τῆς οἰκείας ἀρετῆς Ko; τῆς κασίας τῆς
ἀρετῆς Ku

others. In the he latter, the amount of aloes is increased
by twenty drachms, so that, into one hundred and twenty
drachms of aloes, I put six drachms of each of the others.
It is clear, however, that someone who lacks cinnamon
puts in a double amount of the best cassia. For some cas-   430K
sias reach such a degree of specific goodness as to lack
nothing compared to weak cinnamon, although such cassia falls far short.

But just as, when we lack good bread, we eat the worst,
so too, when faced with a lack of cinnamon, we use the
best cassia. Therefore, in the case of bodies where the
stomach is truly *eukratic* but is inclining toward being
colder, the so-called Diospoliticum medication is suitable.
I am accustomed to compound this in two ways: for those
in whom the stomach is retaining, I mix an equal amount
of the so-called Bernikarian (Macedonian) niter[35] with
cumin, rue and pepper, whereas for those in whom the
stomach is releasing readily, the amount is half. It is agreed
that the phlegm is thinned by this medication and the
flatulent *pneumas* evacuated. The combination of a hot
dyskrasia of the stomach with a cold head is a severe problem and hard to manage. For the phlegm which flows
down to the stomach needs the medication made with
three peppers and catmint, just as it also needs the previously mentioned Diospoliticum medication, whereas the   431K
opening of the stomach is harmed by the heating agents.

[35] This is fine-particled. See Galen, *MM*, X.569K, and *Comp.
Med. Gen.*, XIII.568K.

27 *post* ἡ γαστήρ: ἐστιν Ko; ἐστι ψυχρὰ Ku

μνειν οὖν τῶν οὕτω διακειμένων καὶ ἀπορρύπτειν τὸ φλέγμα πειρᾶσθαι χρὴ διὰ τῶν μὴ θερμαινόντων, ὁποῖόν ἐστι καὶ ὀξύμελι.

χαλεπὴ δὲ καὶ δυσδιάθετός ἐστιν ἐπιπλοκὴ καὶ ἡ τῶν τοιούτων τοῦ σώματος κατασκευῶν, ἐν αἷς ἄτονον μέν ἐστι καὶ ναυτιῶδες τὸ στόμα τῆς κοιλίας, ὃ δὴ καὶ στόμαχον ὀνομάζουσιν, ἴσχεται δὲ ἡ γαστήρ. ὅσα γὰρ αὐτὴν προτρέπει, εὐθέως πάντα ἀνατρέπει τὸν στόμαχον, ὡς ἐπιπολάζειν τε αὐτῷ τὰ σιτία καὶ ναυτίαν ἐργάζεσθαι πολλάκις· οἷς ἐξ ἀνάγκης ἕπεται τὸ μηδὲ πέττεσθαι καλῶς τὴν τροφήν. ἐὰν δὲ πάλιν εὐστόμαχα διδῷς αὐτοῖς ἐδέσματα, καὶ τρίτης καὶ τετάρτης ἡμέρας ἡ γαστὴρ ἐπέχεται. τοῖς τοιούτοις οὖν ἓν εἶδος εὗρον διαίτης ἁρμόζον, ἐν ἀρχῇ μὲν προσφερομένοις λάχανά τε δι᾽ ἐλαίου καὶ γάρου καὶ τἆλλα, ὅσα λαπάττειν εἴωθε τὴν γαστέρα, μετὰ δὲ τὴν αὐτάρκη τροφὴν ἐπιλαμβάνουσί τι τῶν τονούντων τὸν 432K στόμαχον. ἔστι δὲ δὴ ταῦτα μῆλά τε καὶ ἄπια καὶ ῥοιαί· ἔνια γάρ ἐστι τοιαῦτα γένη χωρὶς ὀξύτητος στύφοντα.

αὐτὸς δὲ ὁ ἄνθρωπος ἁπάντων αὐτῶν ἐν μέρει πειραθείς, ὅπερ ἂν ἀβλαβέστατόν τε καὶ ἥδιστον εὑρίσκῃ, τούτῳ χρήσθω.[28] μὴ πολὺ δὲ μηδὲ τούτων ἑκάστου λαμβάνειν, ἀλλ᾽ ὅσον, ὡς ἔφην, ἰάσεται τὴν ἀτονίαν τοῦ στομάχου· τῆς γὰρ τοιαύτης αὐτῶν χρή-

---

[28] post χρήσθω.: μὴ πολὺ δὲ μηδὲ Ko; μὴ πάνυ δὲ Ku

---

[36] It is not altogether clear what the supposed pathology is here. If atonia is understood as "slackness" or "relaxation" (LSJ),

Therefore, for those in such a state, you must attempt to cut and thoroughly cleanse the phlegm through agents that are not heating—oxymel is an example.

Also difficult and hard to manage is a combination of such constitutions of the body in which the opening of the stomach (which they call *stomachus*) is weak and causes nausea, and the stomach is retaining. For those things that stimulate the stomach immediately upset the whole *stomachus* so that the foods in it remain floating on the surface and often create nausea. From these things, failure of proper concoction of the nutriment inevitably follows. If, contrariwise, you were to give them easily digested foods, within three or four days the stomach is held in check. In such cases, I found one kind of regimen to be suitable: in the beginning they are given vegetables with olive oil and fish sauce, and other things that customarily empty the stomach. After sufficient nourishment, they take up one of those things that strengthen the opening of the stomach—apples are such things, and pears and   432K pomegranates, for such classes are astringent without being acidic.

Let the person himself discover what is least harmful and most pleasant of all the things tried by him in turn and use this. He should not take very much of each of these—as I said, only as much as cures the "slackness" of the *stomachus*.[36] I have tried this kind of use of these things

or even "weakness," and *stomachus* as the esophagus, the stomach itself, or the opening of the stomach (whether cardia or pylorus is not specified), it may be that the condition is too ready a passage of material through the pyloric sphincter (which is R. M. Green's interpretation). However, a later passage suggests Galen believes the cardia to be at fault.

σεως ἐπειράθην εὔλυτον ἐργαζομένης τὴν κοιλίαν.
καὶ μᾶλλόν γε τοὐπίπαν οἱ οὕτω διακείμενοι διαχω-
ροῦσιν ἐπιφαγόντες τι τῶν στυφόντων ἢ εἰ μηδ' ὅλως
ἐχρήσαντο. τονωθὲν γὰρ αὐτῶν τὸ στόμα τῆς κοιλίας
ὠθεῖ κάτω τά τ' ἐπιπολάζοντα καὶ σὺν αὐτοῖς τὰ καθ'
ὅλην τὴν ἄνω γαστέρα καὶ τήν γε φορὰν τῶν σιτίων
ὑποδεχόμενα τὰ ἔντερα διαφυλάττει, προωθοῦντα καὶ
αὐτὰ τὸ παραγινόμενον ἐκ τῶν ὑπερκειμένων εἰς αὐτά.
φαίνεται γὰρ ἡ προωστική τε καὶ ἀποκριτικὴ δύναμις, κἂν ἐκ τῶν κατὰ τὴν ἕδραν χωρίων ὁρμήσῃ τι
πρὸς τὴν ἄνω φοράν, ἄχρι τοῦ πλείστου διαφυλάτ-
τουσα, καίτοι παρὰ φύσιν οὔσης ἐν ζῴῳ τοιαύτης
ὁδοῦ. ὅτι δὲ ἀληθές ἐστιν ὃ λέγω, πάρεστι μαθεῖν
ἑκάστῳ τῶν πολλάκις ἡμῖν συμβαινόντων ἀναμνη-
σθέντι. δακνώδης γοῦν ἐνίοτε χυμὸς εἰς τὰ κατὰ τὴν
ἕδραν χωρία παραγενόμενος, ἐρεθίζει μὲν ἡμᾶς ἐπὶ
τὴν ἔκκρισιν αὐτοῦ· κατασχεῖν δὲ αὐτὸν ἀναγκασθέν-
τες, ἐπειδὰν ἐν πολιτικαῖς ὦμεν πράξεσιν, ἀπαλλαγέν-
τες αὐτῶν οὐκέτ' ἀποκρίνομεν, αἰσθανόμεθά τε κἀκ
τούτου πολλάκις τῆς κεφαλῆς ὀδυνηρᾶς γινομένης
ἀνατρεπομένης τε τῆς γαστρός. εὐλόγως οὖν, εἰ καὶ
ὑπακτικοῖς σιτίοις ἐπιπολαζομένου τοῦ στομάχου τὰ
στύφοντα ληφθέντα τόνον ἐμποιήσαντα τοῖς ἄνω
μέρεσι τῆς γαστρός, ἀρχὴν τῆς κάτω φορᾶς ἐργάζεται τοῖς ἐν αὐτῷ περιεχομένοις, ἣν διαδεχόμενα τὰ
ἔντερα μέχρι τοῦ κάτω πέρατος εἴωθε φυλάττειν.

11. Αὐτάρκως οὖν εἰρημένων τούτων ὥρα μεταβαίνειν ἐπ' ἄλλην κατασκευὴν ἀνώμαλον οὐδεμιᾶς τῶν

to make the stomach release easily. And on the whole, those in such a condition excrete more, if they have eaten one of the astringents as well, rather than if they used none at all. For when the opening of their stomach is strengthened, it thrusts downward what is lying on the surface, and with these things, those things that are in the whole stomach above, and the intestines, receive the passage of the foods, preserve them, and also propel forward what has come to them from the structures situated above. For it seems the propulsive and expulsive capacity, even from the places in relation to the fundament, sets in motion something related to the passage above, persisting for a long time, although the path of this is contrary to nature in an animal. That what I say is true is there for each to   433K
learn, if he recalls what often happens to us. At any rate, sometimes a biting humor, when it has come to the places in relation to the fundament, irritates us until we excrete it. If, however, we are forced to retain it, when we are engaged in civic matters, there is a change in these things that we haven't yet excreted, and we often sense from this pains arising in the head when the stomach is upset. It is reasonable, then, if the *stomachus* prevails over aperient foods, astringents that are taken create strength in the parts of the stomach above. This brings about a start of the downward passage for the things contained in it, which the intestines receive and are accustomed to maintain up to the lower limit (anus).

11. Having said enough about these matters, it is time to pass on to another irregular (nonuniform, anomalous)

πρόσθεν ἀπολειπομένην τῇ κακίᾳ, καθ' ἣν οἱ μὲν νε-
φροὶ λίθους ἢ πώρους ἢ ὁπωσοῦν ὀνομάζειν ἐθέλεις
γεννῶσιν, ἡ δὲ τοῦ σώματος ἅπαντος φύσις ἰσχνὴ
τετύχηκεν οὖσα. χρῄζουσι μὲν γὰρ οὗτοι φαρμάκων
καὶ διαιτημάτων λεπτυνόντων, ἐναντιώτατα δ' ἐστὶ
ταῦτα τοῖς ἰσχνοῖς σώμασιν. ὥστε τις τῶν χρωμένων
αὐτοῖς διὰ τοὺς νεφρούς, δυσκινήτων τε καὶ δυσαι-
σθήτων καὶ ὥσπερ ψοφούντων καπυρόν,[29] ὡς αὐτὸς
ὠνόμαζεν, αἰσθανόμενος τῶν δακτύλων, ἐκοινώσατο
μὲν τὰ πρῶτα τοῖς κατὰ τὴν Καμπανίαν ἰατροῖς, ἔνθα
διέτριβεν, οἱ δὲ ὡς[30] καταψυχθέντων αὐτῶν ἐπὶ διαθέσει παράλυσιν ἀπειλούσῃ τοῖς δι' εὐφορβίου καὶ
λιμνήστεως, ἣν "ἀδάρκην" τε καὶ "ἀδάρκιον" ὀνομάζουσιν, ἐχρῶντο φαρμάκοις. ὡς δὲ πολὺ χείρων ἡ
διάθεσις ἐγίνετο καὶ προσανέβαινεν ἀεὶ τὰ συμπτώματα τοῖς ὑπερκειμένοις μέρεσι μετὰ τοῦ σφόδρα
ὀδυνᾶσθαι, κατά τινα τύχην ἐπιδημήσαντί μοι τῇ
Καμπανίᾳ δηλώσας ὁ κάμνων τὰ συμβάντα παρ-
εκάλει βοηθεῖν. ἐγὼ δὲ ὅτι μὲν ἐκ τῶν φαρμάκων ὧν
εἶπεν ὑπερξηρανθεὶς εἰς τὴν τοιαύτην ἧκε διάθεσιν
συνῆκα, ζητῶν δέ, τίνα ἄν τις αὐτῷ δίαιταν εὕροι,
καθ' ἣν ἄνευ τοῦ βλάπτεσθαι τοὺς νεφροὺς ἰάσεται
τὴν ξηρότητα, πτισάνης τε χυλὸν ἐπενόησα καὶ τῶν
ἰχθύων τοὺς πετραίους καὶ πελαγίους, ὅσοι τ' ἄλλοι
μηδὲν ἔχουσι γλίσχρον.

<hr>

[29] καπυρόν Ko; τι καπνηστόν Ku   [30] post ὡς: καταψυχθέντων αὐτῶν Ko; κατισχομένου αὐτοῦ Ku

constitution, which is in no way lacking in badness compared to those previously mentioned. In this, the kidneys generate stones or calculi, or whatever you may wish to call them, although the nature of the body as a whole happens to be dry. These [stones or calculi] require medications and regimens that are thinning, although these are most inimical to thin bodies. As a result, one of those people using these measures for the kidneys, on becoming aware that his fingers were difficult to move and dysaesthetic, and made a cracking sound, as he himself put it, communicated these things first to the doctors in Campania, where he was spending some time. As the fingers were cold due to the condition and he feared paralysis, they used medications made with euphorbium and centaury (*limnēstion*), which they call *adarkēn* and *adarkion*.[37] As the condition became much worse and the symptoms were progressively ascending to the parts above, accompanied by severe pain, and since I happened to be staying in Campania at the time, the patient showed me what had happened and asked for my help. Because I recognized that he had come to such a condition through being overly dried out by the medications which he spoke of, and was looking for someone who might discover a regimen for him that would cure the dryness without harming the kidneys, I thought of the juice of ptisane and fish from the rocks and deep sea, and those other fish that have no viscidity.

434K

[37] According to LSJ, this is a salt efflorescence on the herbage of marshes, referring to Dioscorides, 5.119, and Galen, XII.370K. The Latin is *adarcen* or *adarcum*.

435K   οὕτω δὲ καὶ τῶν πτηνῶν ζῴων ὅσα παραπλήσιον
ἔχει τὴν σάρκα. πολλὰ δὲ τῶν ὀρείων ὀνομαζομένων
ὀρνίθων τοιαῦτα ταῖς κράσεσιν· ὅσα γὰρ ἐν ταῖς
πόλεσι κατακλείσαντες ὑγραῖς καὶ πολλαῖς τροφαῖς
πιαίνουσιν οἱ τὰ τοιαῦτα καπηλεύοντες, ἐναντιώτατα
τούτοις ἐστίν. ἀρίστη δὲ σὰρξ εἰς τὴν τοιαύτην διάθε-
σίν ἐστιν³¹ ἡ τῶν ὀρείων περδίκων, ἐπ' αὐτῇ δὲ τῶν
ἀτταγήνων καὶ ψάρων καὶ κοττύφων καὶ κιχλῶν. ἐὰν
δὲ ἀπορῶμεν ὀρείων ὀρνίθων, ἐκ τῶν ἐν τοῖς ἀγροῖς
τρεφομένων καὶ τῶν κατὰ τοὺς πύργους περιστερῶν
νομάδων λαμβάνειν προσήκει, καθάπερ καὶ τῶν ἐν
αὐτοῖς τοῖς πύργοις νεοττευόντων στρουθίων, οὓς ὀνομάζουσι πυργίτας. ἔστωσαν δὲ ἐν ὑψηλοῖς χωρίοις οἱ
τοιοῦτοι τῶν ἀγρῶν·³² οὕτω γὰρ καὶ ταῖς ἀλεκτορίσι
καὶ τοῖς ἀλεκτρυόσι τοῖς κατὰ τὰς ἐπαύλεις τρεφομέ-
νοις ἔνεστι χρῆσθαι. γάλακτος δὲ τοῦ μὲν τῶν ἄλλων
ζῴων εἴργειν τοὺς οὕτω διακειμένους, μόνῳ δ' ἐπιτρέπειν χρῆσθαι τῷ τῶν ὄνων, ἐπειδὴ λεπτομερέστερόν
ἐστι τοῦτο τῶν ἄλλων. καὶ συνελόντι φάναι, μεταξὺ
τῆς τε λεπτυνούσης διαίτης καὶ τῆς παχυνούσης εἶναι
436K   χρή. τὴν δὲ τούτοις ἁρμόττουσαν τὴν κατὰ μέρος
ὕλην ἐν τοῖς Περὶ τῶν ἐν ταῖς τροφαῖς δυνάμεων ὑπομνήμασιν ἔγραψα καὶ προσέτι τῷ Περὶ εὐχυμίας
τε καὶ κακοχυμίας. διὸ καὶ μηκύνειν ἐν τῷ παρόντι
περιττόν· ἀρκεῖ γὰρ τοῦτο μόνον εἰπεῖν, ὡς καὶ τῆς
ὅλου τοῦ σώματος ἀναθρέψεως ἐπὶ τῶν τοιούτων προνοεῖσθαι χρὴ δι' ἧς ἔμπροσθεν εἶπον ἀγωγῆς, ἡνίκα
ἐμνημόνευσα τοῦ πάντα μᾶλλον ὑπομένοντος παθεῖν

194

So too, some of the winged creatures have a similar 435K
flesh. Many of the so-called mountain birds are like this in
terms of their *krasias*. But those kept enclosed in the cities
that are fattened up with much moist nutriment—birds
such as dealers sell—are absolutely contraindicated for
these patients. The best flesh for such a condition is that
of partridges, and next that of francolins, starlings, black-
birds and thrushes. If, however, we lack mountain birds,
it is appropriate to take those nurtured in the countryside
and wandering doves living in towers, as it is to take spar-
rows that nest in the towers themselves, which they call
"tower birds." Those from the countryside should be in
high places, for in this way too it is possible to use cocks
and hens raised on farms. For people in this condition,
prohibit the use of the milk of other animals, trusting only
the use of the milk of asses, since this is more fine-particled
than the others. In summary, the diet must be midway
between thinning and thickening. I wrote about the mate- 436K
rial suitable for these cases individually in the treatises,
The Powers of Foods, and in addition, the work *On the
Good and Bad Humors of Nutriments.*[38] For this reason,
it is superfluous to waste further time on the present matter. It is enough to say this alone: in such cases it is necessary to take care of the restoration of the whole body, using
the method I spoke about previously, when I mentioned
the person who preferred to submit to anything rather

---

[38] See note 6 above.

---

[31] post ἐστιν: ἡ τῶν ὀρείων περδίκων, ἐπ' αὐτῇ Ko; ὡς τῶν
ὀρνίθων περδίκων, ἐπ' αὐτοῖς Ku

[32] ἀγρῶν Ko; πύργων Ku

ἢ πιττοῦσθαι. παχέος δὲ ὄντος τοῦ λιθιῶντος ἀνθρώπου, θαρρῶν ἂν χρῷο³³ τῇ λεπτυνούσῃ διαίτῃ.

τὰ δ' αὐτὰ νόμιζε καὶ περὶ τῶν ἀρθριτικῶν προσ-
ιέναι. πίνοντες γὰρ ἐκεῖνοι τὰς ἀρθριτικάς τε καὶ ποδαγρικὰς ἀντιδότους οἱ μὲν εὔσαρκοι καὶ παχεῖς καὶ
πιμελώδεις οὐδὲν βλάπτονται, λεπτοὶ δὲ ὑπάρχοντες
ὅλον τε τὸ σῶμα ξηραίνονται καὶ τοὺς πόδας αὐτοὺς
ἐνίοτε καὶ τὰς χεῖρας, εἰ πρότερον ἐπεπόνθεισαν, εἰς
χαλεπωτέραν ἄγουσι διάθεσιν. εἰδέναι μέντοι χρὴ
τὰς ποδαγρικὰς ταύτας ἀντιδότους πολὺ τῶν νεφριτικῶν διαλλαττούσας οὐ μόνον τῷ λεπτύνειν τοὺς πα-
437K   χεῖς καὶ γλίσχρους χυμούς, ἀλλὰ καὶ τῷ θερμαίνειν
τε καὶ ξηραίνειν· ἐπὶ γάρ τοι τῶν νεφριτικῶν οὐδὲ τὴν
ἀρχὴν ὅσα θερμαίνει σφοδρῶς διδόναι χρή. λέλεκται
δὲ τελέως δηλονότι περὶ τῶν τοιούτων ἁπάντων ἐν τῇ
τῆς Θεραπευτικῆς μεθόδου πραγματείᾳ καὶ ταῖς περὶ
τῶν φαρμάκων, διτταῖς οὔσαις καὶ ταύταις, προτέρας
μὲν τῇ τάξει τῆς τῶν ἁπλῶν, δευτέρας δὲ τῆς τῶν
συνθέτων· ⟨ὧν⟩ οὔτε μηδ' ὅλως οὔτ' ἐπὶ πλέον προσήκει μεμνῆσθαι διὰ τὸ καὶ τὰ τοιαῦτα σώματα μεταξὺ
τῶν τ' ἀκριβῶς ὑγιαινόντων καὶ τῶν ἤδη νοσούντων
εἶναι· τὸ γάρ τοι προφυλακτικὸν ὀνομαζόμενον μέρος
τῆς ἰατρικῆς τῶν οὕτω διακειμένων προνοεῖται σωμάτων. οὐδένα γοῦν τῶν ἄμεμπτον ἐχόντων τὴν τοῦ

³³ χρῷο Ko; προσφέροιτο Ku

than be plastered with pitch. However, if the person with a kidney stone is fat, you may use the thinning diet with confidence.

Consider these same things to apply regarding those with arthritis. If those people who drink the antidotes for arthritis and gout are well-fleshed, stout and fatty, they suffer no harm, whereas, in those who are thin, [these medications] dry the whole body and sometimes the feet and hands themselves, and if they had been affected previously, they come to a condition that is more difficult to deal with. Of course, you need to know that these antidotes for gout are very different from those for nephritis, not only in thinning the thick and viscous humors, but also   437K
in heating and drying. Certainly, in those with nephritis, agents which heat strongly must not be given at the beginning. Obviously, I have spoken comprehensively about all such things in my work, *The Method of Medicine*,[39] and in the works on medications, of which there are two: the first in sequence is the one on simple medications and the second, the one on compound medications.[40] It is not altogether appropriate to mention these at greater length, because such bodies lie between those that are perfectly healthy and those that are already diseased. What is called the prophylactic part of medicine is about the care of bodies that are in such a state. Anyway, we do not think it worthwhile for any of those who are faultless in terms of

[39] For nephritis see *MM*, X.917K, and for gout, *MM*, X.513, 803, and 956K.

[40] There are three major works on medications extant: *Simpl. Med.*, XI.369–892K and XII.1–377K; *Comp. Med. Loc.*, XII.378–1003K; and *Comp. Med. Gen.*, XIII.352–1058K.

σώματος ἕξιν οὔτ' ἀξιοῦμεν φάρμακα πίνειν οὔτε
λεπτυνούσῃ διαίτῃ χρῆσθαι· τοὺς δὲ διὰ φυσικὴν
ἀσθένειαν ἤτοι τοῦ σώματος ἅπαντος ἢ μορίων τινῶν
ἑτοίμως βλαπτομένους εἰς τὴν προφυλακτικὴν ὀνομαζομένην ἀγωγὴν ἀξιοῦμεν ἕλκεσθαι, τῆς ὑπόπτου διαίτης ἀφισταμένους, ὑφ' ἧς οὐδὲν ὁρῶμεν βλαπτομένους, ὅσοι καλῶς κατεσκευασμένοι τὰ σώματα φύσει
διαπονεῖν οὐκ ὀκνοῦσι τὰ μέτρια.

438K

12. Τοῖς μέντοι ῥᾳδίως πάσχουσιν, ὡς καὶ τοῖς με-
γάλοις νοσήμασιν ἐμπίπτειν συνεχῶς, οἷον τοῖς ἐπι-
ληπτικοῖς καὶ σκοτωματικοῖς καὶ κεφαλαλγικοῖς ἔτι
τε τοῖς ἄλλοις, ἐφ' ὧν ἡ κεφαλὴ τοῖς ὑποκειμένοις
μορίοις αἰτία γίνεται τῶν συμπτωμάτων ἐν οἷς εἰσιν
ὀδόντες οὖλά τε καὶ κίων καὶ συλλήβδην εἰπεῖν
ἅπαντα τὰ κατὰ τὸ στόμα καὶ τὴν φάρυγγα μόρια
καὶ πρὸς αὐτοῖς ὦτα καὶ ὀφθαλμοὶ καὶ στόμαχος καὶ
κοιλία καὶ τὰ πνευματικὰ μόρια λάρυγξ ἀρτηρία τραχεῖα καὶ πνεύμων καὶ θώραξ, ὁ σκοπὸς τῶν διαιτημάτων κοινὸς αὐτὴν μὲν καὶ πρώτην μάλιστα τὴν
κεφαλὴν ἐργάζεσθαι δυσπαθῆ, μετὰ ταύτην δὲ καὶ
τῶν βλαπτομένων μορίων προνοεῖσθαι γινώσκοντας,
ὡς καὶ τοῦτ' αὐτὸ πάλιν ἑτέρου δεῖται διορισμοῦ,
λέγω δὲ τὸ μὴ πάντα ῥωννύναι τὰ βλαπτόμενα μόρια.
ἄριστος δὲ διορισμὸς ὁ εἰς τὴν χρείαν αὐτῶν ἀναγό-
μενος.

41 In this paragraph, consisting of one very long sentence and

198

bodily state to drink medications or use a thinning diet. However, we do think it worthwhile for those who are easily harmed due to a natural weakness, either of the whole body or of some of the parts, to be attracted to the so-called prophylactic method and to keep clear of the suspect regimen, from which we see none harmed who properly prepare their bodies naturally and do not shirk hard work in moderation.

438K

12. However, in those people who are easily affected so as to be filled continually with major diseases, like the epilepsies, vertigos and headaches, and with others in which the head becomes a cause of symptoms in the parts lying below it, among which are the teeth, gums and uvula, and to speak collectively, all the parts related to the mouth and pharynx, and in addition to these, the ears, eyes, esophagus, stomach, and the parts to do with respiration—the larynx, trachea, bronchi, lungs and thorax—the objective common to all the regimens is primarily and particularly to make the head itself difficult to affect (*dyspathic*), and after this, to care also for the damaged parts, recognizing that these in turn require another distinction—I mean, not all the damaged parts need strengthening. The best distinction is the one that pertains to their use.[41]

one short final sentence, Galen's punctuation is preserved, in part to exemplify his tendency to prolixity and complex sentence structure. The point, simply stated, is that when the head is affected, symptoms / diseases may occur in the head itself but also in the parts below, which may receive abnormal material transmitted from the head. Whether or not these various parts need treatment is determined by an evaluation of their functions.

ὀφθαλμῶν μὲν γὰρ καὶ ὤτων ἡ χρεία μεγάλη, καὶ
439K διὰ τοῦτο προσήκει τῶν ἐκ τῆς κεφαλῆς περιττωμάτων εἰς αὐτὰ φερομένων τὴν ὑφ᾽ Ἱπποκράτους ὀνομαζομένην "παροχέτευσιν" ἐργάζεσθαι, μάλιστα μὲν
ἐπὶ ῥῖνα περισπῶντα τὸ φερόμενον ἐπ᾽ αὐτά, ταύτης
δ᾽ οὐχ ὑπακουούσης, εἰς τὸ στόμα διὰ τῶν ἀποφλε-
γματιζόντων φαρμάκων, ὥσπερ γε κἀπὶ τὴν ῥῖνα διὰ
τῶν πταρμοὺς κινούντων, ὅσα τε πρὸς τὰς ἐμφράξεις
αὐτῶν ἁρμόττει. ὀφθαλμοὺς δὲ τονώσεις τῷ διὰ τοῦ
Φρυγίου λίθου χρώμενος ξηρῷ κολλυρίῳ, τοῖς βλε-
φάροις ἐπάγων τὴν μήλην χωρὶς τοῦ προσάπτεσθαι
τοῦ κατὰ τὸν ὀφθαλμὸν ἔνδον ὑμένος· οὕτως οὖν
πράττουσιν ὁσημέραι καὶ στιμμιζόμεναι αἱ γυναῖκες.

εἰς δὲ τὴν τῶν ὤτων ῥώμην ἀπόχρη μὲν καὶ
τὸ γλαύκιον μόνον, παρατριβόμενον ἐπ᾽ ἀκόνης σὺν
ὄξει, κἄπειτα διὰ μήλης ἐγχεόμενον ἀτρέμα χλιαρὸν
ἢ διὰ τούτου δὴ τοῦ συνήθους ὀργάνου, καλουμένου
δὲ ὑπὸ τῶν ἰατρῶν[34] "ὠτεγχύτου"· κἀπειδὰν ἤδη σοι
δοκῇ τονοῦσθαι καλῶς, ὡς μηδὲν ἐπιρρεῖν αὐτοῖς,
συνεχῶς ἐνστάξεις ναρδίνου μύρου τοῦ ἀρίστου, πρότερον μὲν ἐν Λαοδικείᾳ μόνῃ τῆς Ἀσίας σκευαζομέ-

_____
[34] τῶν ἰατρῶν Ko; πάντων Ku

_____

[42] Hippocrates, On Humors 1, Hippocrates IV, LCL 150,
64–65; Jones renders παροχέτευσις as "deviation."

[43] The term "apophlegmatic" is listed in the 1933 edition of
the OED as: "promoting the removal of phlegm; expectorant."
See note 11 above.

The use of the eyes and ears is of great importance, and because of this, it is appropriate to create what Hippocrates called a "diversion" of the superfluities being carried down to them from the head, particularly drawing off what is being carried to the nose.[42] But if the latter does not accept the superfluities, draw them off to the mouth by means of medications promoting the discharge of phlegm (apophlegmatics),[43] just as, with those going to the nose, through medications that provoke sneezing, and those suitable for obstructions of the nose. You will strengthen the eyes by the use of a dry collyrium made from Phrygian stone,[44] applying the probe to the eyelids without touching the membrane of the eye within. This is what women who apply kohl[45] to the eyelids do every day.

For strengthening the ears, *glaukion*[46] alone is enough, when it is rubbed on a stone with vinegar, and then gently poured in, lukewarm, either with a probe or the instrument customarily called by doctors an "ear syringe." And when it seems to you they are already properly strengthened, so that nothing flows to them, continue to instill the finest oil of spikenard, which was formerly best prepared

[44] Phrygian stone was an aluminous kind of pumice stone used by dyers—see Dioscorides, 5.141.

[45] Kohl is described in *The Chambers Dictionary* (11th edition [2008]) as "a fine powder of native stibnite (formerly known as antimony), black in color, used (orig. in the East) to darken the area around the eyes." It is presumably what is being referred to here.

[46] *Glaukion* was a preparation from the juice of the horned poppy, *Glaucium corniculatum*—see Dioscorides, 3.86, Galen, *Simpl. Med.*, XI.857K.

440K νου καλλίστου, νυνὶ δὲ καὶ κατ᾽ ἄλλας πόλεις. ἔτι δὲ
μᾶλλον τὸ διὰ γλαυκίου καλούμενον κολλύριον ὀνή-
σει τε καὶ ῥώσει τὰ ὦτα καὶ τὸ διὰ ῥόδων καὶ τὸ
κροκῶδες καὶ τὸ νάρδινον καὶ τὸ δι᾽ οἴνου καὶ τῶν
μύρων τὰ ἐν Ῥώμῃ σκευαζόμενα ταῖς πλουσίαις γυ-
ναιξίν, ἃ φουλιᾶτά τε καὶ σπικᾶτα προσαγορεύουσιν.
εἰ δὲ πολὺ συνεχῶς ἐπὶ τὰ ὦτα καταφέροιτο τὸ ἀπὸ
τῆς κεφαλῆς ῥεῦμα καὶ πύον καὶ ἑλκώσεις γεγονυῖαι
τύχοιεν, τὴν μὲν ἕλκωσιν διὰ τοῦ Ἀνδρωνίου φαρμάκου καὶ τῶν ὁμοίων ἐκθεραπευτέον. ἐν αὐτῷ δὲ τῷ
ταῦτα πράττειν ἀντισπαστέον ἐπὶ ῥῖνά τε καὶ στόμα
καὶ μετὰ ταῦτα τοῖς εἰρημένοις χρηστέον.

13. Ὥσπερ δὲ ὑπὸ τῶν ἐκ τῆς κεφαλῆς καταρρεόντων βλάπτεται πολλάκις πολλὰ τῶν ὑποκειμένων,
οὕτω δι᾽ ἧπαρ ἢ νεφροὺς ἢ σπλῆνα νόσοι συμβαίνου-
σιν ἑτέροις μορίοις ἀσθενεστέροις αὐτῶν. ἐδείχθη
γὰρ ἐν τοῖς τῶν φυσικῶν δυνάμεων ὑπομνήμασιν
ἕκαστον μόριον ἑλκτικὴν τῶν οἰκείων αὐτοῦ χυμῶν
441K ἴσχον δύναμιν, οὓς ἀλλοιοῦσα κἀξομοιοῦσα εἰς τροφὴν καταχρῆται, καί τινι πάλιν ἑτέρᾳ δυνάμει τὸ
περιττὸν ἀποκρίνειν ἐφιέμενον εἴς τι τῶν πλησίον. ἂν
μὲν οὖν αὐτὸ τὸ πεπονθὸς εὔρωστον ἔχῃ τὴν ἕξιν τοῦ
σώματος, οὐ δέχεται τὸ πεμπόμενον, ὥστ᾽ ἐν τῷ
πρωτοπαθοῦντι μένον[35] ἐκείνῳ τῷ μορίῳ συνεχῶς
πάσχοντα τὸν ἄνθρωπον ἀπέδειξεν· ἐὰν δ᾽ ἀσθενέστε-
ρον ᾖ τι τοῦ πέμποντος, ἐδέξατο μέν, αὖθις δὲ ὠθεῖ

---

[35] πρωτοπαθοῦντι μένον Ko; πρώτῳ παθόντι μόνον Ku

only in Laodicea in Asia, but now also in other cities. Still   440K
more, the so-called collyrium made from *glaukion* benefits and strengthens the ears, as do those [preparations]
made from roses, crocuses, and spikenard, and with wine
and oils prepared in Rome for rich women, which they call
foliata and *spikata*.[47] If the flux from the head is continuously carried to the ears over a long period of time, and
there happen to be pus and ulcers, you must treat the ul-
ceration thoroughly with the Andronian medication[48] and
similar things. In doing this, you must draw off [the fluxes]
in it to the nose and mouth, and after this, you must use
the things mentioned.

13. Just as many of the parts lying below are often
harmed by what flows down from the head, so too, due to
the liver, kidneys or spleen, diseases come about in other
parts weaker than they are. It was shown in the treatises,
On the Natural Faculties that each part has a capacity for
attracting humors proper to itself, which it uses to change   441K
and assimilate the nutriment, and in turn, by another capacity, to separate and send on the superfluity to one of
the parts adjacent.[49] If the affected part itself has a strong
state of the body, it doesn't receive what is sent, so that it
remains in the part first affected, rendering the person
continuously affected in that part. If, however, the part is
weaker than the one that sent the superfluity, and has re-

[47] See note 33 above.
[48] This was a multicomponent troche compounded by Andron
from pomegranate flowers, oak gall, myrrh, birthwort, vitriol, fis-
sile alum, and Cyprian misu marinated in sweet wine—see
EANS, 80.
[49] *Nat. Fac.,* 3.15 (II.206K ff.).

πρός τι καὶ αὐτὸ τῶν ἀσθενεστέρων, κἀκεῖνο πάλιν
εἰς ἄλλο, μέχρις ἂν εἴς τι τῶν οὐδὲν ἐχόντων ἀσθενέ-
στερον ἑαυτῶν κατασκήψῃ τὸ περιττόν. κατὰ τοῦτον
μὲν οὖν τὸν λόγον ἄλλος ἄλλῳ μέρει συνεχῶς ἐν-
οχλεῖται τῶν ἀμελῶς διαιτωμένων· τοῖς δὲ μὴν μηδὲν
ἀθροίζουσι περιττὸν ἀβλαβὲς ἀεὶ μένει τὸ ἀσθενές.
ἀπόδειξις δὲ τούτου σαφεστάτη τὸ δι᾿ ἐξαμήνου ἢ
πλεόνων ἐνίους ἐνοχλεῖσθαι τοῖς ἀσθενέσι μορίοις.

εἰ γὰρ αὕτη μόνη ἦν ἀσθένεια[36] τῆς βλάβης αἰτία,
διὰ παντὸς ἂν ἔπασχε τὸ ἄρρωστον μόριον, ὡς ἄν γε
διαπαντὸς ἔχον ἐν ἑαυτῷ συνυπάρχουσαν τὴν τοῦ πά-
442K σχειν αἰτίαν. ἐπεὶ τοίνυν οὐ πάσχει διὰ παντός, εὔδη-
λόν ἐστι καὶ ἄλλο τι προσέρχεσθαι τὸ συμπληροῦν
τοῦ παθήματος αὐτοῦ τὴν γένεσιν, ὅπερ οὐδὲν ἕτερόν
ἐστι τοῦ περιττεύοντος ἢ κατὰ τὸ πλῆθος ἢ κατὰ τὴν
ποιότητα. καὶ τοῦτό γε αὐτὸ τὸ περιττεῦον ἢ καθ᾿
ὅλον ἀθροίζεται τὸ σῶμα καὶ καλοῦσι "πληθώραν"
τὴν τοιαύτην διάθεσιν, ἢ ἔν τινι τῶν κυρίων μέν,
ἀσθενῶν δὲ φύσει μορίων, ὅπερ ἤτοι γ᾿ ἐν αὐτῷ μένον
ἢ πρός τι τῶν ἀσθενεστέρων φύσει μορίων ἀποχω-
ροῦν τὴν βλάβην ἐργάζεται. σκεπτέον οὖν καὶ διορι-
στέον αὐτὸ τοῦτο πρῶτον, εἴτε δι᾿ ἑαυτοῦ αἰτίαν συνεχῶς νοσεῖ τὸ μόριον εἴτε δι᾿ ἄλλο τι πρότερον αὐτοῦ
πάσχον. ἐνίοτε μὲν οὖν ἔξωθεν ἡ βλάβη τοῖς ἀσθενέσι
γίνεται μορίοις ἤτοι ψυχθεῖσιν ἢ φλεχθεῖσιν[37] ἢ πλη-
γεῖσιν ἢ κοπωθεῖσιν, ὡς τὰ πολλὰ δὲ ἐπὶ τοῖς διαι-
τήμασι πλῆθος ἢ κακοχυμίαν ἀθροίζουσι. προσέχειν
οὖν χρὴ τῷ μεγέθει τῆς φυσικῆς ἀσθενείας τοῦ μο-

ceived it, that part again impels it on to another part weaker than itself, which in turn sends it on to another, until the superfluity reaches one of the parts that has nothing weaker than itself. On the basis of this argument, then, one part or another is continuously disturbed in those who are careless in respect of regimen, whereas in those who collect no superfluity, the weakness always remains harmless. The clearest demonstration of this is that some people are troubled in the weak parts for six months or more.

If the actual weakness alone were a cause of the damage, the weak part would have been affected continuously, as it would have been the cause of the affection existing and present continually in itself. Accordingly, since it is 442K not affected continuously, it is clear that there is something else which comes to complete the genesis of the affection itself. This is nothing other than what is superfluous, either in quantity or in quality. And this superfluity is either collected in the whole body (they call such a condition *plethora*) or in one of the important parts which is weaker in nature. It then either remains in this part or passes on to one of the parts weaker in nature and causes the damage. What must be considered and distinguished first of all is this—whether the part is a cause of being continually diseased through itself, or through some other part affected before it. Sometimes, the damage to the weak parts arises from an external cause, being cooled, inflamed, struck or fatigued, so that in most instances, due to the regimens, they collect excess or *kakochymia*. You must, then, direct your attention to the magnitude of the

---

36 *post* ἀσθένεια: τῆς βλάβης αἰτία Ko; τὴν βλάβην Ku
37 φλεχθεῖσιν Ko; θλασθεῖσιν Ku

ρίου, διορισμῷ χρώμενον τοιῷδε· λεπτότερον χρὴ διαιτῆσαι τὸν ἄνθρωπον ἢ πρόσθεν ἅμα τῷ δηλονότι
443K καὶ τὰ προσήκοντα γυμνάσια γυμνάζεσθαι, φυλάττεσθαι δὲ καὶ τὴν ἀπὸ τῶν ἔξωθεν αἰτίων βλάβην· εἶτα,
ἐὰν μὲν ἐπὶ τούτοις μήτ᾽ ἰσχνὸς γένηται μήτε διο-
χλῆται τὸ ἄρρωστον αὐτοῦ μόριον, ἐπιμένειν τοῖς δι-
αιτήμασιν, ἐὰν δ᾽ ἤτοι λεπτύνηται πᾶσι τοῖς μέρεσιν
ἢ καί τι πάσχῃ τὸ φαύλως κατεσκευασμένον, ὑπαλλάττειν τὴν τοιαύτην δίαιταν, ἐὰν μὲν πάσχῃ, δυοῖν
θάτερον ἐργαζομένους, ἢ τὴν δίαιταν ἐπὶ τὸ λεπτότερον ἄγοντας ἢ κενώσεσιν ὡραίαις[38] χρωμένους. ἐνίοις
μὲν οὖν ἦρος εἰσβάλλοντος αὐτάρκης κένωσις μία
καθ᾽ ἕκαστον ἔτος γινομένη, τισὶ δ᾽ ἐπὶ τῇδε καὶ δευτέρας φθινοπωρινῆς γίνεται χρεία. πλῆθος μὲν οὖν
ἀθροίζοντος τοῦ σώματος, αἵματος ἀφαιρέσει κενοῦν,
κακοχυμίαν δὲ διὰ καθαίροντος φαρμάκου τὸν ἐπι-
κρατοῦντα χυμόν. ἐὰν δὲ ἐπὶ τῇ προσηκούσῃ διαίτῃ
μηδὲν πάσχῃ τὸ μόριον, ἰσχναίνηται δὲ τὸ σῶμα, τῇ
πιττώσει χρηστέον, ὡς ἔμπροσθεν εἴπομεν, εἰς ἀνά-
θρεψιν ἕξεως λεπτυνομένης.

14. Μοχθηροτάτη[39] δὲ σύστασίς ἐστι σώματος καὶ
ἡ τοιάδε· σπέρμα πολὺ καὶ θερμὸν ἔνιοι γεννῶσιν
444K ἐπεῖγον αὐτοὺς εἰς ἀπόκρισιν, οὗ μετὰ τὴν ἔκκρισιν
ἔκλυτοί τε γίνονται τῷ στόματι τῆς κοιλίας (ὃ καὶ
αὐτὸ καλεῖται "στόμαχος" οὐ μόνον ὑπὸ τῶν ἰατρῶν,
ἀλλὰ καὶ ὑπὸ τῶν ἄλλων ἀνθρώπων, ὥσπερ ὑπὸ τῶν
παλαιῶν ἐκαλεῖτο "καρδία") καὶ τῷ σώματι δὲ παντὶ
καταλύονταί τε καὶ ἀσθενεῖς γίνονται καὶ ξηροὶ καὶ

natural weakness of the part, using the following determination: is it necessary to administer to the person a lighter diet than before, at the same time obviously, as getting him to practice the appropriate exercises, while also guarding against damage from external causes. Then, if after these measures, he neither becomes thin nor is troubled excessively in his weak part, continue the diets. On the other hand, if he does become thin in all the parts, or some constitution affects him adversely, change such a diet. If he is affected, then do one of two things—change to a thinner diet or use evacuation appropriate for the seasons. In some, one evacuation at the start of spring each year is sufficient, whereas in others, there is need of a second evacuation in autumn in addition to this. When the body collects an excess, evacuate by withdrawals of blood; when there is *kakochymia*, evacuate the predominant humor with a purging medication. If, with the appropriate regimen, the part is not affected but the body becomes thin, you must use the application of pitch, as I said before, for restoration of the thinned state.

14. A very distressing state of the body is the following: some men generate semen that is large in amount and hot, impelling them toward its expulsion. After the emission they become relaxed at the opening of the stomach (this is called *stomachus* not only by doctors but also by other men, just as it was called *cardia* by the ancients), and they are debilitated and become weak in the whole body. Those

443K

444K

---

38 ὡραίαις Ko; ἀραιαῖς Ku

39 *post* Μοχθηροτάτη: δὲ σύστασίς ἐστι σώματος Ko; δὲ σώματος ἐστι Ku

λεπτοὶ καὶ ὠχροὶ καὶ κοιλοφθαλμιῶντες οἱ οὕτω διακείμενοι. εἰ δὲ ἐκ τοῦ ταῦτα πάσχειν ἐπὶ ταῖς συνου-
σίαις ἀπέχοιντο μίξεως ἀφροδισίων, δύσφοροι μὲν
τὴν κεφαλὴν γίνονται,[40] δύσφοροι δὲ τῷ στομάχῳ καὶ
ἀσώδεις. οὐδὲν δὲ μέγα διὰ τῆς ἐγκρατείας ὠφελοῦν-
ται· συμβαίνει γὰρ αὐτοῖς ἐξονειρώττουσι παραπλη-
σίας γίνεσθαι βλάβας, ἃς ἔπασχον ἐπὶ ταῖς συνου-
σίαις.

εἷς[41] δέ τις ἐξ αὐτῶν ἔφη μοι δακνώδους τε καὶ
θερμοῦ πάνυ τοῦ σπέρματος αἰσθάνεσθαι κατὰ τὴν
ἀπόκρισιν οὐ μόνον ἑαυτόν, ἀλλὰ καὶ τὰς γυναῖκας,
αἷς ἂν ὁμιλήσῃ. τούτῳ τοίνυν ἐγὼ συνεβούλευσα
βρωμάτων μὲν ἀπέχεσθαι τῶν γεννητικῶν σπέρματος, προσφέρεσθαι δὲ οὐ βρώματα μόνον, ἀλλὰ καὶ
φάρμακα τὰ τούτου σβεστικά (λέλεκται δὲ αὐτῶν ἡ
445K ὕλη κατὰ τὰ Περὶ τροφῶν καὶ τὰ τῶν ἁπλῶν φαρμάκων ὑπομνήματα), γυμνάσια δὲ γυμνάζεσθαι τὰ
διὰ τῶν ἄνω μορίων μᾶλλον, ὁποῖόν ἐστι τό τε διὰ
τῆς σμικρᾶς σφαίρας καὶ τὸ διὰ τῆς μεγάλης καὶ τὸ
διὰ τῶν ἁλτήρων, μετὰ δὲ τὸ λουτρὸν ὅλην τὴν ὀσφὺν
ἀλείφεσθαι τῶν ψυχόντων τινὶ χρισμάτων· ἔστι δὲ δὴ
τὰ τοιαῦτα τὸ καλούμενον ὠμοτριβὲς καὶ ὀμφάκινον,
ἔλαιον ῥόδινόν τε καὶ μήλινον ἐκ τοῦ τοιούτου γεγο-
νὸς ἐλαίου. συνέθηκα δ' αὐτὸς ἐνίοις καὶ παχύτερα τῇ

[40] γίνονται add. Ko
[41] εἷς Ko; ὡς Ku

208

in such a state are dry, thin, pale and hollow eyed. If, due
to suffering these things, they abstain from indulging in
sexual intercourse, they become dysphoric in the head and
in the cardiac orifice of the stomach, and are nauseated.[50]
They are not greatly helped by self-control. For nocturnal
emissions befall them and the harms are similar to those
they suffered due to sexual intercourse.

One of these men told me that he felt the semen to be
biting and very hot at the time of emission—and not only
he himself, but also the women with whom he copulated.
Accordingly, I advised this person to abstain from foods
that generate semen and to take not only foods but also
medications that quench this (I have written about the
material of these in my work, *The Powers of Foods* and   445K
also in the treatises on simple medications);[51] to carry out
the exercises through the upper parts particularly, like the
one with the small ball, the one with the large ball, and
leaping while holding weights; and after a bath, to anoint
his entire loins with one of the cooling unguents—such
things as the so-called *omotribes* and *omphakinos*,[52] oil of
roses and apples, and oil made from this sort of thing. For
some people, I myself have compounded unguents that

---

[50] Here again there is the terminological issue regarding the
meanings of κοιλία, στόμαχος, and καρδία. In the present context, the former is taken to mean the stomach rather than the
abdomen or abdominal cavity generally, while the latter two are
taken to be alternative terms for the upper opening of the stomach (cardia, cardiac orifice).   [51] The former is *Alim. Fac.*,
VI.453–748K, CMG V.4.2, and the latter *Simpl. Med.*, XI.776K ff.
and Books 7 and 8.   [52] Both terms are taken to refer to the
oil made from unripe olives—see note 30 above.

συστάσει χρίσματα πρὸς τὸ μὴ ῥᾳδίως ἀπορρεῖν. ἡ
δὲ σύνθεσις αὐτῶν ἐστι διά τε κηροῦ καί τινος ἄλλου
τῶν ψυχόντων γινομένη. πρῶτον δὲ τὸ καλούμενον
ὑπὸ τῶν ἰατρῶν "κηρέλαιον" ποιήσας, εἶτ᾽ ἐν θυίᾳ μα-
λάξας ταῖς χερσὶν ἱκανῶς ἐπίχει τὸν ψύχοντα χυλὸν
ἐπὶ πλεῖστον ἀναφυρῶν⁴² ὡς ἑνωθῆναι. τὴν δ᾽ ὕλην
τῶν τοιούτων χυλῶν ἐν τῇ τῶν ἁπλῶν φαρμάκων
πραγματείᾳ γεγραμμένην ἔχεις. προχειρότατοι δ᾽ αὐτῶν εἰσι καὶ ῥᾷστοι πορισθῆναι οἵ τε τῶν ἀειζώων
καὶ τοῦ στρύχνου κοτυληδόνος τε καὶ ψυλλίου καὶ
πολυγόνου καὶ τριβόλου καὶ ἀνδράχνης· οὐκ ἀνίησι
446K δὲ αὕτη χυλόν, ἐὰν μὴ κοπτομένης ἐν ὅλμῳ παρεγ-
χέηταί τις ἄλλου ὑγροῦ καὶ λεπτοῦ καὶ ὑδατώδους
τὴν σύστασιν καὶ⁴³ γλίσχρου καὶ παχέος, ὥσπερ ὁ
τῆς ὄμφακός τε καὶ τῶν ῥόδων.

ἀλλ᾽ οὗτοι μὲν ἐν τῷ θέρει, τῶν δ᾽ ἄλλων πολλοὶ
καὶ κατὰ τὰς ἄλλας ὥρας εἰσίν, ὥσπερ ὁ τῆς θριδακίνης τῶν ψυχόντων καὶ αὐτὸς ὑπάρχων. ἀλλὰ καὶ τὸ
λινόσπερμον ἑψόμενον ἐν ὕδατι ψύχοντα χυλὸν ἐργάζεται. γυμναστὴν δέ τινα τῶν ἀθληταῖς ἐπιστατούν-
των ἐθεασάμην μολιβδίνην λεπίδα ταῖς ψόαις ὑπο-
βάλλοντα τοῦ ἀθλητοῦ πρὸς τὸ μὴ ὀνειρώττειν αὐτόν,
καί τινι τῶν οὕτω πασχόντων ἰδιωτῶν ἐδήλωσα, καὶ
χάριν ἔγνω τῇ χρήσει τοῦ δηλωθέντος. ἕτερος δέ τις

⁴² ἀναφυρῶν add. Ko
⁴³ τὴς σύστασιν καὶ Ko; τῆς συστάσεως μὴ Ku

are thicker in consistency so they do not rub off easily. The compounding of these is with wax and one of the other cooling agents. First you make what doctors call "wax oil" (κηρέλαιον);[53] then you soften this in a mortar to a sufficient extent with your hands and pour in as much cooling juice as possible, mixing it in so it is fully incorporated. You have the material of such juices written out in the treatise, *On the Nature and Powers of Simple Medications*.[54] The most readily available of these and the easiest to procure are those of the evergreens, sleepy nightshade, navelwort, fleawort, knot grass, water chestnut and purslane. The last does not produce juice unless it is chopped up in a mortar and some other fluid that is thin and watery in consistency and even thick and viscid like *omphakos* and oil of roses, is poured in.

446K

But these are [only] available in the summer whereas many of the others are also available during the other seasons, like that of lettuce which is itself also one of the cooling agents. But linseed too, when boiled in water, makes a cooling juice. I have seen one of the gymnastic trainers in charge of athletes place lead flakes on the loins of the athlete so he doesn't have nocturnal emissions, and I made this known to one of my own patients affected in this way. He was grateful for being made aware and learning of the use. Someone else, being weaker in the nature

---

[53] This is described in LSJ as "wax-oil, a kind of salve," referring to this passage and to *Comp. Med. Gen.*, XIII.953K and 1006K (where it is mentioned as a term used by younger doctors in conjunction with *omotribes* and *omphakinos*). See also the wax salve mentioned in Pliny, 13.22.43, no. 124, and in Caelius Aurelanus, *Acute*, 2.11.   [54] See note 40 above.

ἀσθενεστέραν ἔχων τὴν φύσιν τῆς σαρκὸς οὐκ ἠνέ-
σχετο τῆς τοῦ μολίβδου σκληρότητος, ᾧ συνεβούλευσα τῶν εἰρημένων ἄρτι βοτανῶν ὑποστρώννυσθαί
τινας, ἀναμιγνύναι δὲ αὐταῖς καὶ ἄγνου κλῶνας ἁπα-
λοὺς καὶ πηγάνου, ᾔσθετό τε παραχρῆμα τῆς ἐξ αὐτῶν ὠφελείας, ὡς χρῆσθαι τοῦ λοιποῦ διαπαντός.
ἀλλὰ καὶ συνεχῶς ἐσθίειν τὸ σπέρμα τοῦ ἄγνου συμ-
447K βουλεύσαντός μου καὶ τούτου χάριν ᾔδει,[44] καθάπερ
καὶ τοῦ πηγάνου.

τῶν μὲν οὖν τοιούτων ἡ ὕλη κατά τε τὰ τῶν ἁπλῶν
φαρμάκων ὑπομνήματα καὶ τὰ τῶν ἐν ταῖς τροφαῖς
δυνάμεων εἴρηται νυνὶ δ᾽ ἀναγκαῖον εἶναί μοι δοκεῖ
προσθεῖναι τῷ λόγῳ φυλάττεσθαι τὰ σφοδρῶς ψύχοντα, καθάπερ ὅσα διὰ μήκονός τε καὶ μανδρα-
γόρου γίνεται χρίσματα· μήτε γὰρ τούτων τι προσφέρειν μήτε, ὅταν ἀκμάζῃ τὰ φυτά, καθάπερ τοῖς
προειρημένοις οὕτω καὶ τούτοις ἔστιν ὑποστορέσμασι
χρῆσθαι.[45] καὶ ῥόδων δὲ ἠξίωσά τινα πειραθῆναι, καὶ
ὠνήσατο καὶ οὗτος ὑποστρωννὺς αὐτὰ χωρὶς τοῦ
βλαβῆναί τι κατὰ τοὺς νεφρούς. αἱ γὰρ σφοδραὶ ψύ-
ξεις τῶν ἐπιτιθεμένων τοῖς κατὰ τὴν ὀσφὺν χωρίοις
ἀδικοῦσι τοὺς νεφρούς. ἐπενόησα δέ τι καὶ ἄλλο τοῖς
οὕτω διακειμένοις χρήσιμον, ὡς ἐκ τῆς πείρας ἐμαρ-
τυρήθην. ταύτην γὰρ ἀεὶ κριτήριον ἔχειν τῶν ἐπινοη-

---

[44] post ᾔδει: ἐμοὶ (Ku) om.
[45] In the Kühn text the two sentences following χρῆσθαι. are
in the reverse order to that given above, which is as in Ko.

of his flesh, could not tolerate the hardness of the lead. I advised him to spread under it some of the herbs mentioned just now, mixing with them also the twigs of the chaste tree[55] and rue. He was immediately aware of the benefit from these so that he used this continuously for the rest of the time. But when I advised him to keep eating the seeds of the chaste tree, he was grateful to know of this, just as he was of the rue.   447K

The material of such things has been stated in the treatises, *On the Nature and Powers of Simple Medications* and *The Powers of Foods.*[56] For the present, what seems to me necessary to add to the discussion is to guard against things that are strongly cooling, as the ointments made from poppy and mandrake are. Do not apply any of these, or the herbs when they are mature, just as you also should not use in this way those previously mentioned that are spread beneath the bedclothes.[57] I thought it was worthwhile for oil of roses to be tried, and when this was spread under the bedclothes, he was benefitted without injury to the kidneys. For the strong cooling effects of things placed in the region of the loins are injurious to the kidneys. I also thought of something else useful for those in such states, based on the evidence from experience. There must always be this criterion of the things thought of, and nothing

---

[55] *Vitex Agnus-castus.* According to LSJ, the branches of this tree were placed in the beds of matrons at the Thesmophoria. See Dioscorides, 1.103.

[56] *Simpl. Med.,* XI.777K, 807K ff., XII.100K; *Alim. Fac.,* VI.550K.

[57] The order of the text here follows Koch.

θέντων χρὴ καὶ μηδὲν γράφειν ὡς χρήσιμον, οὗ[46] τις
αὐτὸς οὐκ ἐπειράθη, πλὴν εἰ προσγράφοιτο τοῦτο[47]
αὐτῷ ἐννοεῖσθαι μὲν αὐτό, πεπειρᾶσθαι δὲ μηδέπω. τί
448K   οὖν ἐστιν, ὃ ἔφην ἐπινοήσαντός μου μεγάλης[48] ὠφελείας ᾐσθῆσθαι τοὺς χρησαμένους, ἤδη σοι φράσω.
παραφυλάττειν ἔφην χρῆναι τοὺς τῇ τοιαύτῃ κατασκευῇ σώματος ἐνοχλουμένους, ἡνίκα μάλιστα φαίνονται πλῆθος ἠθροικέναι σπέρματος ἀποκρίσεως
δεόμενον, ἐν ἡμέρᾳ τινὶ διαιτηθέντας εὐχύμως τε καὶ
μετρίως χρῆσθαι μὲν ἐπὶ τῷ δείπνῳ τρεπομένους εἰς
ὕπνον τῇ συνουσίᾳ· κατὰ δὲ τὴν ἑξῆς ἡμέραν, ὅταν
αὐτάρκως ἔχωσιν ὕπνου, διαναστάντας ἀνατρίψασθαι
σινδόνι, μέχρις ἂν ἔρευθός τι σχῇ τὸ δέρμα· κἄπειτά
τινι δι' ἐλαίου τρίψει συμμέτρως χρησαμένους, εἶτα
μὴ πολὺ διαλείποντας ἄρτον ἄζυμον[49] κριβανίτην
καθαρὸν ἐξ οἴνου κεκραμένου προσενεγκαμένους, οὕτως ἐπὶ τὰς συνήθεις ἔρχεσθαι πράξεις· ἐν δὲ τῷ
μεταξὺ τῆς δι' ἐλαίου τρίψεως καὶ τῆς δι' ἄρτου
προσφορᾶς, εἰ χωρίον ἔχει τι πλησίον ἐπιτήδειον,[50]
ἐμπεριπατῆσαι τούτῳ, πλὴν εἰ κρύος εἴη χειμέριον·
ἄμεινον γὰρ ἔνδον μένειν τηνικαῦτα.

ταύτην μέντοι τὴν διὰ τῆς ἐδωδῆς ῥῶσιν τοῦ στομάχου καὶ γραμματικῷ τινι συνεχῶς ἁλισκομένῳ
σπασμοῖς ἐπιληπτικῶς συνεβούλευσα, καὶ μεγάλως

<hr>

[46] οὗ Ko; ὅτε Ku

[47] post τοῦτο: αὐτῷ ἐννοεῖσθαι μὲν αὐτό, πεπειρᾶσθαι δὲ
μηδέπω. Ko; αὐτὸ, πεπειρᾶσθαι μηδέπω. Ku

should be written about as useful that has not been tried by the person himself, unless he writes in addition that what was thought of has not yet been tested. What it is, then, I said I thought of, that was perceived as being of great benefit to those using it, I shall tell you now. I said it was necessary to watch closely those troubled by such a constitution of the body, especially when they have obviously collected a quantity of sperm requiring emission. On a particular day, having dined moderately on *euchymous* food, they should take themselves off to bed after dinner for intercourse. On the following day, when they have had enough sleep, after getting up, they should be massaged with a muslin cloth until the skin has some redness. And then, using moderate massage with oil and allowing a short interval, provide them with well-leavened bread, pan baked, pure and mixed with wine, and in this way proceed to their customary affairs. Between the massage with oil and the taking of bread, if there is some suitable place nearby, they should walk around in this, unless there is a wintry cold. It is better to remain indoors under these circumstances.

Moreover, I recommended this strengthening of the *stomachus* (cardiac orifice) through food to a certain grammarian who was continually afflicted with epileptic

448K

ὤνητο. μάλιστα δὲ αὐτὸν ὠφελήσειν ἤλπισα πυθόμε-
449K νος παρ' αὐτοῦ τῷ τῶν σπασμῶν συμπτώματι[51] περι-
πίπτειν, ὅταν ἐπὶ πλεῖον ἄσιτος διαμείνῃ, καὶ μᾶλλον[52] ἐὰν ἐν τῷ μεταξὺ λυπηθεὶς ἢ θυμωθεὶς τύχῃ. καὶ
μέντοι καὶ τὴν ἕξιν αὐτοῦ τοῦ σώματος ἑώρων ἰσχνήν,
ἐρωτώμενός τε συνεχῶς ἐκχολοῦσθαι τὸν στόμαχον
ὡμολόγει. κοινὴν δέ τινα συμβουλὴν ἅπασι τοῖς
ταῦτα ἀναγνωσομένοις, ἰδιώταις μὲν τῆς ἰατρικῆς,
οὐκ ἀγυμνάστοις δὲ τὸν λογισμόν, ὑποτίθημι τήνδε·
μή, καθάπερ οἱ πολλοὶ τῶν ἀνθρώπων ὡς ἄλογα ζῷα
διαιτῶνται, καὶ αὐτοὺς οὕτως ἔχειν, ἀλλὰ διὰ τῆς πείρας κρίνειν, τίνα μὲν αὐτοὺς ἐδέσματά τε καὶ πόματα
βλάπτει, τίνες δὲ καὶ πόσαι κινήσεις· ὁμοίως δὲ καὶ
περὶ χρήσεως ἀφροδισίων ἐπιτηρεῖν, εἴτε ἀβλαβὴς
αὐτοῖς ἐστιν εἴτε βλαβερὰ καὶ διὰ πόσων ἡμερῶν
χρωμένοις ἀβλαβής τε καὶ βλαβερὰ γίνεται. καθάπερ γὰρ ἱστόρησά τινας μεγάλως βλαπτομένους,[53]
οὕτως ἑτέρους ἀβλαβεῖς διαμένοντας μέχρι γήρως ἐπὶ
ταῖς χρήσεσιν αὐτῶν. οὗτοι μὲν σπάνιοι καθ' ἑκάτερον γένος, οἵ τε μεγάλως βλαπτόμενοι καὶ οἱ μηδὲν
ἀδικούμενοι· τὸ δὲ μεταξὺ πᾶν ἐν τῷ μᾶλλόν τε καὶ
ἧττον εἰς τὸ πολὺ τῶν ἀνθρώπων ἐκτέταται πλῆθος.
450K ὧν τοῖς πεπαιδευμένοις (οὐ γὰρ δὴ οἱ τυχόντες γε
ταῦτα ἀναγνώσονται) συμβουλεύω παραφυλάττειν,
ὑπὸ τίνων ὠφελοῦνταί τε καὶ βλάπτονται· συμβήσεται γὰρ οὕτως αὐτοῖς εἰς ὀλίγα δεῖσθαι τῶν ἰατρῶν,
μέχρις ἂν ὑγιαίνωσιν.

seizures, and he benefitted greatly. I particularly hoped to
help him when I learned from him that the symptom of 449K
seizures occurred whenever he remained without food
for a long time, and especially if, in the meantime, grief
or anger befell him. And furthermore, I saw the state of
the body itself was thin, and on questioning him, he admit-
ted that the *stomachus* (cardiac orifice) was continuously
charged with bile. As a general recommendation for all
those who read these things, laymen as regards the medical art but not unpracticed in reasoning, I advise the following: Do not, as the majority of men do, eat like irrational animals and be like them, but judge by experience
which foods and drinks harm them, and the kinds and
amounts of movements. Similarly too, they should keep an
eye on the use of sexual intercourse—whether it is harmless for them or harmful to those using it, and after how
many days it is harmless or harmful. As I learned by inquiry, some men are greatly harmed by the use of sexual
intercourse, while others remain unharmed into old age
by the use of sexual intercourse. There are, however, few
in each class—that is, those who are greatly harmed and
those who are not harmed at all. The whole intermediate
range in terms of more and less extends to include the vast
majority of men. My advice to those who are educated (for 450K
ordinary men will not read these things) is to observe
closely those things that benefit and harm them, as in this
way it will come about for them that they will have little
need of doctors whilever they are healthy.

---

51 συμπτώματι Ko; σύμπτωμα τότε Ku

52 post μᾶλλον: ἐὰν ἐν τῷ μεταξὺ λυπηθεὶς ἢ θυμωθεὶς
τύχῃ. Ko; ἐπανιέντων μεταξὺ λύπης ἢ θυμοῦ. Ku

53 post βλαπτομένους: ἐνίους (Ku) om.

15. Τούτων ἤδη γεγραμμένων ἀναγνούς τις τῶν ἑταίρων ὅλον τὸ βιβλίον ἔν τι τῶν ἔμπροσθεν ἀναβληθέντων,[54] ὕστερον δὲ ῥηθησομένων παραλελεῖφθαί μοι τελέως ἔφη, τὸ διὰ τοῦ χυλοῦ τῶν κυδωνίων[55] μήλων φάρμακον, ἐπιτήδειον εἴς τε τὰς ὀρέξεις τοῖς ἀνορέκτοις καὶ τὰς πέψεις τοῖς μὴ καλῶς πέττουσι καὶ συνελόντι φάναι τὴν γαστέρα ῥωμαλεωτέραν ἐργαζόμενον. ἐπαινέσας οὖν τὸν ἀναμνήσαντα προσθήσω τῷ λόγῳ τὴν σκευασίαν αὐτοῦ τοιαύτην οὖσαν· τῶν κυδωνίων μήλων τὰ μείζω τε καὶ ἡδίω καὶ ἧττον στρυφνά, ἃ στρουθία καλοῦσιν οἱ κατὰ τὴν ἡμετέραν Ἀσίαν Ἕλληνες, ἐκ τούτων τοῦ χυλοῦ λαβόντας ξέστας Ῥωμαϊκοὺς δύο, χρὴ μῖξαι μέλιτος ὅτι καλλίστου τὸ ἴσον μέτρον, ὄξους δὲ ξέστην ἕνα καὶ ἡμίσειον· καὶ ταῦτα ἐπ᾽ ἀνθράκων διακεκαυμένων προεψήσαντας μετρίως καὶ προαπαφρίσαντας μῖξαι ζιγγιβέρεως οὐγκίας τρεῖς, πεπέρεως δὲ τοῦ λευκοῦ δύο, καὶ οὕτω πάλιν ἐπὶ τῶν ὁμοίως διακεκαυμένων ἀνθράκων ἑψῆσαι μέχρι μελιτώδους συστάσεως, ἐν οἷα καὶ τὰ στομαχικὰ τῶν φαρμάκων σκευάζεται. τοῦτο τὸ φάρμακον τοῖς ἄτονον ἔχουσι τὸ ἧπαρ ὠφελιμώτατόν ἐστιν. εὔδηλον δέ, ὅτι μάλιστα μὲν αὐτὸ νῆστιν ὄντα προσφέρεσθαι χρὴ πλῆθος, ὅσον ἂν ᾖ μύστρου συμμέτρου τῷ μεγέθει. βλάπτει δὲ οὐδέν, οὐδὲ ἐὰν μετὰ τροφήν τις αὐτὸ λαμβάνῃ. καλῶς δ᾽ ἂν

451K

---

[54] ἀναβληθέντων Ko; ἀναλειφθέντων Ku
[55] κυδωνίων add. Ko

15. After these books had already been written, one of my good friends, who read the whole treatise, said one of the matters previously deferred for later consideration had been completely omitted by me. This was the medication made from the juice of quinces[58] which is useful for the appetite in those who are anorexic and for concoction in those who are not concocting properly, and in a word, for making the stomach stronger. So, having praised the man who brought this to my attention, let me add to the work what the preparation of this is. Taking two Roman pints[59] of juice from the largest, sweetest and least astringent Cydonian apples, which the Greeks in Asia Minor call *strouthia* (quinces), you must mix an equal measure of the best honey and a pint and a half of vinegar. Boil these things moderately over hot charcoals, and after despumation, mix in three ounces of ginger and two of white pepper. Boil this again in the same way over hot charcoals until it has the consistency of honey, as in the preparation of medications good for the stomach (stomachics).[60] This medication is especially beneficial for those who have a weak liver. Obviously it is necessary to administer a quantity of this, particularly in one who is fasting—as much as a moderate spoonful in amount. There is no harm if one takes this after food. It is also good to administer this to

451K

[58] On this medication see Paul. Aegin., 7.11.

[59] One ξέστης was approximately equal to one Roman pint (*sextarius*).

[60] The term "stomachics" is still in use to describe medications good for digestion.

προσφέροιτο καὶ εἰ προηριστηκώς τις εἴη, εἶτα δειπνεῖν μέλλων προσλάβοι. κάλλιστος δὲ καιρὸς ὁ πρὸ δυοῖν ἢ τριῶν ὡρῶν τῆς τροφῆς.

εἰ δὲ στυπτικώτερον αὐτὸ βούλοιο ποιῆσαι, καὶ διὰ τῆς σαρκὸς[56] τῶν κυδωνίων μήλων σκεύαζε. τοῖς μέντοι δύσκρατον ἔχουσι κατὰ θερμότητα τὴν γαστέρα καὶ τοῖς ὁπωσοῦν πληρουμένοις χολῆς ἀφελὼν τό τε ζιγγίβερι καὶ τὸ πέπερι τὸν χυλὸν τῶν μήλων δίδου, μετ' ὄξους τε καὶ μέλιτος ἑψήσας μόνον ἐν τῇ προγεγραμμένῃ συμμετρίᾳ. ὅσαι δὲ μέσαι πώς εἰσι κατὰ τὴν κρᾶσιν γαστέρες, ὡς μήτε χολῶδες ἀθροίζειν
452K περίττωμα μήτε φλεγματῶδες, ἥμισυ τῆς προειρημένης συμμετρίας τοῦ τε ζιγγιβέρεως ἐμβάλλειν καὶ τοῦ πεπέρεως, ὡς εἶναι πεπέρεως μὲν οὐγκίαν μίαν, ζιγγιβέρεως δὲ μίαν καὶ ἡμίσειαν. ὅσαι δὲ δυσκρασίαι κατὰ ψυχρότητα, τέτταρας μὲν τοῦ ζιγγιβέρεως οὐγκίας, τοῦ πεπέρεως δὲ τρεῖς ἢ δύο καὶ ἡμίσειαν. ἔξεστι δέ σοι καὶ κατὰ τὴν μέσην συμμετρίαν σκευάζοντι προσβάλλειν ἐπὶ τῆς χρήσεως τοῦ πεπέρεως.

οὕτω μὲν οὖν προνοητέον ἐστὶ τῶν ἐν διαφέρουσι μορίοις τοῦ σώματος ἐναντίας ἐχόντων κράσεις. ὅσοι δὲ καθ' ἓν ὁτιοῦν ἤτοι γε ὁμοιομερὲς ἢ ὀργανικὸν ἀνώμαλον ἔχουσι κρᾶσιν, ἕτερος ἐπ' αὐτοῖς εἰρήσεται λόγος.

[56] τῆς σαρκὸς Ko; τοῦ χυλοῦ Ku

someone who has previously breakfasted and takes it when about to have the midday meal. However, the best time is two or three hours before food.

If you wish to make this more astringent, prepare it with the flesh of quinces. Of course, for those who have a hot *dyskrasia* of the stomach and those who are in any way at all filled with bile, omit the ginger and pepper, and give the juice of quinces boiled with vinegar and honey alone, in the previously stated proportions. In the case of those stomachs intermediate in *krasis* such that they do not collect bilious or phlegmatic superfluities, put in half the previously stated measures of ginger and pepper—that is, one ounce of pepper and one and a half of ginger. For the cold *dyskrasias*, put in four ounces of ginger and three or two and a half of pepper. You can also add the pepper to what is prepared in the standard proportions at the time of use.

452K

This, then, is how you must care for those who have opposite *krasias* in different parts of the body. In the case of those who have an irregular (nonuniform, anomalous) *krasis* in any one part, either *homoiomerous* or organic, another book will be written.[61]

61 *Inaequal. Intemp.*, VII.733–52—see note 13 above.

ΓΑΛΗΝΟΥ ΘΡΑΣΥΒΟΥΛΟΣ
[ΠΟΤΕΡΟΝ ΙΑΤΡΙΚΗΣ Η
ΓΥΜΝΑΣΤΙΚΗΣ ΕΣΤΙ ΤΟ
ΥΓΙΕΙΝΟΝ]

THRASYBULUS
[ON WHETHER HYGIENE
BELONGS TO MEDICINE OR
GYMNASTICS]

INTRODUCTION

The question to be examined—whether hygiene (healthiness) is part of medicine or gymnastics—is presented by Galen as a theoretical issue. The answer may seem obvious, and to many it would be, but it does also have practical implications which to Galen are of some importance. Basically, the question may be rephrased in greater detail as follows: are the curing of disease, the restoration of health after recovery from disease, the preservation and improvement of the restored health, and the maintenance and improvement of health independent of recovery from disease all part of one overarching medical art, and as a corollary, what part does gymnastics play in these processes and what is its status as an art? In short, are the treatment of those who are diseased, the restoration of health during recovery from disease, the maintenance of health in those who are healthy, and perhaps the improvement of health in such people all ultimately the responsibility of the medical practitioner with the help, as necessary, of those practiced in other, subsidiary disciplines such as gymnastics? The practical relevance is the matter of who manages the restoration, preservation, and improvement of health in people generally, or more precisely, what are the roles of doctors, related health professionals, exercise therapists, and gymnastic trainers?

224

Galen's answer is detailed and unequivocal; it may be summarized as follow:

1. An art is defined and identified by its end (*telos*).
2. The *telos* of medicine as therapeutics is health—in this case its restoration and establishment as a stable state (*hexis*); the *telos* of hygiene is also health, but in this case its preservation or improvement; the *telos* of gymnastics can also be health, but in certain instances it may go beyond health in an adverse direction.
3. Health is defined and identified as either a constitution of the body that allows the bodily functions to be carried out faultlessly and naturally (*kata physin*) or the bodily functions themselves all being normal (again, *kata physin*).
4. Both therapeutics and hygiene are fairly and squarely parts of the medical art with health as its *telos*, or end, and *ipso facto* are the province of the doctor, as recognized from the time of Hippocrates.
5. Gymnastics is actually a part of hygiene and has health as its *telos* (end) unless it shades over into the excesses of wrestling schools and gymnastic trainers when the end is an abnormal bodily state, defective in terms of the normal functions, achieved in the pursuit of mass and strength aimed at overcoming opponents in contests.

It is of interest that the kind of practical situation Galen is addressing in this essay also obtains today. Doctors and related health professionals such as physical and occupational therapists should be the people responsible for restoring, maintaining, and improving health, but the ur-

gency of the treatment of frank disease and its immediate aftermath (therapeutics generally) may give the therapeutic aspect precedence over the hygienic, in Galen's terms, both conceptually and temporally. Particularly in recent times, various exercise specialists, professional trainers, diet therapists, and the like have proliferated, all with health as their ostensible *telos.* As in Galen's time, more extreme measures developed may in fact be counterproductive in terms of health. The considerable intrusion of drugs into the training for athletic activities of various sorts is a particular case in point. Hippocrates' statement, twice quoted in the present essay, remains apposite: "In those who exercise, the peak *euexias* are dangerous" (*Aphorisms* 1.3).

Galen's treatment of the question he poses in *Thrasybulus* might best be described as discursive and repetitious—both recurring features of his writings (as he himself acknowledges from time to time). Nevertheless, his various digressions and theoretical musings are hardly ever without interest. A summary of the forty-seven sections of the present work is given in the General Introduction to the present work, vol. 1 (LCL 535). A short synopsis of this summary follows:

1. A statement of the question; the four key terms that need to be clearly defined (medicine, health, gymnastics, and belonging to / characteristic of); the importance of a knowledge of logical theory. (1–4)
2. Various aspects of the definition and identification of an art and some aspects pertaining to specific arts; the importance of *telos* (end) in defining an art; a division of forms of health. (5–17)

Galen's concluding statement, given in part in the General Introduction, is repeated here in full.

It is not, therefore, unreasonable, when asked what art hygiene is part of, to answer "the medical art." Since the name has been extended further, and no longer signifies the part but the whole art concerning the body, Hippocrates and all the doctors of the present time are rightly so named, for they know the greatest parts of the art itself are two—therapeutic and hygienic. In turn, they know gymnastics is part of hygiene itself, as has also been shown before. Therefore, just as Hippocrates, Diocles, Praxagoras, Philotimos and Herophilus were knowledgeable in the whole art concerning the body, as their writings show, so conversely the followers of Theon and Tryphon practiced the base art concerning athletes, as in turn the works of these men also reveal; they use the terms "preparatory" and "bodily exercise" and again speak of something "partial"

227

and something "complete," and *apotherapy* (restoration), and inquire into whether the athlete must be trained and exercised according to such a course or in another way. And it comes as a surprise to me when I hear those training athletes now laying claim to hygiene as part of their own art. For when hygiene is really not a part of gymnastics but conversely, gymnastics is part of hygiene, why must there be dispute about the base art of these men, which is altogether not a part of the art practiced regarding the body, and which is deemed unworthy not only by Plato and Hippocrates, but also by all other doctors and philosophers?

OTHER RELEVANT WORKS BY GALEN

Various aspects of the subject matter of this work are treated somewhat differently and often in greater detail in other works. The most important of these are as follows:

On the Constitution of the Art of Medicine (I.224–304K): the kinds of arts and medicine's place in the catalog of the arts (sections 1–2); prophylaxis and restoration of health (sections 18–19)

The Art of Medicine (I.305–412K): terminology in health, disease, and "neither" (intro. and sections 1–4); causes of preservation of health (sections 23–24); causes of health in restoration and recuperation (sections 36–37)

On the Best Constitution of our Bodies (IV.737–49K): definitions and causes of benefit and harm to the constitution (κατασκευή)

On Good Condition (IV.750–56K): the Greek titles is περὶ εὐεξία (*On Euexia*), this being a term of particular importance in *Thrasybulus*

Hygiene (VI.1–452K): Galen's major work on the theory and practice of hygiene

On Exercise with a Small Ball (V.899–910K): a short work detailing the benefits of a particular method of exercise, and including also some remarks on the deleterious effects of some other kinds of exercise

On the Differentiae of Diseases (VI.836–80K): the basic definitions of health (sections 1–3)

On the Differentiae of Symptoms (VII.42–84K): basic definitions

The Method of Medicine (X.1–1021K): important considerations on the basic structure of the body and the definitions of health and related matters (Books 1–2)

TERMINOLOGICAL ISSUES

Three groups of terms are considered:

1. τέλος / σκοπός (*telos / scopos*): these two terms are used somewhat interchangeably in the present work, although the former is by far the more important here. *Telos* is a complex term, occupying two-and-a-half columns in LSJ. The sense here is that given in LSJ III.3, 1774, which reads in part: "Philosoph. full realisation, highest point, ideal . . . (b) the end or purpose of action" with reference to *Gorgias* 499E, LCL 166 (W. R. M. Lamb), 442, which has τέλος εἶναι ἁπασῶν τῶν πράξεων τὸ ἀγαθόν. LSJ continues: "hence a final cause, *Metaphysics* 994b9, 996a16 al.; hence simply = τὸ ἀγαθόν, the chief good,"

which Galen also uses in this sense. *Scopos* is used much less frequently by Galen here; it has been rendered "objective." In the translation the first use of either word in any section has the translation ("end" or "objective") with the transliterated Greek in parentheses. In subsequent uses in the same section only the English is given.

2. σχέσις / ἕξις / εὐεξία (schesis / hexis / euexia): these three terms are critical to Galen's discussion and have been largely given in the transliterated form; a translation in parentheses is variably included in parentheses. The terms are the middle three of a five-component linear progression, as Galen describes it in section 7:

> Think of it as a linear progression. There is some damage of functions that occurs in disease, which is different from external damage, while there is uselessness due to weakness, which is in health in relation to *schesis;* but there is another and third thing, which is delivery from weakness without as yet acquisition of strength, which is health in relation to *hexis.* In addition to this is a fourth, which is a kind of excellence of the functions; this is *euexia.* The highest point, and perfection of these functions, is *euexia* up to the peak of these functions.

Thus, *schesis* is understood as a temporary or unstable state (of health). LSJ has: "of temporary, passing conditions opposing those which have become constitutional" and refers to Galen's *Method of Medicine,* X.533, which has: "Since people term the conditions that are readily resolved 'temporary' (*schesis*) and those that are not readily resolved 'established' . . ." (this is in the context of

fevers and not of health). *Hexis*, conversely, is understood as a relatively permanent state (of health). LSJ has: "being in a certain state, a permanent condition as produced by practice (*praxis*), different from *schesis* which is alterable" (III, 595). *Euexia* is in effect a better *hexis*. LSJ has: "good habit of the body, good health" and refers to Hippocrates, *Aphorisms* 1.3, which Galen quotes in this treatise. LSJ also has: "temporary high condition to permanent health" (I.711).

Galen makes explicit the meanings of these terms in relation to health in the following passage:

> Come now to the concept of one of those sick people whom you have seen often, and who, after falling ill severely, has just recovered from being diseased, but is so weak and incapable of movements that he needs others to carry him around. Such a person no longer needs a cure because he is not still sick, but he does need restoration and strength, so he is strong enough for the natural functions and adequate to bear those things that may befall him. Clearly, a person in such a state as he now is, when he first recovers from disease, is not up to enduring heat or cold, sleeplessness, or failure of digestion, or any other of all those things from which he would very easily become sick again. So he would not be safe, having not yet acquired an established healthy condition. But if this were to become difficult to break down and [health] in relation to *hexis*, it would not yet be *euexia* (for then "eu" would be added and it would become *euexia*). Nevertheless it

would be far removed from being a blameworthy and useless condition. For not to be able to carry out the functions of life adequately and to readily come to harm is a blameworthy and useless condition. If, however, someone is neither impeded in his actions nor readily harmed, such is health in accord with *hexis*. This would no longer be blameworthy and useless in terms of the functions of life but would not as yet have acquired praiseworthiness. It would acquire this if, in addition to function no longer being weak, some significant strength were to be acquired. (section 7)

3. analeptic / hygienic / euectic: these are the components of the overall medical art that relate to the three terms considered in (2.) above. The OED (1930) has the following: *Analeptic:* restorative, strengthening. *Hygienic:* belonging or related to hygiene (which in turn is defined as "that department of knowledge or practice which relates to the maintenance of health; a system of principles or rules for preserving or promoting health"). *Euectic:* pertaining to a good habit of the body (illustrated by an unattributed quotation from Jones [1574]: "Three parts of the Arte curative: First Euectick, whose scope is to keep the helthie in the same state").

TEXTS AND TRANSLATIONS

The translation is based on Helmreich's 1893 text in *Claudii Galeni Pergameni Scripta Minora*, vol. 3, 33–100. Diels lists three Greek manuscripts:

Florence, Laurent. plut. 74.3 (12th c.)
Paris, Parisin. 2164 (16th c.)
Venice, Marcian App. cl.5.45 (15th c.)

Helmreich (H) cites the use of the Florence and Paris manuscripts and the Aldine, Chartier, and Kühn (Ku) editions of Galen. There is an English translation by P. N. Singer, *Galen: Selected Works*.

ΓΑΛΗΝΟΥ ΘΡΑΣΥΒΟΥΛΟΣ
[ΠΟΤΕΡΟΝ ΙΑΤΡΙΚΗΣ Η ΓΥΜΝΑΣΤΙΚΗΣ ΕΣΤΙ ΤΟ ΥΓΙΕΙΝΟΝ][1]

806K   1. Οὐκ ἄλλα μέν, ὦ Θρασύβουλε, παραχρῆμα περὶ τοῦ προβληθέντος ὑπὸ σοῦ ζητήματος εἶπον, ἄλλα δὲ συγγράψασθαι τοῖσδε τοῖς ὑπομνήμασιν ἔχω· πάντως γάρ που γιγνώσκεις, ὡς ἀεί τε τὰ αὐτὰ περὶ τῶν αὐτῶν διεξέρχομαι καὶ ὡς εἰς[2] οὐδὲν ἐπιχειρῶ λέγειν, ὧν οὔτε μέθοδον ἔμαθον οὔτ᾽ ἐγυμνασάμην πω κατ᾽ αὐτήν. ἀρχὴ τοίνυν εὑρέσεως οὐ τούτῳ μόνῳ τῷ νῦν

807K προκειμένῳ σκέμματι τὸ γνῶναι, τί ποτ᾽ ἐστὶ τὸ ζητούμενον, ἀλλὰ καὶ τοῖς ἄλλοις ἅπασιν. αὐτὸ δὲ δὴ τοῦτο τὸ γνῶναι διττόν ἐστιν· ἢ γὰρ τὴν ἔννοιαν μόνην τοῦ πράγματος ἢ καὶ τὴν οὐσίαν γιγνώσκομεν. ὅτῳ δ᾽ ἀλλήλων ταῦτα διαφέρει, γέγραπται μὲν ἐπὶ πλέον ἐν τοῖς περὶ ἀποδείξεως, ἵναπερ καὶ τὰς ἄλλας

---

[1] There is some variation in the title—see H's note, p. 33.
[2] εἰς om. Ku

---

[1] Nothing is known of Thrasybulus, to whom this work now

GALEN, THRASYBULUS[1]
[ON WHETHER HYGIENE
BELONGS TO MEDICINE OR
GYMNASTICS]

1. On the inquiry you presented, Thrasybulus, I have not   806K
said anything offhand that is different to what I have written in these notes. You will know full well that in the matters I go over, I always try not to say anything I have not
learned as a method nor yet put into practice in relation
to this. Moreover, a beginning of discovery, not only in the
issue now before us, but also in all the others, is to know   807K
what is being sought at the time. Undoubtedly this same
knowledge is twofold: we either know the concept of
the matter alone, or we also know its essential nature.[2]
Wherein these things differ from each other, I have written
about in greater detail in the work, *On Demonstration*,[3]

lost is dedicated. The pseudo-Galenic work, *Opt. Sect.*, I.106–
223K, is also dedicated to him.   [2] The pairing of the terms
ἔννοια and οὐσία is generally rendered concept / essential substance—for the former, see Plato, *Phaedo* 76c, and Aristotle,
Nicomachean Ethics 179b15; for the latter, see Plato, *Phaedo*
63D, and Aristotle, *Metaphysics* 1017b22, 1031b1.

[3] This was a major Galenic work now lost—see his *Libr. Propr.*
XIX.41K.

ἁπάσας μεθόδους ἐκτιθέμεθα,[3] γένοιτο δ' ἂν καὶ νῦν
δῆλον ἐξ αὐτῆς τῆς χρήσεως.

2. Ὄντος γὰρ τοῦ προβλήματος τοῦδε, πότερον
ἰατρικῆς ἢ γυμναστικῆς ἐστι τὸ καλούμενον ὑγιεινόν,
καὶ σοῦ διὰ ταῦτ' ἀξιώσαντος ἀκοῦσαι τὴν ἐμὴν ὑπὲρ
αὐτοῦ δόξαν, ὅτι πολλάκις ἰατροῖς τε καὶ γυμνασταῖς
ἔφησθα παραγεγονέναι διαμφισβητοῦσιν, ἀποκρίνε-
σθαί μοι πρότερον ἠξίωσά σε, τίνα ποτὲ τὴν ἔννοιαν
ἑκάστου τῶν ὀνομάτων ἔχεις, ἰατρικῆς τε καὶ γυμναστικῆς καὶ ὑγιεινοῦ, μή που σὺ μὲν ὑπὲρ ἄλλων
πραγμάτων ἀκούειν ποθῇς, ἐγὼ δ' ὑπὲρ ἄλλων σοι
διαλέγωμαι, κἄπειτα πρὸς τοὔνομα μόνον ὁ λόγος
808K   ἡμῖν οὐ πρὸς αὐτὸ γίγνηται τὸ πρᾶγμα. σὺ μὲν οὖν
πρὸς τοῦτ' ἀπεσιώπησας οἰόμενος ἐμὲ χρῆναι λέγειν
αὐτὸν ὑπὲρ ἁπάντων ἅπαντα, τὸ δ' ἀληθὲς οὐχ ὧδ'
ἔχει· πολλὰ γὰρ ἂν εἴης οὐχ ἓν προβεβληκώς, πρῶτον
μὲν τί ποτ' ἐστὶν ἰατρική, δεύτερον δὲ τί ποτε γυμναστικὴ καὶ τρίτον ἐπὶ τούτοις τί ποτε τὸ ὑγιεινόν, ἢ
τοῦτο δὴ τὸ μάλιστα προβεβλημένον, ὁποτέρας αὐτῶν εἴη. καίτοι τοῦτ' αὐτὸ τὸ τέταρτον οὐχ ἁπλῶς
ἔφασκον δεῖν οὕτω προβάλλειν, ὁποτέρας εἴη αὐτῶν
τῶν τεχνῶν τὸ ὑγιεινόν, ἀλλ' εὐθὺς προστιθέντας, εἴτ'
ἴδιον εἴτ' οἰκεῖον εἴτε μέρος εἴθ' ὁπωσοῦν ἄλλως ἐθέλοι τις· ἔσεσθαι γάρ τινα κἀνταῦθ' ἀρχὴν τῆς εὑρέσεως τῷ ζητουμένῳ. προβληθέντος γὰρ αὐτοῦ τελέως
ὧδέ πως καὶ νὴ Δί' ἐρωτηθέντος, ἆρά γε τῆς ἰατρικῆς

---

[3] ἐκτιθέμεθα H; ἐξετιθέμεθα, Ku

where I also set down all the other methods. This too should now become clear from the actual use.

2. The problem is this: whether what is called hygiene (healthiness)[4] is part of medicine or part of gymnastics. And the reason you think it worthwhile to hear my opinion on this is because, as you said, you have often been present with doctors and gymnastic trainers when they are disputing [on this point]. But first, I expect you to answer the question as to what concept you have of each of the terms—medicine, gymnastics and hygiene—lest you desire to hear about some matters while I discourse for you about others, and then our discussion is only about the name and not about the matter itself. Since you remain 808K silent on this, I think it behooves me to say everything about all these matters myself. But the truth is not like this; if it were, the matter put forward would not be one thing but many: first, what is medicine; second, what is gymnastics; and third in addition to these, what is hygiene, or particularly the issue of which of the first two hygiene belongs to. And further, I said this fourth question must not be put forward simply in this way: "hygiene belongs to which of these two arts?" Straightway, one must add either "specific," or "characteristic of," or "part of," or whatever else one might wish, for even here there will be some starting point of the discovery in the inquiry. If this is put forward completely thus, and by Zeus, we are asked

---

[4] The various nuances of the term ὑγιεινόν are considered in the General Introduction to the present work, vol. 1 (LCL 535). In this work, it is generally translated "hygiene," although some prefer "healthiness." However, it is taken as the term for the art (techne) of preserving health.

ἔστιν ἢ τῆς γυμναστικῆς οἰκεῖον⁴ τὸ ὑγιεινόν, ἐπειδὰν
ἑκάστου τῶν τριῶν ἐκείνων ὀνομάτων εἴπωμεν τὸ
σημαινόμενον, ἰατρικῆς τε καὶ γυμναστικῆς καὶ ὑγιεινοῦ, δεήσει τέταρτον ἐπ᾽ αὐτοῖς ἐξηγήσασθαι τὸ οἰκεῖον ὅ τι σημαίνει καὶ τίς ὁ κανὼν αὐτοῦ τῆς διαγνώ-
σεως. ἀλλὰ τοῦτο μὲν ἴδιον αὐτοῦ τοῦ προβλήματος
809K ὑπάρχει, τὸ δ᾽ ἰατρικὴν ἐξηγήσασθαι τί ποτ᾽ ἐστὶ καὶ
γυμναστικὴν καὶ ὑγιεινὸν οὐκ ἴδιον μὲν τοῦ προβλήματος ὑπάρχει, ὁμολογεῖσθαι δ᾽ ἀναγκαῖον. ὅθεν,
ἐπειδὴ τοῦθ᾽ ἓν μόνον, ὃ προὔβαλες, σὺ διελθεῖν
ἐβουλήθης τότε καί σε κινδυνεύειν ἐν ταῖς ἀποκρίσε-
σιν ἑώρων ὀκνοῦντα, τῶν τινα γεγυμνασμένων ἐν λο-
γικῇ θεωρίᾳ φιλοσόφων ἐπιφανέντα πως τοῖς λόγοις
κατὰ τύχην ἠξίουν ἀποκρίνασθαί μοι. προθύμως δ᾽
ἐκείνου τοῦτο ποιήσαντος καὶ πάντ᾽ ὀρθῶς ἀποκρι-
ναμένου ῥᾳδίως, ὡς οἶσθα, διεπεράνθη τὸ πρόβλημα.

3. Καὶ μὴν κατάδηλος ἦσθα καὶ τότε μὲν εὐθὺς
ἱκανῶς χαίρων ἐπὶ τῇ τῶν λόγων μεθόδῳ· καὶ γὰρ
θᾶττον ἢ κατὰ τὴν προσδοκίαν εὑρέθη τὸ ζητούμενον·
ἐξ ὑστέρου δὲ προσκείμενος ἀεὶ λιπαρῶς ἐβιάσω τε
καὶ κατηνάγκασας οὐ πάνυ τί με πρόθυμον ὄντα γρά-
ψασθαι τοὺς λόγους· οὔτε γὰρ ἐν τουτὶ τὸ πρόβλημα
μόνον ἀκριβῶς ἐρευνηθὲν ᾤμην χρῆναι διασῴζειν ἐν
γράμμασιν οὔθ᾽ ὁμοίως τῷδε τὰ λοιπὰ πάντα διεξέρ-
810K χεσθαι σχολὴν ἦγον. ὅθεν, ὅπερ αὐτὸς ἐπ᾽ ἐμαυτοῦ
πράξας ἔτυχον, ἱκανὸν ᾤμην ἔσεσθαι καὶ πρὸς ὑμᾶς

⁴ οἰκεῖον add. H

238

whether hygiene belongs to medicine or gymnastics, whenever we speak of the signification of each of those three terms—medicine, gymnastics and hygiene—it will require a fourth term in addition to these to explicate "belongs to"—what it signifies and what the measure of its recognition is. But this is specific to the issue itself, whereas the detailed explication of what medicine is, and 809K gymnastics, and hygiene is not specific to the issue, although it is agreed to be necessary. From which, since you only wanted to go over this one thing alone which you put before me, and at that time I saw you were hesitant to take a risk in the answers, I asked one of the philosophers practiced in logical theory who were present somehow by chance in the discussions to answer me. This he did zealously, and when he answered everything correctly, the issue was easily brought to a conclusion, as you know.

3. And it was quite clear at the time also that you immediately derived sufficient pleasure from the method of the discussions, for what was being sought was discovered quicker than expected. Later, however, you were always pressing me importunately, forcing and constraining me to write down the discussions very much against my wishes. I didn't think it was necessary to preserve in writing what was accurately discovered on this one issue alone, when I could not spend the time to go over all the remaining matters similar to it. Wherefore, I followed my own 810K practices and thought it would be enough for you, my

τοὺς φίλους, δεῖξαι τὴν ὁδόν, ᾗ χρώμενος ἄν τις οὐ
τοῦτο μόνον ἀλλὰ καὶ τἆλλα πάντα διαιροῖτο προβλήματα.

4. Τὴν γὰρ λογικὴν ὀνομαζομένην παρὰ τοῖς φιλο-
σόφοις θεωρίαν ὅστις ἂν ἱκανῶς ἀσκήσῃ, πᾶν οὗτος
ὁμοίως δυνήσεται μεταχειρίζεσθαι ζήτημα· τὸ δ' ἄνευ
ἐκείνης ἀναγιγνώσκειν ὑπομνήματα προβλημάτων
οὐδὲν ἄλλ' ἢ χρόνον ἀπολλύειν ἐστὶν οὔτε κρίνειν
εἰδότα, τίνα μὲν ἐν αὐτοῖς ἀληθῶς εἴρηται, τίνα δὲ
ψευδῶς, οὔτε πάντων τῶν γεγραμμένων μνημονεύειν
δυνάμενον. ἀλλ' ἐπειδὴ κατέστην ἅπαξ εἰς τὸ λέγειν
περὶ τοῦ προβλήματος, ὅθενπερ ὀλίγον ἔμπροσθεν ὁ
λόγος ἔδειξεν, ἄρχεσθαι δεῖ.

5. Καὶ νὴ Δι' εἴ τις ἐρωτηθείς, ὅ τί ποτ' ἐστὶν ἰατρική, φαίη τέχνην εἶναι θεραπευτικὴν μὲν νοσούντων, φυλακτικὴν δ' ὑγιαινόντων, ἄντικρυς ἂν οὗτος
δοκοίη τὸ ζητούμενον εἰληφέναι μέρος αὐτῆς ἀπο-
811K φαίνων τὸ ὑγιεινόν· ὥσπερ αὖ καὶ ὅστις ἰατικὴν[5] ἀρ-
ρωστούντων μόνον εἶναι λέγει τὴν ἰατρικήν, ἑτέρῳ
τρόπῳ καὶ οὗτος ἐξ ἑτοίμου λαμβάνει τὸ ζητούμενον
ἀφαιρῶν αὐτῆς τὸ ὑγιεινόν. οὕτω δὲ καὶ τὴν γυμναστικὴν εἴ τις ὑγιείας φυλακτικὴν εἶναι φήσειεν, ἐκ
προχείρου τὸ ζητούμενον λήψεται, ὥσπερ γε καὶ ὃς
ἂν εὐεξίας αὐτὴν ἀποφαίνηται δημιουργόν, ἑτέρῳ
τρόπῳ καὶ ὅδε τὸ ζητούμενον ὡς ὁμολογούμενον θή-
σεται. χρὴ γάρ, εἴθ' ὁρισμόν τις τῆς τέχνης εἴθ' ὑπο-

---

[5] ἰατικὴν H; ἰατρικὴν μὲν Ku

friends, to show the path which someone might use to resolve not only this issue alone, but also all the others.

4. With respect to what is called logical theory among philosophers, whoever would train himself sufficiently will be able to handle any inquiry similarly, whereas to read notes on issues without that theory is nothing other than to waste time and not to judge what is read—that is, what is said truly and what falsely in them. Nor is a person able to remember all the things written. But since I set myself to speak about the issue definitively, for this reason the discussion I indicated a little earlier needs to begin.

5. And, by Zeus, if someone when asked what medicine is, were to say it is an art that treats those who are sick and preserves those who are healthy, this person would seem to have understood the matter being sought outright, declaring hygiene to be a part of this art. On the   811K other hand, someone may say healing the sick alone is the medical art, and in another way immediately takes the matter sought, which is hygiene, to be separate from this art. In this way too, he will say that, if the gymnastic art is what preserves health, he will take what is sought from what is at hand, as he would declare this to be productive of *euexia* (a good bodily state),[5] and will in another way establish the matter sought as something agreed upon. Otherwise it is necessary for someone who would attempt

[5] Galen considers this term in detail in section 7, and it is the subject of his short essay *Bon. Habit.*, IV.750–56K. The three key terms, *hexis, schesis* and *euexia*, are considered briefly in the General Introduction to the present work; an English translation is given in parentheses after the first use in each section.

γραφὴν ἐγχειρήσειεν ἀποδιδόναι, μὴ τὸ ζητούμενον ἀναποδείκτως τοῦτον ἀναιρεῖν ἢ τίθεσθαι ἀλλ᾽ ἑτέρωθέν ποθεν ἐξ ὁμολογουμένων ἀρξάμενον ἀποδεικνύναι πειρᾶσθαι.

ἆρ᾽ οὖν ἄμεινον ἰατρικὴν μὲν εἶναι λέγειν, ἧς τέλος ἡ ὑγίεια, γυμναστικὴν δ᾽, ἧς τέλος ἡ εὐεξία, καὶ ταύτας τὰς ὑπογραφὰς ἀρχὰς τίθεσθαι τῆς ζητήσεως; ἀλλά τοι καὶ οὕτως ἐκ προχείρου μὲν τῆς γυμναστικῆς ἀφαιρησόμεθα τὸ ὑγιεινόν, ἐπὶ δ᾽ αὖ τῆς ἰατρικῆς ἔτ᾽ ἄδηλόν τε καὶ ζητούμενον ἀπολείψομεν· εἰ γὰρ δὴ τέλος ἡ ὑγίεια τῆς τέχνης ἐστὶ ταύτης, τάχ᾽ ἂν ἔχοι
812K λαβὴν ὁ λόγος εἰς διορισμόν, οὐ φυλακὴν τῆς οὔσης ἀλλὰ ποίησίν τε καὶ γένεσιν τῆς οὐκ οὔσης τέλος εἶναι τιθέμενος αὐτῆς. εὐπορία γὰρ εἰς ἑκάτερον ἐντεῦθεν ἐπιχειρεῖν ἐπαγωγαῖς χρωμένους,[6] εἰ μὲν τῆς αὐτῆς τέχνης τό τε ποιῆσαί τι πρότερον οὐκ ὂν καὶ τὸ φυλάξαι σῶον, ἐπειδὰν γένηται, δεικνύναι βουληθείημεν, οἰκοδομικῆς τε καὶ ναυπηγικῆς καὶ τεκτονικῆς τε καὶ χαλκευτικῆς μνημονεύοντας,[7] εἰ δ᾽ ἄλλης μὲν τὸ δημιουργεῖν, ἄλλης δὲ τὸ διαφυλάττειν σῶον, ὑφαντικῆς τε καὶ ῥαπτικῆς καὶ πρὸς ταύταις σκυτοτομίας τε καὶ νευρορραφίας· οὕτω γὰρ ὀνομάζουσι τὴν τὰ πεπονηκότα τῶν ὑποδημάτων ἐπανορθουμένην. ἄλλου μὲν γὰρ εἶναι δοκεῖ τεχνίτου ποιεῖν ἱμάτιον, ἄλλου δ᾽ ἠπήσασθαι ῥαγέν, ὥσπερ αὖ καὶ ὑπόδημα δημιουργῆσαι μὲν τοῦ σκυτοτόμου, πονῆσαν δ᾽ ἐπανορθώσασθαι τοῦ νευρορράφου. δέδεικται δ᾽ ἡμῖν ἐν

to expound a definition or outline of the art not to deny or affirm this matter under investigation without demonstration. On the contrary, he should attempt to give a demonstration, beginning from what is agreed upon.

Would it, then, be better to say that health is the end (*telos*) of the medical art while *euexia* is the end (*telos*) of the gymnastic art, and place these outlines as starting points of the inquiry? But certainly in this way, we are casually removing hygiene from the gymnastic art, while leaving it still unclear whether the matter being investigated is of the medical art, for if health is the end of this art, perhaps the discussion might take as a definition, not the preservation of what exists, but would assume the end of this to be the creation and genesis of what does not exist. A solution here for each is to attempt to use dialectical arguments, should we wish to show that the production of what did not previously exist, and its preservation when it is sound, belong to the same art—we might mention housebuilding, shipbuilding, carpentry and metalwork [as cases in point]. On the other hand, if we wished to show that production belongs to one art and preservation of what is sound to another, we might mention [as cases in point] weaving and stitching, and in addition to these, shoemaking and shoemending, for people term thus the restoration of shoes that are worn out. For it seems to fall to one craftsman to make a cloak and to another to repair it when it is torn, just as it also is the task of the shoemaker to make shoes and the shoemender to restore those that are worn. I have, however, shown, in the work *On Dem-*

812K

---

6 χρωμένους H; χρωμένον Ku
7 μνημονεύοντας H; μνημονεύοντες Ku

τοῖς περὶ ἀποδείξεως, ὡς οὐ χρηστέον ἐπαγωγαῖς
εἰς ἀποδείξεις ἐπιστημονικάς· ὥσθ᾽ ὅστις ἐν ἐκείνοις
813K ἐγυμνάσατο, καταφρονήσει μὲν τῆς τοιαύτης ὁδοῦ,
ζητήσει δ᾽ ἑτέραν βελτίω, ὁ μὴ γυμνασάμενος δὲ
θάτερον τῶν μερῶν ἑλόμενος, ὁπότερον ἂν βουληθῇ,
δι᾽ ὅλης ἡμέρας ἐρίζειν ἕξει.

6. Καὶ γὰρ αὖ καὶ ζήτημ᾽ ἄλλο πρὸς τοῖς εἰρημένοις
οὐ σμικρὸν ἀναφυήσεται (τῆς μὲν ἰατρικῆς ὑγιείας
ποίησιν, οὐ σωτηρίαν τε καὶ φυλακὴν ὑποθεμένοις τὸ
τέλος, τῆς γυμναστικῆς δὲ τὴν εὐεξίαν)·[8] ἀναγκασθή-
σεται γάρ τις, οἶμαι, καθάπερ ἐπὶ τῆς ὑγιείας οὕτω
κἀπὶ τῆς εὐεξίας ἑτέραν μὲν τὴν φυλακτικὴν αὐτῆς
τέχνην ἀποφαίνειν, ἑτέραν δὲ τὴν δημιουργικήν. ἀλλ᾽
εἰ τοῦτο, δύο ἄλλας τέχνας ἐξ ἀνάγκης ζητήσομεν,
ἑτέραν μὲν τῆς ἰατρικῆς, τὴν τῆς ὑγιείας φυλακτικήν,
ἑτέραν δὲ τῆς γυμναστικῆς, τὴν τῆς εὐεξίας διασω-
στικήν. καὶ μὲν δὴ καὶ διττῆς οὔσης τῆς εὐεξίας, ὡς
ἐν ἄλλοις ἀποδέδεικται, ποτέρας αὐτῶν ἡ γυμναστικὴ
δημιουργός ἐστιν, χαλεπὸν ἔσται διελεῖν, ἆρά γε τῆς
κατὰ φύσιν ἢ τῆς ἀθλητικῆς· ἢ δῆλον, ὡς δύο ἄλλας
ἡμῖν τέχνας ἀναγκαῖόν ἐστι ζητεῖν; καὶ δὴ καὶ αἱ
σύμπασαι τὸν ἀριθμὸν ἐξ γενήσονται, τρεῖς μὲν αἱ
814K δημιουργοῦσαι τὰ τέλη, τρεῖς δ᾽ αἱ φυλάττουσαι.
τριῶν γὰρ ὑποκειμένων τελῶν, ὑγιείας, εὐεξίας τῆς
κατὰ φύσιν, εὐεξίας τῆς τῶν ἀθλητῶν, εἰς τοσοῦτον
ἀνάγκη τὸν ἀριθμὸν ἐξήκειν τὰς τέχνας. καὶ μὴν εἴ-

---

[8] From τῆς μὲν to εὐεξίαν precedes Καὶ γὰρ in Ku

onstration, that dialectical arguments must not be used for scientific demonstrations. As a consequence, anyone who has become practiced in those writings will despise such 813K a path and will seek another that is better. Anyone who is not so practiced may choose whichever of the two paths is preferred, and argue about it all day long.

6. For surely another inquiry in turn, and by no means a small one, will arise about the things said, for those who suppose the end (*telos*) of the medical art is the creation of health and not its preservation and maintenance, while that of the gymnastic art is *euexia.* Such a person will be compelled, I think, to give in the case of *euexia,* as in the case of health, an account of one art preservative of *euexia* and another productive of it. But if this is so, we shall, of necessity, seek two other arts—one of medicine preservative of health and another of gymnastics preservative of *euexia.* Furthermore, since *euexia* is also twofold, as has been shown elsewhere,[6] it will be difficult to determine which of these the gymnastic art is productive of—whether it is that which is natural (in accord with nature) or that of the athlete, so it is clear there are necessarily two other arts for us to seek. Furthermore, the number of arts will become six in all; there will be three productive of the ends and three preservative of them. Since there are three 814K proposed ends—health, *euexia* in accord with nature, and *euexia* of athletes—of necessity the arts would reach such a number. And if, because the *euexias* differ from each

[6] The reference is to "ordinary" *euexia* and the *euexia* of athletes—on this see particularly the work cited in the previous note, of which there is an English translation by Singer (*Galen: Selected Works*).

περ τῷ διαφέρειν τὰς εὐεξίας ἀλλήλων τε καὶ τῆς
ὑγιείας διαφόρων δεήσεται[9] τῶν τεχνῶν, ἀνάγκη
πᾶσα καὶ τῆς ὑγιείας οὔσης διττῆς, ἑτέρας μὲν τῆς
καθ᾽ ἕξιν, ἑτέρας δὲ τῆς κατὰ σχέσιν ὀνομαζομένης,
διττὰς εἶναι καὶ τὰς τέχνας. οὐ γὰρ δὴ πλέονι μὲν ἡ
εὐεξία τῆς καθ᾽ ἕξιν ὑγιείας, ἐλάττονι δὲ τῆς κατὰ
σχέσιν ὑπερέχει.

7. Γνοίης δ᾽ ἂν ἐναργῶς ἑκάστης αὐτῶν ἐπιμελῶς
ἀνασκεψάμενος τὴν φύσιν, ἀνασκέψῃ δὲ τόνδε τὸν
τρόπον. ἡκέτω τις εἰς ἔννοιάν σοι νῦν ὧν ἐθεάσω πολλάκις ἀρρώστων ἰσχυρῶς μὲν διανοσησάντων, ἀπηλ-
λαγμένων δ᾽ ἄρτι τοῦ νοσεῖν, οὕτως[10] ἰσχνός τε καὶ
ἀδύνατος εἰς τὰς κινήσεις, ὡς ἑτέρων δεῖσθαι τῶν
μετακομιούντων. οὗτος ἰάσεως μὲν οὐκέτι χρῄζει μηκέτι γε νοσῶν, ἀναθρέψεως δέ τινος καὶ ῥώμης, ἵν᾽
815K ἰσχυρός θ᾽ ἅμα καὶ πρὸς τὰς κατὰ φύσιν ἐνεργείας
γένηται καὶ φέρειν ἱκανὸς τὰ προσπίπτοντα. δῆλον
γὰρ ὡς ὁ οὕτω διακείμενος, ὡς διάκειται νῦν, ὁπότε
πρότερον ἀπήλλακται τοῦ νοσεῖν, οὔτε θάλπος οὔτε
κρύος οὔτ᾽ ἀγρυπνίαν οὔτ᾽ ἀπεψίαν οὔτ᾽ ἄλλο τῶν
πάντων οὐδὲν ἐνεγκεῖν ἱκανός ἐστιν, ἀλλ᾽ ἐκ τοῦ ῥᾴ-
στου νοσήσει πάλιν, ὡς ἂν οὐκ ἀσφαλῆ τε καὶ μήπω
πεπηγυῖαν κεκτημένος τὴν ὑγιεινὴν κατάστασιν. ἀλλ᾽
εἴπερ δύσλυτος γένοιτο καὶ καθ᾽ ἕξιν, οὔπω μέν ἐστιν
εὐεξία, (τὸ γὰρ εὖ προσλαβοῦσα τότ᾽ ἂν εὐεξία γένοιτο,) τοῦ μέντοι μεμπτή τις εἶναι καὶ ἄχρηστος διάθεσις ἀπήλλακται. τὸ γὰρ μήτ᾽ ἐνεργεῖν δύνασθαι
κατὰ τὸν βίον ἱκανῶς ἑτοίμως τε βλάπτεσθαι μεμπτὴ

other, and from health, there will be need of different arts. Of necessity, health as a whole, being twofold, will need one art in relation to what is called health in relation to *hexis* (a stable or permanent state) and one in relation to what is called health in relation to *schesis* (an unstable or temporary state), so there are also these two arts, for certainly *euexia* does not exceed the health in relation to a *hexis* by more and that in relation to *schesis* by less.

7. You would know clearly the nature of each of these after a careful and close examination. [carried out] in this way: Come now to the concept of one of those sick people whom you have seen often, and who, after falling ill severely, has just recovered from being diseased, but is so weak and incapable of movements that he needs others to carry him around. Such a person no longer needs a cure because he is not still sick, but he does need restoration and strength, so he is strong enough for the natural func-   815K
tions and adequate to bear those things that may befall him. Clearly, a person in such a state as he now is, when he first recovers from disease, is not up to enduring heat or cold, sleeplessness, or failure of digestion, or any other of all those things from which he would very easily become sick again. So he would not be safe, having not yet acquired an established healthy condition. But if this were to become difficult to break down and [health] in relation to *hexis,* it would not yet be *euexia* (for then "eu" would be added and it would become *euexia*). Nevertheless, it would be far removed from being a blameworthy and useless condition. For not to be able to carry out the functions of life adequately and to readily come to harm is a blame-

---

9 δεήσεται H; δεήσονται Ku   10 οὕτως H; οὗτος Ku

καὶ ἄχρηστος διάθεσις. εἰ δὲ μήτ᾽ εἰς τὰς πράξεις
ἐμποδίζοιτο μήθ᾽ ἑτοίμως βλάπτοιτο, τοιαύτη δ᾽ ἐστὶν
ἡ καθ᾽ ἕξιν ὑγίεια, τοῦ μὲν μηκέτ᾽ εἶναι μεμπτὴ καὶ
ἄχρηστος εἰς τὰς κατὰ τὸν βίον ἐνεργείας ἀπήλλακται, τὸ δ᾽ ἐπαινετὸν οὔπω κέκτηται, κτήσαιτο δ᾽
816K ἄν, εἰ καὶ πρὸς τῷ¹¹ μηκέτ᾽ ἀσθενῶς ἐνεργεῖν ἔτι καὶ
ῥώμην τιν᾽ ἀξιόλογον προσλάβοιτο.

μέση γὰρ ἡ καθ᾽ ἕξιν ὑγίεια ταῖς ἐνεργείαις ἐστὶ
τῆς τε κατὰ σχέσιν ὑγιείας καὶ τῆς εὐεξίας. ἡ μὲν
γὰρ κατὰ σχέσιν ἀρρώστους ἔχει τὰς ἐνεργείας, ἡ δ᾽
εὐεξία ῥωμαλέας, ἡ δ᾽ αὖ καθ᾽ ἕξιν οὔπω μὲν εὐρώ-
στους, ἀρρώστους δ᾽ οὐκέτι. κἂν τῷδε δῆλον, ὡς, εἰ
καὶ μέση τέτακται, πλέον τῆς κατὰ σχέσιν ὑγιείας
χρήσιμος εἰς πάσας τοῦ βίου τὰς πράξεις ἐστὶν ἡ
καθ᾽ ἕξιν, οὐ μὴν ἤδη τὸ κατ᾽ ἀρετὴν ἔχουσα· μόνη
γὰρ τοῦθ᾽ ὑπάρχει τῇ εὐεξίᾳ. καί μοι νόει κατὰ στί-
χον τινὰ τὸ μὲν βλάβην ἐνεργείας, ὅπερ ἐν τῷ νοσεῖν
γίγνεται, ἕτερον δ᾽ ἔξω μὲν τῆς βλάβης, ἄχρηστον δ᾽
ὑπ᾽ ἀσθενείας, ὅπερ ἐν τῷ κατὰ σχέσιν ὑγιαίνειν,
ἄλλο δὲ τρίτον ἀσθενείας μὲν ἀπηλλαγμένον, εὐρω-
στίαν δ᾽ οὔπω κεκτημένον, ὅπερ ἐν τῷ καθ᾽ ἕξιν, ἐφ᾽
ᾧ τέταρτον οἷον ἀρετή τίς ἐστι τῶν ἐνεργειῶν ἡ εὐεξία. [καὶ ἡ ἀκρότης τε καὶ ἡ τελειότης αὐτῶν ἡ ἐπ᾽
ἄκρον ἐστὶν αὐτῶν εὐεξία.]¹² ἡ τοίνυν εἰς ἕξιν ὑγιεινὴν
817K ἄγουσα τέχνη τὸν ἄνθρωπον ἑτέρα δηλονότι τῆς ἰα-

worthy and useless condition. If, however, someone is neither impeded in his actions nor readily harmed, such is health in accord with *hexis*. This would no longer be blameworthy and useless in terms of the functions of life but would not as yet have acquired praiseworthiness. It would acquire this if, in addition to function no longer being weak, some significant strength were to be acquired.   816K

In terms of functions, health in relation to *hexis* (a stable state) is midway between health in relation to *schesis* (an unstable state) and *euexia*, for health in relation to *schesis* has weak functions whereas *euexia* has strong functions; in health in relation to *hexis*, the functions are not yet strong but are no longer weak. And it is clear from this also that, if it is placed in the middle, health in relation to *hexis* is of greater use than health in relation to *schesis* in terms of all the activities of life, although it is not already excellence. This is for *euexia* alone. Think of it as a linear progression. There is some damage of functions that occurs in disease, which is different from external damage, while there is uselessness due to weakness, which is in health in relation to *schesis;* but there is another and third thing, which is delivery from weakness without as yet acquisition of strength; this is health in relation to *hexis*. In addition to this is a fourth, which is a kind of excellence of the functions; this is *euexia*. The highest point, and perfection of these functions, is *euexia* up to the peak of these functions. Accordingly, the art which leads the person to *hexis* is clearly different from that which cures   817K

---

11 τῷ H; τὸ Ku   12 *On the sentence in parentheses, included in Ku, H has:* verba καὶ ἡ ἀκρότης usque εὐεξία ab interpolatore profecta esse ratus uncis inclusi.

σαμένης αὐτὸν ὑπάρξει, συγκεχωρημένου γ' ἅπαξ
τοῦ τε δεῖν ἐπὶ διαφόροις τέλεσι διαφόρους εἶναι τὰς
τέχνας τοῦ τε δύο καθ' ἕκαστον ἑτέραν μὲν τὴν δημι-
ουργοῦσαν, ἑτέραν δὲ τὴν διαφυλάττουσαν. ᾧ καὶ
δῆλον, ὡς ἡ περὶ τέλους ἔννοια περιείληφεν ἅπασαν
τὴν νῦν ἡμῖν προκειμένην ζήτησιν, καὶ διὰ τοῦθ' ὅσοι
τῶν ὁρισμῶν ἀπὸ τοῦ τέλους τὴν σύστασιν ἔχουσιν,
οὐ σμικρὰς γεννῶσιν ἀπορίας.

8. Ἴσως οὖν ἄμεινον ἐπ' αὐτὴν ἀφικόμενον τὴν
οὐσίαν τῆς τέχνης ἀρχὴν τῆς ζητήσεως ἐκείνην ποιήσασθαι. τί[13] οὖν ἐστιν ἰατρική; εἴποι ἄν τις εἶναι ἐπιστήμην ὑγιεινῶν τε καὶ νοσερῶν. ἀλλὰ κἂν τούτῳ
δόξει προχείρως εἰλῆφθαι τὸ ὑγιεινὸν ὡς μέρος δηλονότι τῆς ἰατρικῆς ὑπάρχον. ὅθεν οἶμαι τῶν τἀναντία
τιθεμένων ἔνιοι μόνων τῶν νοσερῶν ἐπιστήμην εἶναι
συγχωρήσουσι τὴν ἰατρικήν. ἀλλ' οὗτοί γε πρῶτον
μὲν ἀγνοοῦσιν, ὡς ἔστι μία τῶν ἐναντίων ἁπάντων
ἐπιστήμη καὶ ὅστις γιγνώσκει τὰ νοσερά, τοῦτον
ἀνάγκη πᾶσα μὴ ἀγνοεῖν τὰ ὑγιεινά.

τουτὶ μὲν ὡς μεῖζον ἢ κατ' ἐκείνους παρείσθω,
λεγέσθω δ' ἤδη τὸ δεύτερον ὧν ἀγνοοῦσιν, ᾧ τάχ' ἂν
ἴσως ἀκολουθήσειεν, ὅτι μὴ περιλαμβάνεται κατὰ τὸν
ὅρον τῆς ἰατρικῆς ἐξ ἀνάγκης τὸ ζητούμενον· ἔνεστι
γάρ τινι διαστειλαμένῳ τὴν ὁμωνυμίαν, εἶτα δείξαντι
σωμάτων μὲν τῶν ἐχόντων ὑγίειαν καὶ σημείων τῶν
δηλούντων καὶ αἰτίων τῶν ποιούντων ἐπιστήμην εἶναι
τὴν ἰατρικήν, οὐ μὴν τῶν γε φυλαττόντων αὐτήν,
ἀντιλαμβάνεσθαι κατὰ τἀναντία τοῦ προβλήματος.

818K

250

him [of a disease], if it is agreed once and for all that for
different ends there is a need for different arts, and that
for each end there must be two arts; one which is productive and one which is preservative. It is clear from this also
that the concept concerning an end has encompassed the
whole inquiry now before us, and because of this, those
definitions that derive their composition from the end create quite a few difficulties.

8. Perhaps it would be better, then, to come to the very
essence of the art, making that the beginning of the inquiry. What, then, is a medical art? Someone might say it
is knowledge of those things that are healthy and those
that are morbid. But even in this, it will seem that hygiene
is readily taken to be an obvious part of the medical art.
From this, I think, that some who hold opposite views only
assent to knowledge of morbid matters being of the medical art. But these people, in the first place, do not know
that one science includes all the opposites, and someone
who knows those things that are morbid, of necessity is not
ignorant of all things that are healthy.

As this might be too much for them, let it pass, and let
me speak now about the second of the matters they do not
know, but which perhaps they might follow: that what is
sought is not necessarily included in the definition of the
medical art; it is possible for someone who distinguishes
the equivocal sense to then show that the medical art is a
knowledge of the bodies possessing health, of signs indi-
cating it, and of causes producing it, but not of those things
preserving it, taking an opposing position on the issue. In

818K

---

13 τί H; τίς Ku

ἐοίκασιν οὖν οἱ τὰ τοιαῦτα σοφιζόμενοι μήτε τὸ ζητούμενον ἀκριβῶς ἐπίστασθαι μήθ᾽ ὅτι καὶ σημεῖον
ὑγιεινόν τι λέγομεν, ὃ τοῖς νοσοῦσιν ἐπιφαινόμενον
ὑγιείας μελλούσης ἐστὶ γνώρισμα, μήτ᾽ αἴτιον, ὃ τοῖς
νοσοῦσι προσαγόμενον ὑγιείας ἐστὶ ποιητικόν, ὥσπερ
ἀμέλει τὰ βοηθήματα σύμπαντα, μήθ᾽ ὡς καὶ τὸ δε-
δεγμένον ὑγίειαν σῶμα καλοῦμεν ὑγιεινόν, ὧν πάντων ἐπιστήμην ὁ ἰατρὸς ἔχει μηδὲν κοινωνούντων τῷ
προβλήματι πλὴν ὀνόματι.

καὶ γὰρ εἰ τῆς τῶν ὑγιαινόντων φυλακῆς ὁ ἰατρὸς
ἐπιστάτης ἐστὶν ἢ γυμναστοῦ τὸ τοιοῦτον ἴδιον, ἐξ
ἀρχῆς ἡμῖν προὔκειτο ζητεῖν, ὅπερ, ὡς[14] ὀλίγον ἔμπροσθεν ἐδείχθη, τὴν κρίσιν ἐν τῇ τοῦ τέλους ἔχει
ζητήσει. πότερον γὰρ ἁπλῶς ὑγίεια τὸ τέλος ἐστὶ τῆς
ἰατρικῆς οὐδὲν διαφέρον εἴτε ποιοῦντός τινος αὐτὴν
οὐκ οὖσαν εἴτε φυλάττοντος οὖσαν ἢ τὸ ποιῆσαι μὲν
μόνον ἰατρικῆς ἐστι, τὸ φυλάξαι δὲ τῆς γυμναστικῆς
αὐτῆς; οὕτω δὲ καὶ περὶ τῆς γυμναστικῆς, ἆρά γ᾽ εὐεξία τὸ τέλος ἐστὶν ἢ ὑγίεια ἢ ποίησις ὁποτέρας τούτων ἢ φυλακή.

9. Κινδυνεύσομεν γάρ, ὡς ὁ λόγος ἔδειξεν, ἐὰν
ὑπερβῶμεν τὴν μίαν, ἑπτὰ ποιῆσαι περὶ τὸ σῶμα τὰς
τέχνας, τὴν μὲν πρώτην καὶ σαφεστάτην καὶ σχεδὸν
μόνην ἀναμφισβήτητον ἰωμένην τὰ νοσήματα, δύο δ᾽
ἄλλας, τὴν μὲν ἐκ τῆς κατὰ σχέσιν ὑγιείας εἰς τὴν
καθ᾽ ἕξιν ἄγουσαν, τὴν δ᾽ ἐν ταύτῃ φυλάττουσαν καὶ

819K

---

[14] ὅπερ, ὡς H; ὥσπερ Ku

respect of such things, the Sophists seem neither to have accurately understood what is being sought, nor that what we call a sign that is healthy, which is a sign of impending health displayed by those who are diseased, nor what is an effective cause producing health in those who are diseased, as of course all remedies are, nor that we also call the body that has received health, "healthy," all of which the doctor has a knowledge of, apart from their having a share in the issue of the name.

For truly, the question of whether the doctor is a supervisor of the preservation of those who are healthy, or such a thing is specific to the gymnastic trainer, which is what was set before us to investigate at the beginning, has 819K its resolution in the investigation of the end (*telos*), as was shown a little earlier. For if health absolutely is the end of the medical art, does it make any difference if it is the creation of this when it doesn't exist, or the preservation of it when it does, or is creation of it alone the province of the medical art, while the preservation is the province of the gymnastic art itself? In this way too, regarding the gymnastic art, there is the question of whether *euexia* or health is the end, or the creation of both of these is, or their preservation.

9. As the discussion has shown, we shall be in danger of creating seven arts concerning the body, if we go past one. The first, which is the clearest and alone almost beyond dispute, is curing diseases. There are, however, two others: the art of bringing someone from health in relation to *schesis* (an unstable state) to health in relation to *hexis* (a stable state), and the preservation of the latter. Then

δύο ἄλλας ὁμοίως περὶ τὴν εὐεξίαν, δημιουργικὴν μὲν τὴν ἑτέραν, φυλακτικὴν δὲ τὴν ἑτέραν, ἔτι τε πρὸς ταύταις ἄλλας δύο περὶ τὴν ἀθλητικὴν εὐεξίαν. ἁπλῶς μὲν γὰρ τὴν ἑτέραν τὴν κατὰ φύσιν ὀνομάζομεν εὐ-

820K εξίαν, οὐχ ἁπλῶς δὲ τὴν οὐ φύσει, τὴν ἀθλητικήν, ἀλλ᾽ ἀεὶ μετὰ προσθήκης, ὥσπερ καὶ Ἱπποκράτης ὁτὲ μὲν ὡδί πως λέγων· "διάθεσις ἀθλητικὴ οὐ φύσει, ἕξις ὑγιεινὴ κρείσσων," ὁτὲ δ᾽ αὖ πάλιν· "ἐν τοῖσι γυμναστικοῖσιν αἱ ἐπ᾽ ἄκρον εὐεξίαι σφαλεραί" . . .[15] ἀλλ᾽ ἐν τοῖς ἀθλητικοῖς τε καὶ γυμναστικοῖς σώμα-σιν· ἀκούειν γάρ σε χρὴ γυμναστικὰ σώματα λέγε-σθαι νῦν οὐ τὰ τῶν ὁπωσοῦν γυμναζομένων, οἷον ἤτοι σκαπτόντων ἢ ἐρεσσόντων ἢ ἀμώντων ἤ τι τῶν ἄλ-λων, ὅσα κατὰ φύσιν ἀνθρώποις ἔργα, πραττόντων, ἀλλ᾽ οἷς αὐτὸ τοῦτ᾽ ἔστιν ἀγώνισμα τὸ γυμνάζεσθαι καταβλητικὴν τῶν ἀντιπάλων ἰσχὺν ἐπασκοῦσιν.

διὰ τί μὲν οὖν ἡ τοιαύτη διάθεσις οὐ φύσει, δι᾽ ἑτέρων ἐξηγήμεθα· τὸ δ᾽ οὖν ἐκ τοῦ λόγου χρηστὸν εἰς τὰ παρόντα τοῦτ᾽ ἔστιν αὐτὸ τὸ νῦν εἰρημένον, ὡς ἑπτὰ γενήσονται περὶ τὸ σῶμα τέχναι τὴν μίαν ὑπερ-βᾶσιν ἢ καὶ νὴ Δι᾽ ἐννέα. τί γὰρ οὐ χρὴ δύο ἄλλας τέχνας τίθεσθαι, ἑτέραν μὲν τῆς ἄκρας εὐεξίας δημι-

821K ουργόν, ἑτέραν δὲ τῆς αὐτῆς ταύτης φυλακτικήν; ἐπὶ μὲν γὰρ τῆς τῶν γυμναστικῶν εὐεξίας ὅτι χρὴ φεύ-γειν τε καὶ δεδιέναι τὴν ἀκρότητα, πρὸς Ἱπποκράτους

---

[15] On this, H has the following: post σφαλεραί nonnihil intercidisse videtur; cf. Galen, IV.752K, lines 5–10.

there are similarly two others regarding *euexia:* one is the production of this and the other its preservation. Further, in addition to these, there are two others concerning the *euexia* of the athlete. For one state we call simply *euexia* in accord with nature, whereas the other, is not simply natural, but always has the addition of "the athletic." Thus, Hippocrates on one occasion said, speaking thus: "An athletic condition is not natural; a healthy state is better."[7] And at another time again: "In those who exercise the peak *euexias* are dangerous."[8] [. . . but in the athletic and gymnastic bodies.][9] It is necessary for you to understand "gymnastic bodies" as now used, not as referring to any kind of exercise whatsoever, like digging, rowing or reaping, or any of the other things which are done by people as natural activities, but to those whose practice is for the purpose of the actual strength to throw their opponents in a contest.

We have explained elsewhere why such a condition is not natural. What is useful from that discussion for present matters is this, which is now being stated: that there will be seven arts concerning the body, if we go beyond one, and, by Zeus, there will be nine. For must we not assume two other arts—one for creating the peak of *euexia* and the other for preserving this same *euexia?* In the case of the *euexia* of gymnastics, that we must avoid and be fearful of the peak was most clearly stated by Hippocrates.

820K

821K

[7] Hippocrates, *Nutriment* 34, LCL 147 (W. H. S. Jones), 354–55.

[8] Hippocrates, *Aphorisms* 1.3, LCL 150 (W. H. S. Jones), 98–99.

[9] See H's note, p. 43 on the text here.

εἴρηται σαφέστατα· τῆς δ᾽ ἁπλῶς λεγομένης εὐεξίας
τῆς κατὰ φύσιν οὐχ ὅπως φυλάττεσθαι προσήκει τὸ
ἄκρον ἀλλὰ καὶ παντὶ τρόπῳ σπουδάζειν. οὕτως οὖν
αἱ σύμπασαι γένοιντ᾽ ἂν ἐννέα περὶ τὸ σῶμα τέχναι
καὶ τούτων ἑπτὰ μέν, ἃς ἂν καὶ ἐπαινέσειέ τις, αἱ δύο
δ᾽ αἱ λοιπαὶ κακοτεχνίαι δηλονότι καθάπερ καὶ ἡ κομ-
μωτική. ταύτας μὲν οὖν, εἰ βούλει, παραλίπωμεν, ἐπ-
έλθωμεν δ᾽ αὖθις τὰς ἑπτά, τὴν μὲν πρώτην ἁπασῶν,
ἣν ἰᾶσθαι τὰς νόσους ἐλέγομεν, ἄλλας δ᾽ ἐφεξῆς
αὐτῇ, δύο μὲν περὶ τὴν καθ᾽ ἕξιν ὑγίειαν, δύο δὲ περὶ
τὴν εὐεξίαν καὶ δύο ἄλλας περὶ τὴν ἄκραν εὐεξίαν.
ἤδη μὲν οὖν κἀκ τούτων[16] εὔδηλον, ὡς, εἰ μή[17] τις θείη
περὶ τὸ σῶμα τέχνην ἓν ἔχουσαν δηλονότι καὶ τὸ
τέλος, ἀναγκαῖόν ἐστι τοῦτον μέχρι τῶν ἑπτὰ προ-
ϊέναι καὶ ὡς οὐδὲν ἄλλο τὸ τέλος τοῦτ᾽ ἔστι παρὰ τὴν
ὑγίειαν.

10. Ἄμεινον δ᾽ ἴσως ἀκριβέστερον ἐπεξελθεῖν τῷ
λόγῳ πρῶτον μὲν τοῦτ᾽ ἀναμνήσαντας, ὃ μηδεὶς
822K   ἀγνοεῖ, τὰς μὲν κακοτεχνίας τὸ φαινόμενον ἀγαθὸν
ἐκποριζομένας ἑκάστῳ τῶν ὄντων, τὰς τέχνας δὲ τὸ
κατ᾽ ἀλήθειαν ἐν αὐτοῖς ὑπάρχον, ἐφεξῆς δὲ τούτῳ
λέγοντας, ὡς, εἴπερ ἡ κομμωτικὴ καλουμένη νόθου
κάλλους ἐστὶ δημιουργός, αὕτη μὲν ἂν εἴη κακο-
τεχνία τε καὶ κολακεία, τέχνη δέ τις ἑτέρα συστήσε-
ται περὶ τὸ γνήσιόν τε καὶ ὄντως ἀληθινὸν κάλλος,
ὅπερ ἐστὶν εὐχροιά τε καὶ εὐσαρκία καὶ συμμετρία
τῶν μορίων, ἃ τῇ κατὰ φύσιν εὐεξίᾳ συμβέβηκεν, ἣν
καὶ δι᾽ ἄλλου μὲν λόγου τῆς γυμναστικῆς διωρισά-

On the other hand, the peak of what is simply called *euexia* in accord with nature is something which it is not only appropriate to preserve as it is, but also to pursue zealously in every way. So then, there are nine arts in all concerning the body; seven of these may be praised, while the remaining two are clearly bad arts, just as the art of embellishment (the cosmetic art) also is. Let us, then, leave these aside, if you will, and come again to the seven. The first of all of these, we said, cures diseases; in respect of the others subsequent to this, there are the two pertaining to health in relation to *hexis,* two pertaining to *euexia,* and two others pertaining to peak *euexia.* It is already quite clear from these that, unless someone were to assume there to be one art pertaining to the body and one end (*telos*), it is necessary to go up to seven, and that nothing else is the end of these apart from health.

10. Perhaps it would be better to proceed with the discussion more strictly, after first calling to mind this, which no one is unaware of—that the base arts provide an 822K apparent good in each of the things they exist for, whereas there are arts that are true in these. Following this, let me say that, if the so-called cosmetic art is productive of a counterfeit beauty, it would be a base art and a form of flattery, while some other art will be contrived concerning legitimate and genuinely true beauty, which is a good complexion, good flesh, and proportion of the parts, these being qualities that are present contingently in the *euexia* in accord with nature, which I also define in another work on

---

16 κἀκ τούτων H; καὶ τοῦτο Ku
17 om. μή Ku

μεθα, καὶ νῦν δ᾽ ἂν ἴσως ἐπὶ κεφαλαίων οὐδὲν ἂν εἴη χεῖρον ὑπὲρ αὐτῶν εἰπεῖν ἀρχὴν τῷ λόγῳ τήνδε ποιησαμένους.

11. Ἤτοι γὰρ τὸ κατὰ φύσιν ἐνεργεῖν ἑκάστῳ τῶν μορίων ὑγιαίνειν ἐστὶν ἢ τοῦτο μὲν ἐξ ἀνάγκης ἔπεται τῇ κατὰ φύσιν τοῦ σώματος κατασκευῇ, τὸ δ᾽ ὑγιαίνειν αὐτὸ τοῦτ᾽ ἔστι τὸ κατὰ φύσιν κατεσκευάσθαι. τούτων ὁπότερον ἂν ἐθελήσῃς ὑποθέμενος ὑγίειαν, οὐδὲν γὰρ εἰς τὰ παρόντα διαφέρει, τῶν ἑξῆς λόγων οὕτως ἄκουε προσέχων τὸν νοῦν. πότερα τοῦ κατεσκευάσθαι κατὰ φύσιν ἕκαστον τῶν μορίων ἢ τῆς ἐνεργείας αὐτοῦ χρῄζομεν; ἐμοὶ μὲν πάντες δοκοῦσιν οὐδ᾽ ἂν ὅλως θελῆσαι μόριον ἔχειν οὐδὲν ἀργὸν ἔργου τινός, οὔτ᾽ οὖν ὀφθαλμοὺς μὴ βλέποντας οὔτε ῥῖνας ὀσμᾶσθαι μὴ δυναμένας οὔτε σκέλη μὴ βαδίζοντα οὔτ᾽ ἄλλο τῶν πάντων οὐδὲν ἢ μηδ᾽ ὅλως ἐνεργοῦν ἢ κακῶς ἐνεργοῦν. οὐδενὸς γὰρ ἁπάντων, ὧν χρῄζομεν, ἀτελοῦς χρῄζομεν οὔτ᾽ οἰκίας οὔθ᾽ ὑποδήματος οὔτε σκίμποδος οὔθ᾽ ἱματίου, ἀλλ᾽ ἅμα τε χρῄζομεν καὶ τελείου χρῄζομεν. οὕτως οὖν οὐδὲ τοῦ περιπατεῖν ἀσθενῶς καὶ ἀρρώστως οὐδὲ τοῦ βλέπειν ἢ ἀκούειν ἀμβλέως οὔτ᾽ ἄλλου τῶν πάντων οὐδενὸς ἐλλιποῦς ὀρεγόμεθα. τίς γὰρ εὔχροιαν ἢ εὐσαρκίαν ἢ κάλλος ἁπλῶς τοῦ σώματος ἢ ῥώμην ἐλλιπῆ σχεῖν εὔχεται; καὶ μὴν εἴπερ οὐκ ἐλλιποῦς ἐνεργείας ἀλλὰ τελέας χρῄζομεν, οὐδὲ τῆς κατασκευῆς τοῦ σώματος, ἀφ᾽ ἧς

gymnastics,[10] although now it would perhaps be no bad thing to speak about these in summary, making this a start to the discussion.

11. Either to function in accord with nature in each of the parts is to be healthy, or this necessarily follows the constitution of the body in accord with nature, while to be healthy is this itself, in relation to being constituted in accord with nature. You may wish to assume health is either one of these two things—it makes no difference for our present purposes—and to focus your attention on understanding the following discussions in this way.[11] Do we need the constitution of each of the parts to be in accord with nature, or their function? Everyone seems to me to wish that no part at all should have any failure of some action—the eyes not to be able to see, the nose to smell, the legs to walk, nor any other of all the parts not to function at all or to function badly. Of all the things we need, we need none of them to be imperfect—neither a house, nor a sandal, nor a bed, nor a cloak; what we need, we also need to be perfect at the same time. In this way, then, we do not desire to walk feebly or weakly, nor to see or hear indistinctly, nor for any other of all the functions to be deficient. For who would boast of having a complexion, or flesh, or simply beauty of the body, or strength that is deficient? And so, if we need perfect and not deficient functions, we shall need the constitution of the body with

823K

---

10 Galen may be referring here to his short treatise, *Bon. Habit.*, IV.750–56K (English trans., Singer, *Galen: Selected Works*) or possibly the relevant sections in his *De sanitate tuenda*, included in the present work.

11 This issue is addressed at length by Galen in *MM*, X.41–63K.

ἐνεργοῦμεν, ἐλλιποῦς δεησόμεθα. τούτων δ᾽ ἦν θάτε-
824K ρον ὑγίεια· δῆλον οὖν, ὡς οὐδεὶς ἀτελοῦς ὑγιείας ἀλλ᾽
ὡς ἔνι μάλιστα τελειοτάτης ἅπαντες χρῄζομεν.

12. Εἰ μὲν οὖν ἄλλο τι τὴν εὐεξίαν οἴεταί τις εἶναι
παρὰ τὴν τελείαν ὑγίειαν, ἄλλην μὲν τέχνην ὑγιείας,
ἄλλην δ᾽ εὐεξίας ζητείτω· εἰ δ᾽ ἓν καὶ ταὐτόν ἐστιν
ἄμφω, μίαν ἀνάγκη καὶ τέχνην εἶναι. πῶς οὖν ἓν καὶ
ταὐτόν ἐστιν εὐεξία τε καὶ ἡ παντελὴς ὑγίεια; πρῶτον
εἴπερ τὴν ταύτης αἰτίαν κατασκευὴν τὴν αὐτὴν δήπου
καὶ τῆς εὐεξίας εἶναι πεπιστεύκαμεν, ἄμφω ταὐτὸν
ἔσται· δεύτερον δ᾽ ἐξ αὐτῆς τῆς οὐσίας. εὐεξία μὲν
γὰρ οὐδὲν ἄλλ᾽ ἐστὶν ἢ εὖ ἔχουσα ἕξις, ἡ δ᾽ ἕξις διάθεσίς ἐστι μόνιμος, ὥστε, οὗπερ ἕξις, τούτου καὶ εὐεξία, πρός τι δ᾽ ἄμφω. λέγεται δ᾽ οὖν τις ἕξιν ἔχειν ἐν
γραμματικῇ καὶ ἄλλος ἐν ἀριθμητικῇ καὶ ἄλλος ἐν
γεωμετρίᾳ καὶ ἄλλος ἐν ἀστρονομίᾳ καὶ καθ᾽ ἕκαστον
τῶν ὄντων, ὅταν ἡ διάθεσις δύσλυτος ᾖ, ἕξις ὀνομάζεται. διαφερέτω δὲ μηδὲν ἔν γε τῷ παρόντι διάθεσιν
825K ὀνομάζειν ἢ σχέσιν. εἴπερ οὖν, οὗπερ ἕξις, τούτου καὶ
εὐεξία, τινὸς δ᾽ ἡ ἕξις, δῆλον, ὅτι[18] καὶ ἡ εὐεξία τινὸς
καὶ τοῦ αὐτοῦ γε, οὗ καὶ ἡ ἕξις.

ἀλλ᾽ ἡμῖν νῦν οὐ περὶ τῆς γεωμετρικῆς ἕξεως ἢ
μουσικῆς ἢ γραμματικῆς, ἀλλὰ περὶ τῆς ὑγιεινῆς ὁ
λόγος ἐστίν. οὐκοῦν, ὅταν εἴπωμεν εὐεξίαν, οὐ γραμ-
ματικὴν ἢ μουσικὴν ἢ γεωμετρικὴν ἀλλ᾽ ὑγιεινὴν λέγομεν. εὐθὺς δ᾽ αὐτὸ τοῦτο παρορῶσιν οἱ πολλοὶ καὶ

[18] δῆλον, ὅτι H; δηλονότι, Ku

which we function not to be deficient. Health is one or other of these things [mentioned above]. It is clear, then, that no one needs imperfect health; we all need first and foremost the most perfect health.

12. If, then, *euexia* is thought to be something else beside complete health, let us seek one art for health and another for *euexia*. If, however, both are one and the same, it is also necessary for there to be one art. How, then, are *euexia* and complete health one and the same? The first point is, of course, if we were to have the belief that the constitution causative of complete health and of *euexia* were the same, both will be the same. The second point is from the actual essence. For [the term] *euexia* is nothing but εὖ followed by ἔχις (*hexis*—state), while *hexis* is a stable condition, so that *hexis* and *euexia* are of this, and are both relative to something. There is, then, one *hexis* in grammar, another in arithmetic, another in geometry, and another in astronomy, and in each of these instantiations, whenever the condition is difficult to break down, it is termed *hexis* (a stable state). Let it make no difference in the present context whether we term this *diathesis* (condition) or *schesis* (unstable or temporary state). If, then, *hexis* is of something, and *euexia* is also of this, it is clear that what *hexis* is of, *euexia* is also of this—that *hexis* and *euexia* are, in fact, of the same thing.

But our discussion is not now about the *hexis* of geometry, or of music or grammar, but about the *hexis* of health. Therefore, when we say *euexia*, we are not speaking about grammar, music or geometry, but about health. The majority, however, openly disregard this very thing and think

824K

825K

261

νομίζουσι τὸ τῆς εὐεξίας ὄνομα τῷ τῆς ὑγιείας ὁμοίως
λέγεσθαι. καίτοι τὸ μὲν τῆς ὑγιείας ὄνομα διαθέσεώς
τινός ἐστι, τὸ δὲ τῆς εὐεξίας οὐχ ἁπλῶς τὴν διάθεσιν,
ἀλλὰ τὸ κατ᾽ αὐτὴν ἄριστόν τε καὶ μόνιμον ἐνδείκνυ-
ται· ἀρίστη γάρ ἐστιν ἕξις ἐκείνης τῆς διαθέσεως, ἣν
ὑγίειαν ὀνομάζομεν, ἡ εὐεξία. οὔκουν (αὖθις γάρ που
καὶ αὖθις ἀνάγκη ταὐτὸν εἰπεῖν, εἰ μέλλοι τις ἐξαιρή-
σεσθαι τῶν πολλῶν ἄγνοιαν παλαιάν,) οὔτε διαθέσεως οὔτε κατασκευῆς οὔτ᾽ ἐνεργείας ἐστὶ τὸ τῆς εὐεξίας ὄνομα δηλωτικόν, ὥσπερ οὖν οὐδὲ τὸ τῆς ἕξεως,
826K   ἀλλὰ τοῦτο μὲν αὐτοῦ μόνου τοῦ δυσλύτου τε καὶ
μονίμου τῆς διαθέσεως, ἡ δ᾽ εὐεξία πλέον οὐδὲν αὐτῷ
προστίθησι τοῦ εὖ. σύγκειται γοῦν ἐκ τοῦ τῆς ἕξεως
ὀνόματος τὸ εὖ προσλαβόντος, ὅπερ εἰώθαμεν ἐπαι-
νοῦντες ἅπασιν ἐπιφέρειν, ὅσα καλῶς καὶ κατὰ τὴν
οἰκείαν ἀρετὴν διάκειται.

τίνος οὖν ὀρεγόμεθα καὶ τί τῷ σώματι διὰ παντὸς
ὑπάρχειν βουλόμεθα; πότερον διάθεσιν ὑγιεινὴν αὐτὸ
τοῦτο μόνον ἴσχειν ἢ τοῦτό γε παραπλήσιον τῷ ζητεῖν, ἆρά γ᾽ οἰκίας μοχθηρᾶς καὶ μελλούσης ὅσον
οὔπω καταπεσεῖν χρῄζομεν ἢ καὶ πρὸς τὴν χρείαν, ἧς
ἕνεκα γέγονεν, ἄριστα διακειμένης καὶ ὡς ἔνι μάλιστα πολυχρονιωτάτης; οὐδεὶς γάρ ἐστιν οὔτ᾽ ἐλλιπῶς
οὔτ᾽ ὀλιγοχρονίως ὑγιαίνειν βουλόμενος. ἀλλ᾽ οὐδ᾽
ἐστί τις ὅλως τέχνη σκοπὸν ἔχουσα τὸ χεῖρόν τε καὶ
ὀλιγοχρονιώτερον, ἀλλ᾽ ὅπερ ἂν ἐκ τῆς αὐτῆς ὕλης
ἄριστόν τε μέλλῃ καὶ πολυχρονιώτατον ἔσεσθαι,
τοῦτο ταῖς ποιητικαῖς ἁπάσαις τέχναις ὁ σκοπός. οὐ

the term *euexia* is said similarly to the term health. However, the term health is of some condition, whereas that of
euexia is not simply the condition, but indicates that this
is excellent and stable in relation to this (health), for there
is an excellent *hexis* of that condition, which we term
health. *Euexia* is not, therefore, (for it is necessary to say
this again and again if we are going to put an end to the
antiquated ignorance of the majority) indicative of a condition, or constitution, or function, just as the term *hexis*
is not. The last is only of a stable condition that is difficult   826K
to break down; *euexia* adds nothing more to this than the
eu (εὖ). Anyway, it is a compound of the term ἕξις (*hexis*)
with εὖ attached besides, which we are accustomed to apply to all things that are praised—things which are in a
state that is good and in relation to their characteristic
excellence.

What, then, do we desire? What do we wish the body
to be continuously? Is it to possess this healthy condition
in itself alone, or is this in fact similar to asking whether
we need a bad house—one that is on the point of falling
down—or one that, in terms of the use for which it has
been built, is in an excellent state, and is as long-lasting as
it can possibly be? For no one would wish for health that
is deficient and of short duration. But there is no art at all
which has as its objective the bad and the more short-
lasting. The objective (*scopos*) in all the productive arts is
what is going to be the best and most long-lasting from its
own material. The end (*telos*) is nothing other than the

μὴν οὐδὲ τέλος ἄλλο τι παρὰ τὴν ἐπιτυχίαν ἐστὶ τοῦ
827K σκοποῦ καὶ δῆλον ἐκ τῶν εἰρημένων, ὡς καὶ τέλος ἕν
ἐστι τῆς περὶ τὸ σῶμα τέχνης καὶ σκοπὸς εἷς. εἴτ᾽ οὖν
ἀρτιότητά τις ἐθέλοι τὸν σκοπὸν τοῦτον ὀνομάζειν
εἴτ᾽ εὐεξίαν εἴθ᾽ ὑγίειαν εἴτε τὴν κατὰ φύσιν κατασκευὴν τοῦ σώματος εἴτε τὴν κατὰ φύσιν ἐνέργειαν
εἴτε διάθεσιν ἢ κατασκευήν, ἀφ᾽ ἧς ἐνεργοῦμεν τελέως
ταύτας τὰς κατὰ φύσιν ἐνεργείας, οὐδὲν εἴς γε τὰ
παρόντα δέομαι διαιρεῖν.

13. Ἕν γάρ μοι πρόκειται δεῖξαι τὸ πᾶσαν τέχνην
καὶ σκοποῦ καὶ τέλους ἐφίεσθαι. τέλος δ᾽ ἐν ἑκάστῳ
τῶν ὄντων ἕν, ὅπερ οὐδὲν ἄλλ᾽ ἐστὶν ἢ τὸ κατ᾽ ἐκείνην
τὴν οὐσίαν ἀγαθόν. ἀμπέλου γοῦν οὐκ ἄλλο μὲν τελειότης, ἄλλο δὲ τὸ ἀγαθόν, οὐ μὴν οὐδ᾽ ἄλλο τι
παρὰ τοῦτο πρόκειται σκοπεῖσθαι τῇ περὶ τὰς ἀμπέ-
λους τέχνῃ, καθάπερ οὐδὲ τῇ περὶ τὰς ἐλαίας, ἀλλὰ
καὶ ἥδε τὴν τελειότητα[19] τῆς ἐλαίας φύσεως ἔχει τὸν
σκοπόν. ἐν δή τι τῶν ὄντων ἐστὶ τὸ τἀνθρώπου σῶμα
καὶ τίς ἐστι καὶ τοῦδε πάντη τελειότης, ἣν καὶ παροῦ-
828K σαν φυλάττειν καὶ ἀποῦσαν ἀνασώζεσθαι πρόκειταί
τινι τέχνῃ. ταύτης δ᾽ εἴπερ τι κατωτέρω ποιήσομεν
ἕτερον τέλος, ὁ προειρημένος ἡμᾶς ἐκδέξεται λόγος.
ἔσται γὰρ ἡ μὲν τῆς εὐεξίας, ἡ δ᾽ ἄκρας εὐεξίας,
ἑτέρα δὲ τῆς ὑγιεινῆς ἕξεως καὶ τετάρτη τις ἄλλη
τέχνη σχέσεως ὑγιεινῆς δημιουργὸς ἔξωθέν τε τούτων
ἄλλη πέμπτη τῆς ἀθλητικῆς εὐεξίας. εἰ δὲ καὶ δημι-

success of the objective (*scopos*), and it is clear from what
is being said that there is one end and one objective of the   827K
art pertaining to the body. Whether, then, someone wishes
to term this objective either soundness, or *euexia,* or
health, or the constitution of the body in accord with nature, or the function in accord with nature, or condition
or constitution from which we function perfectly in respect of these natural functions is not something we need
to determine for our present purposes.

13. One thing lies before me to show—that every art is
directed at an objective (*scopos*) and an end (*telos*). In
each of the things that exist, there is one end, which is
nothing other than the excellence in relation to that essence. Anyway, the perfection of a grapevine is nothing
other than the excellence [of this], and there is nothing
else proposed to consider apart from this in the art concerned with grapevines, just as there is not in the art concerned with olive trees; this also has as its objective the
perfection of the nature of an olive tree. Now the body of
the person is one of the existing things, and what constitutes the complete perfection of this, which is to be preserved when present and restored when absent, is pro-   828K
posed as a certain art. If, however, we should make
something lesser another end, the previously stated argument would show us this. For one will be either *euexia,* or
peak *euexia,* while another will be healthy *hexis,* and another fourth art the production of a healthy *schesis,* and
apart from these, another and fifth will be the production
of an athletic *euexia.* And if we are going to entrust the

---

19 *post* τὴν τελειότητα: τῆς ἐλαίας φύσεως H; τῆς τῆς
ἐλαίας φύσεως Ku

ουργεῖν μὲν ἄλλαις τὰ τέλη, φυλάττειν δ' ἄλλαις ἐπιτρέψομεν, οὐ ταύτας μόνον ἀλλὰ καὶ σὺν αὐταῖς
ἄλλας τοσαύτας ἐπιζητήσομέν τε καὶ συστησόμεθα
περὶ τὸ σῶμα τέχνας.

14. Ὧι καὶ δῆλον, ὡς οὔτε πολλὰ τὰ τοῦ σώματος
ἀγαθὰ χρὴ ποιεῖν οὔτ' ἄλλην μέν τιν' αὐτοῦ δημιουργόν, ἄλλην δὲ φύλακα. τῷ μὲν δὴ σκοπεῖσθαι δυναμένῳ διὰ κεφαλαίων τὸ σύμπαν οὐδὲν ἔτι προσδεῖ·
τῷ δὲ τοῦτο μὴ δυναμένῳ δρᾶν ἢ καὶ πάντ' ἐξαλεῖψαι
τῆς διανοίας, ὅσα τοῖς κακῶς ὑπειλημμένοις ἕπεται,
δεομένῳ παμπόλλων ἔθ' ἡγοῦμαι δεῖν. ὅτι τε γὰρ οὐχ
ἕν ἐστι τὸ τοῦ σώματος ἀγαθόν, ἀλλ' ὑγίειά τε καὶ
ῥώμη καὶ κάλλος ὅτι τε δυνατὸν ἑτέραν μὲν εἶναι δη-
829K   μιουργὸν τούτου τέχνην, ἑτέραν δὲ φυλακτικὴν ὅσα τ'
ἄλλα τούτοις ἑπόμενα κακῶς ἐγνώκασιν ἔνιοι, διελέγ-
ξαι χρή.

καὶ πρῶτόν γε, ὅτι τοῦ σώματος ἓν ἀγαθόν ἐστι,
διελθεῖν ἄμεινον πρώτως τε καὶ κυρίως λεγόμενον, ἐφ'
ὃ πάντα τὴν ἀναφορὰν ἔχει. τὰ δ' ἄλλα τὰ λεγόμενα
σώματος ἀγαθὰ τὰ μὲν ὡς μόρια ἐκείνου λέγεται, τὰ
δ' ὡς αἴτια, τὰ δ' οἷον καρπός τις. ὥσπερ γὰρ τὸ
κάλλος ἥ τ' εὔχροια καὶ ἡ εὐσαρκία καὶ ἡ συμμετρία
καθάπερ καὶ ἄλλα τινὰ συμπληροῖ, τί κωλύει καὶ τὸ
τοῦ σώματος ἀγαθὸν ἐξ ὑγιείας καὶ ῥώμης καὶ κάλλους συμπληροῦσθαι; τί δὲ κωλύει πάλιν αὐτοῦ μὲν
τοῦ σώματος ἀγαθὸν εἶναι τὴν ὑγίειαν, οἷον δὲ καρ-

production of the ends to some arts, and their preservation
to others, we shall seek after and establish not only these
arts, but also with them such other arts as concern the
body.

14. From this it is also clear that we must not create
many "goods" for the body, nor any one art productive of
this and another preservative of it. For someone who is
able to consider the whole from a summary account, there
is no need of anything further. On the other hand, for
someone who is not able to do this, or wipe from his thinking all those things that follow faulty assumptions, there is
still need of a great deal, I believe. That there is not one
good for the body, but there is health, strength and beauty,
and that it is possible for there to be one art productive of   829K
this and another preservative, and those other things that
follow from them, as some people wrongly think, must be
refuted.

First, in fact, it would be better to go over, primarily
and precisely, the statement that there is one "good" for
the body, to which everything has reference. With respect
to the other so-called "goods" of the body, some are named
as parts of that [good], some as causes, and some as a fruit,
as it were. For as the good is made up of good complexion,
good condition[12] and due proportion, and certain other
things, what is to prevent the good of the body also being
made up of health, strength and beauty? Contrariwise,
what is to prevent the good of the body itself being health,

---

[12] The term here is *eusarkia*—see Aristotle, *History of Animals*, 493b22, LCL 437 (A. L. Peck), 52–53, and Hippocrates,
Joints 53, 149 (E.T. Withington), 324–25.

πόν τινα τὸ κάλλος αὐτοῦ καὶ τὴν ἐνέργειαν ὑπάρχειν; τί δὲ κωλύει τὴν μὲν ἐνέργειαν εἶναι τὸ πρῶτον
ἀγαθὸν τοῦ σώματος, αἰτίαν δ᾽ αὐτοῦ τὴν ὑγίειαν; οὐ
γὰρ ἐξ ἄλλων μέν τινων ὑγιαινὸν ἀκριβῶς ἔσται τὸ[20]
σῶμα, δι᾽ ἄλλων δ᾽ ἰσχυρὸν ἢ καλόν·[21] αὐτὸ μέλλει
ποιήσειν, εὐθὺς δὲ τοῦτο καὶ ἀκριβῶς ὑγιεινόν.

830K   15. Ὥστε καὶ διὰ τοῦτο μία τέχνη περὶ τὸ σῶμα.
τὰ γὰρ αὐτὰ πράττοντες ἰσχυροί θ᾽ ἅμα κατὰ τὰς
ἐνεργείας ἐσόμεθα καὶ καλλίους ὀφθῆναι καὶ ὑγιεινότεροι καὶ εὐεκτικώτεροι, καθάπερ εἰ καὶ σφαλείημέν
τι περὶ τὸ σῶμα, καὶ τῶν ἐνεργειῶν τὴν ῥώμην κατα-
λύσομεν καὶ τῷ κάλλει λυμανούμεθα καὶ τὴν εὐεξίαν
καθαιρήσομεν καὶ τὴν ὑγίειαν μειώσομεν· ἅπαντα
γὰρ ταῦτα συναυξάνεταί τε καὶ συμμειοῦται προσηκόντως. ἡ γάρ τοι κατὰ φύσιν ἐνέργεια δεῖται τῆς
κατὰ φύσιν τοῦ σώματος κατασκευῆς, ὑφ᾽ ἧς γίγνεται, λόγον αἰτίας ἐχούσης πρὸς αὐτήν. ὥστ᾽ οὐκ ἐνδέχεται πρότερον τὸ μὲν ἕτερον αὐτῶν εἶναι, τὸ δ᾽
ἕτερον μὴ παρεῖναι. καὶ μὲν δὴ καὶ συναυξάνεται καὶ
συμμειοῦται ταῦτ᾽[22] ἀμφότερα. βελτίω μὲν δὴ γενό-
μενα τὸ μὲν εὐεξία, τὸ δὲ ῥώμη κέκληται. τὸν αὐτὸν
γὰρ καὶ ἡ ῥώμη πρὸς τὴν ἐνέργειαν ἔχει λόγον, ὅνπερ
ἡ εὐεξία πρὸς τὴν ὑγίειαν· ἑκάτερον γὰρ ἑκατέρῳ γίγνεται καὶ ὥσπερ ἡ εὐεξία τινός ἐστιν, οὕτω καὶ ἡ
831K   ῥώμη. τῆς μὲν γὰρ κατὰ φύσιν εἴτε κατασκευῆς εἴτε

[20] τὸ add. H
[21] H includes the following note here: post καλόν nonnulla

while beauty of it and function are like some kind of fruit? What is to prevent the function being the primary "good" of the body, while health is a cause of this? For the body will not be perfectly healthy through other factors than those due to which it is strong or beautiful; if they are going to do this, it will also immediately be perfectly healthy.[13]

15. Consequently, because of this, there is one art concerning the body. For in respect of these same things, actions will be strong at the same time in terms of the function, and the body would be seen as more beautiful, healthier and more *euectic*. If, however, we commit some error concerning the body, we shall dissipate the strength of the functions, harm beauty, do away with *euexia*, and reduce the health. For fittingly all these things are increased together and diminished together. Certainly, function according to nature needs the constitution of the body, from which it arises, to be in accord with nature, since it has the ground of cause regarding this [function]. As a consequence, it is not possible for one of these to be first and another not be present. For surely both of them will be increased together and diminished together. Indeed, when *euexia* becomes better, it is called strength. For strength has the same causal relationship to function as *euexia* has to health; each arises with the other, and just as *euexia* is of something, so too is strength. *Euexia* is a

830K

831K

[13] There is some issue about the text here, so the translation is somewhat tentative—see H's note on the facing page.

vocabula intercidisse videntur; conicias ἃ γὰρ ἰσχυρὸν ἢ καλόν αὐτὸ μέλλει ποιήσειν, εὐθὺς τοῦτα (p. 50).
[22] ταῦτ᾽ add. H

διαθέσεως ὀνομάζειν ἐθέλοις ἀρετή τις ἡ εὐεξία, τῆς
δ᾽ ἐνεργείας ἡ ῥώμη.

ταὐτὸν δ᾽ ἀρετή τε καὶ τελειότης ἐστὶ καὶ τὸ καθ᾽
ἕκαστον τῶν ὄντων ἀγαθόν, ὅπερ αὐτοῦ πρώτως τε
καὶ ἁπλῶς ἀγαθὸν ὀνομάζεται. καὶ μὲν δὴ καὶ χείρω
γίγνεται διὰ τῶν αὐτῶν ἀμφότερα. καὶ ἡ μὲν ὑγιεινὴ
διάθεσις, ὥσπερ ἔμπροσθεν εἴπομεν, εὐεξία προσαγο-
ρεύεται, εἰ δ᾽ αὖ τις ἐνεργεῖ ἀσθενῶς,[23] ῥώμης ἀρρω-
στία τε καὶ ἀσθένεια. καὶ δὴ καὶ κάλλος μὲν τοῖς
προτέροις, αἶσχος δὲ τοῖς δευτέροις ἐξ ἀνάγκης ἕπεται. πάντ᾽ οὖν ταῦτα καὶ συναύξεται καὶ συμμειοῦται
καὶ τελειοῦται καὶ καθαιρεῖται πάνθ᾽ ἅμα καὶ τὸ βλά-
πτον ὁτιοῦν ἓν ἐξ αὐτῶν εὐθὺς καὶ τἆλλα σύμπαντα
βλάπτει τό τ᾽ ὠφελοῦν ὡσαύτως ἅπαντ᾽ ὠφελεῖ. καὶ
δῆλον, ὡς καὶ διὰ τοῦτο μίαν ἀνάγκη τέχνην εἶναι
περὶ σύμπαντα ταῦτα.

τί δὴ τούτων ἐστὶ τὸ πρῶτόν τε καὶ ἁπλῶς ἀγαθὸν
τοῦ σώματος, οὐδὲν μὲν ἐπείγει[24] τό γε νῦν εἶναι λέγειν, ἵνα δὲ μηδὲ τοῦτ᾽ ἐνδέῃ, προσθήσω. σώματος
832K ἀγαθὸν ἁπλῶς καὶ πρῶτον, οὗ μάλιστα δεόμεθα, τελειότης τῆς ἐνεργείας ἐστίν, ὅπερ δὴ ῥώμην τε καὶ
ἰσχὺν ὀνομάζουσιν ἐλλειπτικῶς ἑρμηνεύοντες· ἐχρῆν
γὰρ αὐτοὺς οὐ ῥώμην ἁπλῶς ἀλλ᾽ ἐνεργείας ῥώμην
οὐδ᾽ ἰσχὺν ἁπλῶς ἀλλ᾽ ἐνεργείας ἰσχὺν εἰπεῖν. ἑξῆς
δὲ τούτῳ δεύτερον οὐχ ἁπλῶς οὐδὲ καθ᾽ ἑαυτὸ τοῦ
σώματος ἀγαθόν, ἀλλ᾽ ὅτι γε τὸ πρῶτόν τε καὶ καθ᾽

---

[23] *After listing three alternatives for this phrase, H has:*

certain excellence (*arete*)—you may wish to term it of the constitution in accord with nature or of a condition—while strength is of the function.

Excellence and perfection are the same, and are the good in relation to each existing thing, which is called primarily and absolutely a good of this. Furthermore, both become worse due to the same things. And the healthy condition, as we said before, is called *euexia,* whereas, in turn, someone who functions weakly is called weak in terms of strength and asthenic. And in particular, beauty necessarily follows the first of these while ugliness follows the second. All these, then, are increased together and diminished together, and all are brought to perfection or done away with at the same time. What injures one of these in any way, immediately also injures all the others, while what benefits one, similarly benefits all. And it is clear also from this that there is necessarily one art pertaining to them all.

Which of these is the primary and absolute good of the body is not a pressing matter to be stated now, so although I shall add it, this is not because it is lacking. The absolute and primary good of the body, which we particularly need is perfection of function, which those explaining it term inadequately "strength" (*rōmē*) and "power" (*ischus*). It behooves them to say not simply "strength," but strength of function and not simply "power," but power of function. Next in order to this, the second is not simply or of itself a good of the body, but because the good that is primary

832K

Corruptum esse hunc locum hiatu quoque monemur; suspicor legendum: ἡ δ᾽ αὖ τῆς ἐνεργείας ἀσθένεια (p. 51). *This is what has been translated.*   24 ἐπείγει H; ἐγείρει Ku

271

ἑαυτὸ τοιούτου δεῖται πάντως εἰς γένεσιν, ἡ τῆς ὑγιείας ἐστὶν εὐεξία, ἣν καὶ αὐτὴν πάλιν ἐλλειπτικῶς
ὀνομάζοντες ἀφορμὴν σοφισμάτων παρέχουσιν· δέον
γὰρ εὐεξίαν ὑγιείας εἰπεῖν, οὐχ οὕτως ἀλλ' ἁπλῶς εὐεξίαν ὀνομάζουσιν. ἕπεται δ' ἐξ ἀνάγκης τῇδε τὸ κάλλος ἄλλο τι γένος· ἕπεται ἐκείνοις, τρίτον ἀγαθὸν σώματος, ὥστ' οὔθ' ὁμογενῆ τὰ τοῦ σώματος ἀγαθὰ
καθάπερ οὐδὲ τὰ τῆς ψυχῆς οὔθ' ὁμοίως λέγεται ἅπαντα, ἀλλὰ τὸ μὲν ὡς πρῶτόν τε καὶ καθ' ἑαυτό, τὸ δ'
ὡς αἴτιον ἐκείνου, τὸ δ' ὡς ἐξ ἀνάγκης ἑπόμενον.

τίνος οὖν τούτων τῶν τριῶν ἡ περὶ τὸ σῶμα τέχνη
833K πρώτως[25] ἐστὶ δημιουργός; ἆρά γε τῆς ὑγιείας ἢ τῆς
ἐνεργείας ἢ τοῦ κάλλους; ὅτι μὲν γὰρ ἐξ ἀνάγκης
ὠφελήσει τε καὶ ποιήσει τὰ τρία, κἂν ἓν ἐξ αὐτῶν
ὁτιοῦν ὠφελῇ, πρόδηλον ἤδη γέγονεν· εἴτε γὰρ τὴν
ἐνέργειαν, ἐξ ἀνάγκης καὶ τὴν ὑγίειάν τε καὶ τὸ κάλ-
λος· ἡ μὲν γὰρ οὐχ οἷά τ' ἐστὶ χωρὶς τῆς ποιούσης
αὐτὴν αἰτίας γενέσθαι, τὸ κάλλος δ' ἐξ ἀνάγκης ἕπε-
ται· εἴτε τὴν ὑγίειαν, εὐθὺς καὶ τὴν ἐνέργειαν καὶ τὸ
κάλλος· ἄμφω γὰρ ἦν ταύτης ἔγγονα. καὶ μὴν εἰ τὸ
κάλλος ποιεῖ, πάντως που καὶ τὴν ὑγίειαν προπεποίη-
κεν, εἰ δὲ τοῦτο, καὶ τὴν ἐνέργειαν.

16. Ἀλλὰ τί τὸ πρώτως ἐστὶν ὑπὸ τοῦ τεχνίτου
γιγνόμενον, ὁ λόγος ἐπόθει θηρᾶσαι. καὶ δὴ φαίνεται
σαφῶς ἤδη διὰ τῶν εἰρημένων ἡ μὲν τέχνη τὴν ὑγίειαν ἐργαζομένη, ταύτῃ δ' ἐξ ἀνάγκης ἐνέργειά τε καὶ
τὸ φυσικὸν κάλλος ἀκολουθεῖ. ὅπου γὰρ πρῶτον ἐνεργ-
γῶν ὁ τεχνίτης ἵσταται, τοῦτ' ἔστιν αὐτοῦ τὸ τέλος,

and of itself altogether needs such a thing for genesis; this is the *euexia* of health, which again people named inadequately, thus providing the starting point for sophistical deliberations. For although there is a need to say *euexia* of health, they do not name it in this way, but simply term it *euexia*. It follows this of necessity that beauty is some other class. It follows those goods and is a third good of the body, so that the goods of the body are not of the same class, just as those of the soul are not all similarly stated. Rather, there is what is primary and of itself, there is what causes that, and there is what necessarily follows.

So which of these three is the art pertaining to the body primarily the producer of? Is it of health, or of function, or of beauty? That it will necessarily benefit and create the three, if it benefits one of these in any way whatsoever, has already become clear. Thus, if it benefits function, it will necessarily also benefit health and beauty. Function is not possible apart from the cause producing it, while beauty follows of necessity. If it benefits health, it immediately also benefits function and beauty, for both are offspring (progeny) of this. And further, if it creates beauty, it assuredly has previously somehow created health, and if health, also function. 833K

16. But what the argument is primarily anxious to chase down is what is done by the practitioner of the art. And indeed, it already seems clear from the things said, that the art is productive of health; through this, function and natural beauty follow of necessity. For what the practitioner establishes as primary among its effects is its end

---

²⁵ πρώτως H; πρώτη Ku

ἵσταται δ᾽ ἐν τῷ τὴν κατὰ φύσιν ἐργάσασθαι διάθε-
834K σιν, ἀφ᾽ ἧς ἐνεργοῦμεν, ὅπερ ἦν ὑγίεια· ταύτην δ᾽
ἐργασάμενος οὐδὲν ἔτι πονεῖ περὶ τὴν ἐνέργειαν ἢ τὸ
κάλλος, ἕπεται γὰρ ἐξ ἀνάγκης ἐκεῖνα, κἂν ὁ τεχνίτης
μὴ θέλῃ, καὶ κωλῦσαί γ᾽ αὐτὰ τῷ τεχνίτῃ παντάπασιν ἀδύνατον ἅπαξ γε τὴν ὑγίειαν ἐργασαμένῳ, αὐτὴν μέντοι τὴν ὑγίειαν ἐπ᾽ αὐτῷ κωλύειν ἐστίν. εἰ δέ
γέ πη διαφθείρειεν αὐτήν, οὐκ ἔστιν ἐπ᾽ αὐτῷ τὴν
κατὰ φύσιν ἐνέργειαν ἢ τὸ κάλλος ἐκπορίσαι τῷ
σώματι. τὴν οὖν περὶ τὸ σῶμα τέχνην, ἥτις ἂν ᾖ, περὶ
μὲν τὴν ὑγίειαν ἁπάντων πρώτην τε καὶ καθ᾽ ἑαυτὴν
πραγματεύεσθαι φατέον, ἐφεξῆς δὲ κατά τι συμβεβηκὸς τήν τ᾽ ἐνέργειαν καὶ τὸ κάλλος, οὐ μὴν ἀτελές γέ
τι τούτων οὐδὲν οὐδ᾽ ἐλλιπές, ἀλλὰ τέλεά τε καὶ
πλήρη καὶ ἄκρα. ὅπως μὲν οὖν ἐστι πολλὰ τοῦ σώματος ἀγαθὰ καὶ πῶς ἓν ὅτι τε μὴ πάντων ὡσαύτως ἡ
περὶ τὸ σῶμα τέχνη πεφρόντικεν, ἀλλὰ τοῦ μὲν καθ᾽
αὑτό, τοῦ δὲ κατὰ συμβεβηκός, αὐτάρκως δέδεικται.

17. Διὰ τί δ᾽ οὐκ ἐνδέχεται κατ᾽ οὐδεμίαν ὕλην
ἄλλην μὲν τέχνην τοῦ τέλους δημιουργόν, ἑτέραν δ᾽
835K εἶναι τὴν φυλάττουσαν, ἤδη καὶ ταῦτα πειράσομαι
δεικνύναι. καὶ πρῶτον μὲν ὡς οὐδὲ τὸ παράδειγμα
αὐτοῖς ὁμολογεῖ τὸ σφέτερον, ἐπιδείξω, μετὰ ταῦτα δ᾽
ὡς οὐδ᾽ ἡ τῶν πραγμάτων αὐτῶν φύσις. εἴπερ γὰρ
ἄλλης μέν ἐστι τέχνης ὑπόδημα ποιήσασθαι, τὸ δ᾽
ἠπήσασθαι τοῦτο παθὸν ἑτέρας, οὔπω τὴν τρίτην ἐνταῦθα τέχνην ἐπέδειξαν ἡμῖν τὴν φυλακτικήν. ὡσαύτως δὲ κἀπὶ τῶν ἱματίων, εἴπερ ἑτέρας μέν ἐστι τέ-

(*telos*), while in this he sets the creation of the condition in accord with nature from which we function—that is, health. If he creates this, he has nothing further to work on concerning function or beauty, for those necessarily follow, even if the practitioner doesn't wish it; it is altogether impossible for him to prevent these consequences once he has definitively created health. Nevertheless, it is possible for him to prevent health itself. On the other hand, if health is in fact destroyed in some way, it is not possible for him to provide function in accord with nature or beauty for the body. One must say, then, that the art pertaining to the body, whatever it may be, is first of all about health, and is engaged with this in itself. Next to that, function and beauty are contingent, although neither of these is without an end and a deficiency; there are goals, completions and end points. Thus, it has been adequately shown how there are many "goods" of the body, and how the art which has care of the body is one that is not like all the others; it is of this *per se* but of other things *per accidens*.

17. I shall now attempt to show why it is not possible for there to be one material art productive of the end (*telos*) but another that is preservative. And first I shall show that their own example does not agree with these same things, and after this, that it is not the nature of the matters themselves. Thus, if the making of shoes belongs to one art while the repair of what has suffered belongs to another, this doesn't yet show us that there is here a third art which is the preservative. Similarly, in the case of cloaks, if repair is of one art and production of another, on

834K

835K

χνης ἠπήσασθαι, δημιουργῆσαι δ᾽ ἑτέρας, ἡ τρίτη
κἀνταῦθα τίς ἐστιν ἡ φυλακτική, λέγειν οὐχ ἕξουσιν.
ἐπὶ μὲν γὰρ τῶν ἡμετέρων σωμάτων καὶ ὅλως τῶν
ὑπὸ φύσεως διοικουμένων ἡ μὲν οἷον δημιουργὸς²⁶ ἡ
φύσις ἐστὶν ἀνάλογον ὑφαντικῇ τε καὶ σκυτοτομικῇ,
τὸ πονῆσαν δ᾽ ἐπανορθοῦται γεωργός τε καὶ ἰατρός,
ἀνάλογον αὖ καὶ οἵδε τῷ ῥάπτοντι τὰ ἱμάτια καὶ τῷ
τὰ παλαιὰ τῶν ὑποδημάτων ἐπανορθουμένῳ. τὸ δὲ
φυλάττειν ἱμάτιον ἢ ὑπόδημα καταθέμενον ἐπὶ τῆς
οἰκίας, ὅπως μὴ κλαπείη πρός τινος ἢ ὑπὸ μυῶν
καταβρωθείη, τάχα μὲν οὐδὲ τέχνης ἐστὶν οὐδεμιᾶς
836K   ἀλλ᾽ ἐπιμελείας μόνης. εἰ δ᾽ ἄρα καὶ τέχνην τινὲς εἶναι βούλονται φυλακτικὴν αὐτῶν, αἱ τοιαῦται τέχναι
τοῖς ἀνθρώποις ἥ στρατηγική τ᾽ εἰσὶ καὶ ἡ πολιτικὴ
καὶ πρὸς ταύταις ἥ τε τῶν θυρωρῶν, εἰ βούλει, καὶ
τῶν ἄλλων φυλάκων. ἵνα γὰρ μήτ᾽ ἐπὶ τοῖς πολεμίοις
ὦμεν ἢ ὅλως τοῖς πονηροῖς ἀνθρώποις μήτ᾽ ἐπὶ τοῖς
θηρίοις, οἰκίας τε καὶ πόλεις οἰκοδομησάμεθα καὶ
τείχη περιεβαλόμεθα καὶ στρατηγοὺς καὶ ἄρχοντας
ἀπεδείξαμεν, ἀλλ᾽ οὐ τοιαύτην τιν᾽ ἐζητοῦμεν τέχνην
ὑγιείας φυλακτικήν, ἀλλ᾽ ἥτις οἶμαι περὶ τὸ σῶμα τοῦ
ἀνθρώπου πραγματευομένη τι, μὴ κατὰ συμβεβηκὸς
ἀλλὰ καθ᾽ αὑτό, σῶον αὐτὸ καὶ ὑγιαῖνον ἀποδείκνυ-
σιν. ἑκάστη δὲ τῶν νῦν εἰρημένων οὐκ αὐτὴ καθ᾽
αὑτὴν ὑγιείας ἐστὶ φύλαξ, ἀλλ᾽ ἐπειδὴ συμβέβηκεν
τῷ μὴ σφαγέντι μηδ᾽ ὑπὸ θηρίου καταβρωθέντι φυλάττεσθαι τὴν ὑγίειαν, ὥσπερ οὖν καὶ ὅλην τὴν ζωήν,
διὰ τοῦτο κατά τι συμβεβηκός, οὐ πρώτως οὐδὲ κατὰ

what the third and preservative is here, they will have nothing to say. In the case of our own bodies, and in general of those things governed by Nature, then Nature is, as it were, the demiurge (craftsman), analogous to the weaver and shoemaker, while a farmer and a doctor restore what is worn out, and are in turn analogous to one who sews torn cloaks and repairs old shoes. However, the preservation of a cloak or shoes situated within the household, so they are not stolen by someone or eaten by mice, is perhaps not a matter for any art, but only of being careful. But if there are also those who wish there to be an art 836K preservative of these things, there are such arts among men as military strategy, politics and, in addition to these, the art of guarding the gates, if you like, and the other guardings. We build cities, and houses, and walls around them, and we appoint generals and magistrates so we do not come into contact with enemies and rogues generally, or wild animals, but this is not the kind of art, preservative of health, we were looking for. Rather, this would be, I think, an art which takes some matter in hand concerning the body of the person—and not *per accidens* but *per se* (not contingently but in and of itself), making this body sound and healthy. Each of the arts now mentioned is not a guardian of health *per se,* but because it is *per accidens* that someone who is not slain or devoured by a wild animal preserves his health, just as he does his whole life, due to this, each of such arts is preservative of health, but does

---

26 δημιουργὸς H; δημιουργικὴ Ku

277

τὸν ἴδιον λόγον, ἑκάστη²⁷ τῶν τοιούτων τεχνῶν ὑγιείας γίγνεται φύλαξ.

837K   18. Ἡ πρώτη οὖν ὑγίειαν φυλάττουσα κατὰ τὸν ἑαυτῆς λόγον ἐν τίσιν ἕξει τὴν πραγματείαν; ἐμοὶ μὲν δοκεῖ μήτ' ἐν τοῖς οὐκ ἐξ ἀνάγκης ὁμιλοῦσι τῷ σώματι μήτ' ἐξ ἀνάγκης ἐν τοῖς οὐδὲν διατιθεῖσιν, ἀλλ' ἐν οἷς διατρίβει πάντως τὸ σῶμα, κἂν ἡμεῖς μὴ βουλώμεθα, καὶ τούτων τοῖς ὠφελεῖν ἢ βλάπτειν δυναμένοις, ἐν τούτοις τε καὶ περὶ ταῦτα πραγματεύεσθαι. ξίφεσι μὲν οὖν καὶ θηρίοις καὶ κρημνοῖς καὶ βρόχοις οὐκ ἐξ ἀνάγκης περιπίπτει τὸ σῶμα, ἀλλὰ τῷ περιέχοντι ἀέρι κατὰ διττὸν τρόπον ἐξ ἀνάγκης ὁμιλεῖ καὶ πάντη περικεχυμένῳ καὶ διὰ τῆς εἰσπνοῆς ἑλκομένῳ. καὶ μὲν δὴ καὶ ὕπνος καὶ ἐγρήγορσις ἡσυχία τε καὶ κίνησις ἐκ τῶν τοιούτων ἐστίν· ἀνάγκη γὰρ ἢ ὑπνοῦν ἢ ἐγρηγορέναι ἢ ἡσυχάζειν ἢ κινεῖσθαι, ὥσπερ οὖν καὶ λιμώττειν ἢ ἐσθίειν καὶ διψῆν ἢ πίνειν ἤ τινα μεταξὺ τούτων ἔχειν κατάστασιν. αἱ δὲ τῆς κοίτης διαφοραὶ καὶ τῶν ἱματίων οὐκ ἀναγκαῖαι πᾶσαι· κλίνη γὰρ ἐλεφαντόπους οὐδὲν οὔτ' ὠφελεῖ τὴν ὑγίειαν οὔτε βλά-

838K   πτει, κατὰ ταὐτὰ δὲ καὶ σκίμπους εὐτελὴς ἢ ἱματίοις εὐτελέσιν ἢ πολυτελέσιν ἢ τοῖς σκεύεσιν ὑαλίνοις ἢ χρυσοῖς ἢ ἀργυροῖς ἢ ξυλίνοις χρῆσθαι καὶ παῖδας ἔχειν εὐμόρφους ἢ αἰσχροὺς τοὺς ὑπηρετουμένους ἢ μηδ' ὅλως ἔχειν ἀλλ' ἑαυτῷ διακονεῖσθαι. ταῦτα μὲν γὰρ οὐδὲν οὔτ' ὠφελεῖν οὔτε βλάπτειν ἡμᾶς πέφυκεν οὔτε καθ' ἑαυτὰ καὶ πρώτως οὔτε κατὰ συμβεβηκός. ἀὴρ δὲ θερμὸς ἢ ψυχρὸς ἐδέσματά τε καὶ πόματα καὶ

278

this incidentally and contingently, and not primarily in relation to its specific rationale.

18. In what things, then, will the primary art preserving health have its subject matter by reason of itself? It seems to me it will not be in those things that are not associated with the body of necessity, or in those that do not change anything of necessity. It will be in those things that continually wear away the body, even if we do not wish this to be so. And of these, it is in those that are able to help or harm; it is on these and the things around them that we must work diligently. Thus, swords, wild animals, beetling cliffs and nooses do not befall the body of necessity. But the body associates with the ambient air of necessity in a twofold manner by being everywhere surrounded by it and by drawing it in through inspiration. Furthermore, sleep and waking, rest and movement are among such things, for sleep, waking, rest and movement are necessary, just as there is a condition of being hungry or eating, being thirsty or drinking, or something in between these. On the other hand, all the varieties of beds and cloaks are not necessary; an ivory-footed bed does nothing to help or harm health, and in the same way, neither does a cheap pallet, nor the use of cheap or expensive cloaks, or glass, gold, silver or wooden vessels, and the same applies to a child who has comely or ugly servants, or has none at all, but looks after himself. These things neither benefit nor harm us naturally, either of themselves and primarily, or contingently (*per accidens*). However, hot or cold air,

837K

838K

---

27 ἑκάστη H; ἑκατέρα K

ἠρεμία καὶ κίνησις ἐγρήγορσίς τε καὶ ὕπνος ἐξ ἀνάγκης ὠφελεῖ καὶ βλάπτει κατὰ τὴν ἑαυτῶν δύναμιν·
ἱμάτιον δὲ τρύχινον ἐν χειμῶνι καὶ βαρὺ καὶ πνιγῶ-
δες ἐν τῷ θέρει βλάπτει μὲν οὐκ[28] ἐξ ἀνάγκης ἀλλὰ
κατὰ συμβεβηκός, ὅτι[29] κρύει καὶ θάλπει τιμωρεῖται.

19. Ταῦτ᾽ οὖν, ὅσα καθ᾽ αὑτὰ δύναμιν ἔχει βλάβης
τινὸς ἢ ὠφελείας, ὁ τῆς ὑγιείας προνοούμενος ἐπισκο-
πεῖται. πῶς καὶ τίνα τρόπον; ἐπειδὰν μὲν ἱκανῶς ἤδη
τὸ σῶμα κενώσεως ἔχῃ καὶ κίνδυνος ἤκῃ βλάβης αἰ-
σθητής, τρέφεσθαι κελεύων, ἐπειδὰν δὲ ξηραίνηται
839K   μᾶλλον τοῦ μετρίου καὶ κίνδυνος ᾖ κἀντεῦθεν ἤδη
βλαβῆναι, πίνειν ἀξιῶν· οὕτω δὲ καὶ γυμνάζων μέν,
ἐπειδὰν ῥῶσαι μὲν βούληται τὴν διοικοῦσαν ἡμᾶς
δύναμιν, ἐκκαθῆραι[30] δὲ τοὺς κατὰ λεπτὸν πόρους,
ἡσυχάζειν δὲ προστάττων, ἐπειδὰν ἤτοι κάμνειν πρὸς
τῶν γυμνασίων αἴσθηται ἢ διαφορεῖσθαι τὸ σῶμα
πέρα τοῦ προσήκοντος, ἀλλὰ καὶ τὴν γαστέρα λα-
πάττων μέν, ἢν ἴσχηται, κατέχων δέ, ἢν ἐκταράττη-
ται, καὶ τἆλλα δὴ πάντα κατὰ τὸν αὐτὸν τρόπον ἕκαστα πραγματευόμενος, ἐν κεφαλαίῳ τῷδε προσέχων
τὸν νοῦν, ἐπειδὰν μὲν ἀκριβῶς[31] ὑγιεινῶς ἔχῃ τὸ
σῶμα, μηδὲν νεωτερίζειν, ἐπειδὰν δὲ κατά τι τῆς ἀκριβοῦς ἐξίστηται συμμετρίας, εὐθὺς ἀντεισάγειν τὸ λεῖπον, πρὶν μεγάλην γενέσθαι τὴν εἰς τὸ παρὰ φύσιν
ἐκτροπήν, ὥσπερ ἂν εἰ τὰς κρόκας ἐκρεούσας ἱματίου
τινὸς οὐχ ἅμα πάσας οὐδ᾽ ἀθρόας ἀλλὰ καθ᾽ ἑκάστην

28 οὐκ add. H   29 ὅτι H, τῷ τε Ku   30 H lists the

foods, drinks, rest, movement, wakefulness and sleep necessarily benefit and harm in relation to their own capacity (potency), whereas a ragged cloak in winter and a heavy and stifling one in summer harm contingently and not of necessity, because they aid the cold and heat respectively.

19. The one who provides for health will therefore keep a close watch on those things which have power within themselves to do some harm or be of benefit. How and in what manner? When the body has already been sufficiently evacuated and danger of perceptible harm comes, he orders the person to be nourished; when the body has been dried beyond moderation, and there is already here danger of being harmed, he deems it worthwhile to drink. In this way also, when he wishes to strengthen the capacity that governs us, and for the pores to be cleaned out thoroughly, he orders exercises. He prescribes rest when he perceives the body to be tired out by the exercises or is dissipated beyond what is appropriate. But also, he empties the stomach when it is retaining, whereas he holds it back when it is upset, and manages all the other things, each in the same way. In summary, when the body is completely healthy, he directs his attention to avoiding anything new. However, when there is some departure from precise moderation, he immediately introduces what is lacking, before the deviation to a contrariety to nature becomes great. It is just as if threads have fallen out of a cloak—the person attending does not effect a cure of all the threads at the same time and all at once, but

839K

---

following variations in this verb: ἐκκαθῆραι, ἐκκαθαῖραι, ἐκκαθᾶραι *(see p. 56)—he uses the first, Ku the third.* 31 *post* ἀκριβῶς: ὑγιεινῶς ἔχῃ H; ἔχῃ καὶ ὑγιῶς Ku

ἡμέραν μίαν ὁ τοῦτ᾽ ἔργον πεποιημένος ἰῶτο παρεδρεύων, ὡς λανθάνειν τοὺς πολλοὺς τὴν ἐπανόρθωσιν ὑπὸ σμικρότητος. τοιοῦτον γάρ τινα χρὴ καὶ τὸν τοῖς
840K ὑγιαίνουσιν ἐφεστῶτα τεχνίτην εἶναι σμικρᾶς διαφθορᾶς καὶ βλάβης αἰσθητικόν θ᾽ ἅμα καὶ ἐπανορθωτικόν.

εἰ δέ γ᾽ ἀπαθὲς ἔμενε πάντη τὸ σῶμα καὶ τοιοῦτον, οἷόνπερ ὁ ποιήσας αὐτὸ τεχνίτης ἀπέλιπεν, οὐκ ἂν ἐδεῖτο τοῦ διὰ παντὸς ἐπανορθουμένου. νυνὶ δ᾽ ἐπειδὴ διαρρεῖ τε καὶ φθείρεται, δεῖταί τινος ἐπιστάτου παρεδρεύοντος, ὃς καὶ γνωριεῖ τὸ κενούμενον ὁποῖόν τ᾽ ἐστὶ καὶ ὁπόσον ἰάσεταί τε παραχρῆμα τοιοῦτόν τε καὶ τοσοῦτον ἕτερον ἀντεισάγων. ἀπορρεῖ τὸ κατὰ φύσιν ὑγρόν· ἐπάρδειν χρὴ τὴν ἴσην ὑγρότητα πόμα παρέχοντα. διαφορεῖται τὸ θερμόν· ἀντεισάγειν χρὴ τοσοῦτον ἕτερον. ἐκκενοῦται τὸ ξηρόν·[32] ἤδη καιρὸς τρέφειν· ἑνὶ λόγῳ τὸ διαφορούμενον καὶ ἀπολλύμενον ἐπανορθοῦσθαι κατὰ βραχὺ πρὸ τοῦ τοῖς πολλοῖς γενέσθαι κατάφωρον.

20. Ὥσπερ οὖν εἰ δύο τινὲ πίθω νοήσαις[33] τετρημένω πολλαχόθι, μεστὼ μὲν ἄμφω τὰ πρῶτα, κενουμένω δ᾽ ὁμοίως ὑπὸ τῶν ἐκροῶν, εἴη δὲ τῷ μὲν ἐπιστατῶν τις ἀεὶ προσεδρεύων· ἐπαντλῶν ἴσον ἑκάστοτε
841K τῷ κενουμένῳ, θατέρῳ δέ, πρὶν ἂν[34] ἱκανῶς ἐκκενωθῇ, μηδεὶς παρείη, τηνικαῦτα δ᾽ ἐξαίφνης ἐπιστάς τις ἀθρόως πληρώσειεν, εὔδηλον, ὡς τὸν μὲν ἕτερον τῶν

carries out the task every single day, so the restoration, by being small, escapes the notice of the majority. The person attending to those who are healthy must also be such a craftsman, able to perceive and at the same time repair 840K small losses and harm.

If, however, the body remained unaffected in every way, and just as it was left by the craftsman who made it, there would be no need for continual restoration. But now, since it fades away and is destroyed, there is need of some overseer to attend to it—someone who knows what kind of thing is emptied out and how much, and will immediately effect a cure, introducing in its place something else of the same kind and amount. For example, if the fluid in accord with nature flows away, it is necessary to restore the equal amount of fluid by providing drinks. If the heat is lost, it is necessary to introduce in its stead another similar amount. If what is dry is emptied out, it is already time to nourish. In a word, what is destroyed and lost must be restored little by little before the loss becomes apparent to the majority.

20. It is, then, as if you were to imagine two wine jars perforated in many places. At first both are full, but are being emptied in a similar manner by the outlets. For one, however, there is an overseer who is always in attendance, pouring in an amount equal to what is emptied each time, whereas for the other, no one is at hand before a consider- 841K able amount is emptied out. But then, suddenly, someone is on the spot and fills it all at once. It is clear that the

---

32 ξηρόν H; στηρεόν Ku. *H has the following note:* ξηρόν C στηρεόν coniecit Crassus in ed. Iunt., quod recepit Ch (*p. 57*).
33 νοήσαις H, ὄντε Ku   34 ἂν *add.* H

πίθων οὐδέποτε ἐκκενοῦσθαί τε καὶ πληροῦσθαι, τὸν
δ' ἕτερον ἄμφω ταῦτα πάσχειν οἱ πολλοὶ τῶν θεω-
μένων ἐροῦσιν, οὕτως ἔχει κἀπὶ τῶν ὑγιαινόντων τε
καὶ νοσούντων σωμάτων. ἐπανορθοῦται μὲν ἄμφω
ταῦτα μία τέχνη καθ' ἕνα τρόπον ἀντεισάγουσα τὸ
λεῖπον, ἡ διαφορὰ δ' ἐν τῷ ποσῷ τῆς ἐπανορθώσεως,
οὐκ ἐν τῷ ποιῷ.

21. Καὶ μὴν ἡ κατὰ τὸ ποσὸν ἐν τοῖς γιγνομένοις
διαφορὰ μιᾶς ἐστι τέχνης· εἰ δέ γ' ἄλλης ἔσεσθαι
μέλλοι, πάντως χρὴ παραλλάξαι τῷ ποιῷ. εἰ δ', ὅτι
τοῦ μὲν ἔφθασεν ἐκρυῆναι πολλά, πρὶν ἀφικέσθαι τὸν
τεχνίτην, ὁ δ' ἕτερος ἀεὶ παρεδρεύοντος, εὐτυχήσας
αὐτῷ μεστὸς διαφυλάττεται, τούτου χάριν ἡγῇ διττὸν
εἶναι τὸν τρόπον τῆς ἐπανορθώσεως, οὐκ ὀρθῶς ἔγνω-
κας· οὐ γὰρ ἄλλη μὲν τέχνη μίαν ἀντεισάγει κρόκην
842K   ἐκρυεῖσαν, ἄλλη δὲ τρεῖς ἢ τέτταρας ἢ πεντακοσίας
οὐδ' ἄλλη μὲν τέχνη σμικρόν τι τοῦ τοίχου πονῆσαν
ἐπανορθοῦται, μεῖζον δ' αὐτοῦ πάθος ἑτέραν ἐπιζητεῖ
τέχνην· καθόλου γὰρ ᾧ τρόπῳ³⁵ γέγονεν ἕκαστον τῶν
ὄντων, τούτῳ καὶ φθειρόμενον ἐπανορθοῦται. κρόκαι
στήμοσι διαπλακεῖσαι ποιοῦσιν ἱμάτιον. ἆρ' οὖν ἢ
παθεῖν ἔξω τούτων ἱμάτιον ἢ ἰαθῆναι δυνατόν; δεῖ δὴ
πάντως, εἴ γε πάσχει τι πάθος, ἤτοι στήμονα παθεῖν
ἢ κρόκην ἢ συναμφότερα ἴασίν τε μίαν εἶναι τῆς κρό-
κης³⁶ ἀεὶ διαπλεκομένης τῷ στήμονι μιμήσει³⁷ τῆς γενέσεως.

³⁵ καθόλου γὰρ ᾧ τρόπῳ H; ἐπὶ γὰρ τῷ τρόπῳ ᾧ Ku

majority of those observing will say that one of the wine
flasks is never emptied and filled, whereas both these processes affect the other. This is what obtains in the case of
healthy and diseased bodies. One art corrects both these,
restoring what is deficient in one way. The difference is in
the amount (quantity) of the correction and not in the kind
(quality).

21. And further, the difference in the amount in the
things occurring is of one art, whereas if it is going to be
of another art, there must have been, at all events, a
change in the quality. If, however, because in one case
much is lost by flowing away before the craftsman comes,
while in the other case, he is always in attendance, and it
is fortunately kept full by him, you would think, on account of this, that there is a twofold manner of correction,
but you would not have understood properly. For it is not
customary to introduce one art to replace one thread that
has fallen out, and another to replace three or four, or five   842K
hundred, or one art to restore a small amount of damage
in a wall, and another art for greater damage to it. For in
general, the manner in which each of the existing things
came to be is the manner by which what is destroyed is
restored. Horizontal threads (woof) when plaited with
vertical threads (warp) make a cloak. How, then, is it possible for a cloak, affected by external things, to be re-
paired? It is altogether necessary, if it does suffer some
damage, that this either affects the warp, or the woof, or
both, and the one cure is always when the woof is interwoven with the warp, imitating the original creation.

---

36 τῆς κρόκης H; τὴν κρόκην Ku—see H's note, p. 59.
37 στήμονι μιμήσει H; στήμονιμιμή Ku

22. Ἀλλ᾽ αἱ κατὰ μέρος ἐνέργειαι τοὺς πολλοὺς
ἐξαπατῶσιν οὐ δυναμένους θεάσασθαι τὸ καθόλου·
μεθόδου γὰρ ἤδη λογικῆς τὸ τοιοῦτον, ἣν οὔτ᾽ ἔμαθον
οὔτ᾽ ἤσκησαν οἱ τολμῶντες ἑκάστης ἡμέρας εἰς τὰ
προβλήματα λέγειν. ἄμεινον γὰρ ἦν, εἴπερ ᾔδεσαν
ἐκείνην, διδάξαι τοὺς μαθητὰς ἅπαξ, οὐ μυρία προβλήματα διέρχεσθαι. ὁ μὲν γὰρ ἐκείνην μαθὼν οὐδὲν
ἔτι δεῖται τῶν μυρίων ἑαυτῷ γε πάντα καλῶς διαιρεῖ-
843K σθαι δυνάμενος, ὁ δ᾽ ἀγνοῶν ἔτι τὴν μέθοδον πρὸς τῷ
μὴ γιγνώσκειν, εἰ κακῶς εἴρηται τὰ μυρία, παμπόλλων ἔτι προβλημάτων, ὧν οὐκ ἀκήκοεν, ἐπιδεής ἐστιν.

ἀλλὰ τοῦτο μὲν ὁδοῦ τι πάρεργον ἡμῖν εἰρήσθω
πρός γε τοὺς ἑκάστης ἡμέρας εἰς τὰ προβλήματα
λέγοντας τοῖς μαθηταῖς, οὔτε δ᾽ αὐτοὺς εἰδότας, ὅ τι
λέγουσιν, ἔστ᾽ ἂν ἀμαθεῖς θ᾽ ἅμα καὶ ἀγύμναστοι[38]
θεωρίας ὑπάρχωσι λογικῆς, ἐξαπατῶντάς τε τοὺς μαθητὰς ἀγνοίᾳ τοῦ κριτηρίου. καὶ γὰρ οὖν κἀνταῦθα
πότερον ταῖς κατὰ μέρος ἐνεργείαις ἢ ταῖς καθόλου
διακριτέον ἐστὶ τὰς τέχνας ἀπ᾽ ἀλλήλων, ἢ τοῦτο μὲν
οὐχί, τοῖς σκοποῖς δὲ καὶ τοῖς τέλεσιν ἢ μηδὲ τούτοις,
ἀλλὰ ταῖς ὕλαις τε καὶ τοῖς ὀργάνοις καὶ ταῖς ἀρχαῖς
καὶ τῇ θεωρίᾳ, μηδ᾽ ὅλως σκεψάμενοι ληροῦσιν ὑπὲρ
ὧν οὐκ ἴσασιν. ἄμεινον οὖν ἴσως ἐστὶ καὶ ἡμᾶς, ἐπει-

---

[38] post ἀγύμναστοι: θεωρίας ὑπάρχωσι λογικῆς, ἐξαπατῶντάς τε τοὺς μαθητὰς ἀγνοίᾳ add. H (e codice L restitui—
note, p. 59).

22. But the individual actions deceive the majority who are unable to grasp what is general. Such a thing is already part of logical method, which those who are bold enough to speak about *problēmata*[14] every day, neither learned nor practiced. It would be better if they knew that method and taught their pupils definitively, rather than going through countless problems. For one who has learned that method no longer requires countless problems in it to be able to distinguish all things properly, whereas one who is still ignorant of the method, in addition to not knowing if he states the countless issues wrongly, is still deficient in the very many *problēmata* he has not heard of. 843K

But let this be said as something secondary on our path—something aimed at those who speak every day to their students on the *problēmata* without themselves knowing what they are talking about. Being, as they are, untutored and unpracticed in logical theory, they deceive their students through ignorance of the standard. For surely even here what must be decided is whether the arts are to be distinguished from one another by the actions individually or in general, or not by this, but by the aims (*scopos*) and ends (*telos*), or not by these, but by the materials, instruments, principles and theory, lest on the whole they talk nonsense in their considerations about those things they do not know. Perhaps, then, it is also

[14] The term *problēma,* used in a general sense in the opening sections, is taken as a technical term in logic—see Aristotle, *Topics* 104b1–4, which has: "A *problēma* is a dialectical speculation / investigation leading either to choosing or avoiding, or to truth and knowledge, either by virtue of itself or as secondary to some other of such *problēmata*" (see also 101b28).

δήπερ ἅπαξ ὑπέστημέν σοι τὸν ἆθλον τοῦτον ἐκτελέσαι, διελθεῖν τι καὶ περὶ τῶν ῥηθέντων ἁπάντων ὡς οἷόν τε διὰ βραχυτάτων ἀρξαμένους αὖθις ἀπὸ τῶν κατὰ μέρος ἐνεργειῶν.

23. Ἔστιν οὖν τις ἐνέργεια κατὰ μέρος, ᾗ χρώμενοι
844K τὰ βλέφαρα τῶν ὀφθαλμῶν ἀναρράπτομεν, ἑτέρα δὲ τῇδε μηδὲν ἐοικυῖα, δι' ἧς ὑποχύματα παράγομεν, ἄλλαι δὲ τρίτη καὶ τετάρτη μήτ' ἀλλήλαις τι μήτε ταῖσδε προσεοικυῖαι, καθ' ἃς ὀστοῦν κατεαγὸς ἐκ μὲν τῆς κεφαλῆς ἐκκόπτομεν, ἐν ἄλλῳ δὲ μέρει τοῦ σώματος, οἷον βραχίονι καὶ πήχει, κατατείναντές τε καὶ διαπλάσαντες ἐπιδοῦμεν. ὧν ἁπασῶν ἐνεργειῶν ἀποκεχώρηκε πάμπολυ κήλης χειρουργία καὶ ταύτης ἡ τῶν κιρσῶν καὶ πασῶν ὁμοῦ τῶν εἰρημένων ἡ τοῦ κατὰ τὴν κύστιν λίθου. καὶ τούτων μὲν ἔτι τὰ πλεῖστα μετὰ σμίλης ἐνεργοῦμεν· ὑπαλεῖψαι δ' ὀφθαλμὸν ἢ ἄρθρον ἐμβαλεῖν ἢ καταπλάσαι τι μέρος ἢ καθετῆρι χρῆσθαι καλῶς ἢ σικύαν κολλῆσαι γίγνεται μὲν ἄνευ σμίλης, ἀποκεχώρηκε δὲ καὶ ἀλλήλων πάμπολυ καὶ τῶν προειρημένων, ὥσπερ γε καὶ τὸ φλέβα τεμεῖν καὶ ἀρτηρίαν διελεῖν καὶ ἀποσχάσαι τὸ δέρμα καὶ παρακεντῆσαι τοὺς ὑδεριῶντας ἀλλήλων διενήνοχε καὶ συμπάντων τῶν προειρημένων. καὶ αὗται μὲν ἔτι καὶ πρὸς ταύταις ἕτεραι μυρίαι χειρουργίαι τινὲς ἐνερ-
845K γοῦνται δι' αὐτῶν τῶν χειρῶν.[39] τῶν δὲ φαρμακευ-

---

[39] τῶν χειρῶν add. H

better for me, since I promised once and for all to achieve this prize for you, to go over, as briefly as possible, all the things said, beginning again from the individual actions.

23. There is, then, one individual action which we use to lift up the eyelids by suturing, and another, nothing like this, by which we couch cataracts, and others, a third and fourth, not like each other or the previous two, by which we cut away shattered bone from the head, while in another part of the body, such as the upper arm and forearm, we bind [broken bones], having stretched them out and molded them. Surgery for a hernia[15] is far removed from all these actions, and that for varicose veins from this and all those mentioned here, and from surgery for bladder stone. And yet we still perform most of these with a knife. However, applying a salve to an eye, reducing a joint, applying a poultice to some part, using a catheter properly or a cupping glass, and gluing together occur without a knife and are very far removed from each other and from those procedures previously mentioned. So too are cutting a vein, dividing an artery, scarifying the skin and carrying out paracentesis on people with dropsy (ascites); these differ from each other and from all those previously mentioned. And yet these and countless others besides them are surgical procedures carried out by the hands themselves. However, the use of pharmaceuticals is another

844K

845K

---

[15] The term κήλη has several meanings that may be included under the general heading of tumor or swelling. Here it is taken as referring specifically to inguinal (and perhaps femoral) hernias—see, for example, Hippocrates, *Airs, Waters, Places* 7, which has: "Hernias occur particularly in children and varicose veins and ulcers in the lower part of the leg in men."

ὄντων ὅλως ἕτερον τὸ γένος ἀλλήλων τε καὶ προσέτι
τῶν κατὰ χειρουργίαν· οὐδὲν γὰρ ἔοικεν ἑλλεβόρου
πόσις καὶ ἐδωδὴ σιτίων οὐδ' ἀσιτία κλυστῆρι καὶ τρίψις λουτρῷ. τούτων δ' ἐπὶ πλέον ἀποκεχώρηκεν ἕλκος
πλῦναί τε καὶ καθῆραι καὶ φάρμακον ὑγρὸν ἐπιθεῖναι
καὶ αὖθις ξηρὸν αἰωρηθῆναί τε καὶ περιπατῆσαι καὶ
παλαῖσαι καὶ σκαμμωνίαν ἢ μελίκρατον πιεῖν.

οἷς ἅπασιν οὐδὲν ἔοικεν ἡ τῶν σφυγμῶν τε καὶ τῆς
θερμασίας διάγνωσις, καίτοι δι' ἄκρων τῶν χειρῶν
ἐνεργοῦμεν, καθάπερ οἶμαι καὶ ἡ τῶν ἐν τοῖς ἀφ-
ισταμένοις μέρεσι περιεχομένων ὑγρῶν ἢ τῶν ὄγκων
τῶν ὑδερικῶν ἢ γαγγλίων ἢ ἀθερωμάτων ἢ μελικηρί-
δων ἢ στεατωμάτων· οὐ γὰρ ταῖς αὐταῖς ἐπιβολαῖς τε
καὶ κινήσεσι τῶν δακτύλων καὶ μεταγωγαῖς καὶ θλί-
ψεσιν ἀλλ' ἐνίοτε πάμπολυ διαφερούσαις ἐπὶ τῶν εἰρημένων χρώμεθα. κινδυνεύομεν γοῦν ὅλης ἡμέρας
οὐδὲν ἄλλο λέγειν ἢ τῶν ἐνεργειῶν τὰς διαφοράς·
οὕτω πάμπολυ τὸ πλῆθος αὐτῶν ἐστιν.

24. Ἀλλ' ἤδη κἀκ τούτων ἱκανῶς ἄν τις γνωρίζοι,
πηλίκον ἁμαρτάνουσιν, ὅσοι ταῖς κατὰ μέρος ἐνεργείαις ἡγοῦνται χρῆναι διακρίνειν τὰς τέχνας. οὐδεὶς
γοῦν οὕτως ἠλίθιος οὐδ' ἔμπληκτός ἐστιν, ὃς ἀφαιρή-
σεται μὲν τῆς ἰατρικῆς τὰς εἰρημένας ἐνεργείας, ἄλλην δέ τινα καθ' ἑκάστην αὐτῶν ἐπιστήσει τέχνην

class altogether; these differ from each other, and besides, from those things relating to surgery. For there is no similarity between a drink of hellebore and eating meat and grains, or between fasting and the use of a clyster, or between a massage and a bath. Still further removed from these is washing and cleansing an ulcer (wound), applying a moist medication, and then a dry one, and passive exercises, walking around, wrestling, and drinking scammony or melikraton.

Diagnosis of the pulses and of the [bodily] heat is nothing like all these, and yet we still carry it out through the tips of the fingers, just as, I think, we also do [in the diagnosis] of fluids contained in protuberant parts, or swellings (tumors), or dropsies, or ganglia, or *atheromata, melikerides,* or *steatomata.*[16] But we do not use the same applications and movements of the fingers, and transpositions and compressions; they sometimes differ very greatly in the case of the things mentioned. Anyway, we run the risk of talking about nothing else for a whole day apart from the differences of the actions, so great are they in number.

24. But already someone would realize sufficiently 846K from these things, how greatly mistaken those are who think they must distinguish the arts by their individual actions. Anyway, no one is so foolish or stupid that he would take away the actions mentioned from the medical art and set up some other art for each of them—either

---

[16] On *atheromata, melikerides,* and *steatomata,* see Johnston and Horsley, *Galen: Method of Medicine,* 1.cxii–cxxv (Introduction 9 [Diseases and Symptoms]), and 3.513 and 527. See also Paulus Aeginata (trans. Francis Adams), 6.36 and 6.39 (on ganglia).

ἤτοι κηλοτομικήν, ὡς νῦν ὀνομάζουσί τινες, ἢ λιθο-
τομικὴν ἢ παρακεντητικήν. εἰ γὰρ καὶ ὅτι μάλιστα
τόνδε μέν τινα κηλοτόμον εἶναί φασι, παρακεντητὴν[40]
δὲ τόνδε, λιθοτόμον δὲ τόνδε, πάντας γοῦν ἰατροὺς
αὐτοὺς ὀνομάζουσιν, ὥσπερ οἶμαι καὶ τοὺς ἀπὸ μορίων τινῶν ὠνομασμένους, ὧν ἐξαιρέτως προνοοῦνται.
καὶ γὰρ οὖν καὶ τούτους ὀφθαλμικούς τε καὶ ὠτικοὺς
καὶ ὀδοντικοὺς ἰατροὺς ὀνομάζουσιν, ἑτέρους δ' ἀπὸ
τῆς ὕλης προσηγόρευσαν ἤτοι διαιτητικοὺς καὶ φαρ-
μακευτικοὺς ἢ καὶ νὴ Δία βοτανικούς· εἰσὶ δ' οἳ καὶ
οἰνοδότας καὶ ἑλλεβοροδότας ἰατρούς τινας ἐκάλεσαν
ἐκ τοῦ πολλάκις αὐτοὺς θεάσασθαι ταῖς τοιαύταις
ὕλαις χρωμένους.

847K   ἔχουσι γὰρ οἶμαι λογικὰς ἀρχὰς ἅπαντες ἄνθρωποι φύσει καὶ γιγνώσκουσιν οἱ μὲν μᾶλλον, οἱ δ' ἧττον, ὡς ἔστι μέν[41] τι καὶ ταὐτὸν ἐν ταῖς ἐνεργείαις,
ἔστι δέ τι καὶ διαφέρον. οὗ μὲν οὖν ἕνεκα γίγνονται,
ταὐτόν [ὑγεία παρὰ πᾶσιν ὁ σκοπός],[42] ὁ τρόπος δ'
οὐχ εἷς ἁπασῶν, ἀλλ' εἰς ἀριθμὸν ἐξήκει πάμπολυν.
εἰ μὲν δή τις, ὥσπερ καὶ Πλάτων ἑκάστου τὸν τρόπον
τε καὶ καθόλου σκοπὸν μίαν ἐπιστήσας τέχνην, εἶτ'
εἰς εἴδη τε καὶ διαφορὰς τέμνων αὐτὴν αὖθις ἐκείνων
τῶν τμημάτων ἕκαστον ὀνομάζει τέχνην, οὕτω καὶ αὐτὸς ἐθελήσειε διαιτητικήν τινα καὶ φαρμακευτικὴν
καὶ χειρουργικὴν ὀνομάζειν τέχνην, οὐκ ἂν ἔχοιμι
τούτῳ γ' οὐδὲν ἐγκαλεῖν. οὕτω δὲ κἂν εἰ τῶν εἰρη-

herniotomy (as some now call it), or lithotomy, or paracentesis. For even if they say this person is particularly a herniotomist, and this person a paracentesist, and this person a lithotomist, they at least call them all doctors. It is, I think, also the same with respect to those named for certain parts who are considered exceptional [in relation to that part]. They are then called ophthalmic, or otological, or dental doctors, while others are named from the material they use—dieticians, pharmacists or, by the gods, herbalists. And there are some who were called winegiving and hellebore-giving doctors because they were seen to use such materials often.

All people have, I think, rational principles by nature, 847K and they know, some more and some less, that there is something the same in the actions and something different. What is the same is the purpose for which they occur, which is the objective (*scopos*) of health in all instances, although the manner is not the same in all cases, but comes to a considerable number. And just as Plato posited one art with a manner for each and a general objective,[17] and then, dividing this into kinds and *differentiae,* calls each of those divisions an art, so too, should he also have wished to name a dietetic, pharmaceutical and surgical art, I would have no quarrel with this. So even if he were

---

[17] See, for example, Plato, *Republic* I, 346A, and sections 33 and 36 below.

---

[40] κηλοτομικόν εἶναί φασι, παρακεντητικὸν Ku—see H's note, p. 61.   [41] μέν H; ἕν Ku   [42] H has the following note on this inclusion: verba ὑγεία παρὰ πᾶσιν ὁ σκοπός ab interprete addita esse ratus inclusi (p. 62).

293

μένων ἕκαστον τέμνων, οἷον τὴν δίαιταν εἴς τε τὰ
προσφερόμενα καὶ κενούμενα καὶ ποιούμενα καὶ ἔξωθεν προσπίπτοντα, καθ' ἕκαστον αὐτῶν ἰδίαν τινὰ
τέχνην ἐπιστήσειεν, οὐδ' ἂν τούτῳ τι μεμψαίμην. οὐδὲ
γὰρ οὐδ' εἰ πάλιν αὐτὰ ταῦτα κατατέμνοι πολυειδῶς
ἄχρι τῶν κατὰ μέρος, οὐκ ἂν κωλύσαιμι λέγειν αὐτόν,
848K ὡς ἐν τοῖς προσφερομένοις ἄλλη μέν ἐστι τέχνη φαρμάκων, ἄλλη δ' ἐδεσμάτων, ἄλλη δὲ πομάτων, καὶ
καθ' ἕκαστον αὐτῶν ἐδέσματος τοῦδε καὶ τοῦδε καὶ
πόματος τοῦδε καὶ τοῦδε καὶ φαρμάκου τοῦδε καὶ
τοῦδε συγχωρήσαιμ' ἂν ἰδίαν εἶναι λέγειν τὴν τέ-
χνην· εἰ δ' οἴεται ταύτας ἀλλήλων διαφέρειν τὰς τέχνας, ὡς ἀριθμητικήν, εἰ τύχοι, ῥητορικῆς ἢ ταύτην
οἰκοδομικῆς τε καὶ τεκτονικῆς, οὐκ ἂν ἔτι συγχωρή-
σαιμι. ταύταις μὲν γὰρ[43] οὐδείς ἐστι σκοπὸς κοινός,
ἐκείναις δὲ ταῖς ὀλίγον ἔμπροσθεν εἰρημέναις εἰς κοινὸς ἁπάσαις πρόκειται σκοπός, ὑγίεια.

ὥσπερ οὖν ἐπὶ τῆς ῥητορικῆς, μιᾶς οὔσης τέχνης,
ἄλλην μέν τινα προοιμίου, διηγήσεως δ' ἄλλην καὶ
πίστεών γε καὶ τῶν καλουμένων ἐπιλόγων ἄλλην εἶναι συγχωρήσω τέχνην, εἴ μοι μόνον ἕν γε τοῦτο φυλάττοιτο, μιᾶς εἶναι τέχνης αὐτά, εἴτ' εἴδη βούλει
καλεῖν εἴτε μέρη, κατὰ τὸν αὐτὸν οἶμαι τρόπον ἔχει
κἀπὶ τῆς ἰατρικῆς· ἄλλην μέν τινα τέχνην ἐρῶ χει-
ρουργικήν, ἄλλην δὲ διαιτητικήν, ἄλλην δὲ φαρμα-
κευτικήν, εἰ τοῦθ' ἕν μοι φυλάττοιτο μόνον, ὡς εἰς
ταύταις ἁπάσαις σκοπός, δι' ὃν ἀναγκάζονται μιᾶς
849K εἶναι τέχνης μόρια. τὰ γοῦν ἀνομοιότατα συνάγειν

to take each of the divisions mentioned, and set up a specific art for each of these, for example, dividing the dietetic art into those things administered, those things
evacuated, those things done and those things befalling
externally, I would not censure him for this. Nor, if he
were in turn to cut up these same divisions in various ways
as far as the individual activities, would I prevent him saying this—that in those things administered, there is one   848K
art for medications, another for foods, and another for
drinks. And I would agree if he were to speak of a specific
art for each of the foods individually, and for each of the
drinks and each of the medications. If, however, he were
to think that these arts differ from one another, like for
example arithmetic differs from rhetoric, or this from
house building and carpentry, I would not still agree. For
there is no common objective for these arts, whereas for
those mentioned a little earlier, one common objective lies
before them all—health.

Therefore, in the case of rhetoric which is one art, let
us agree there is another art for the preamble (*proem*),
another for the statement of the case, another for the
proofs, and another for the so-called perorations, providing this one thing only is preserved for me; that they are
of one art, whether you wish to call them kinds or parts.
The same form, I think, holds in the case of the medical
art; I shall say there is another, surgical art, and another
dietetic art, and another pharmaceutical art, provided this
alone is preserved for me—that there is one objective for
all these, from which they are compelled to be parts of one   849K

---

43 *post μὲν γὰρ*: οὐδείς ἐστι σκοπὸς κοινός, H; οὐχ εἶς ἐστι
σκοπὸς, Ku

οὗτος φαίνεται καὶ συνδεῖν καὶ ἀναγκάζειν εἰς μίαν
ἅπαντα τέχνην συντελεῖν. ἐκκόπτει τις ἢ ἀποκόπτει
τὸ σεσηπός, ἄλλος δ' ἀνατρέφει τε καὶ σαρκοῖ τὸ
κοῖλον. ἐναντία μὲν οὕτως καὶ πάντη[44] διαφέροντα δόξει τὰ πράγμαθ' ὑπάρχειν. αἵ τε γὰρ ἐνέργειαι διαφέρουσι πολὺ καὶ τὸ γιγνόμενον ὑπ' αὐτῶν ἐναντίον.
ὁ μὲν γὰρ καὶ τῶν ὄντων ἀφαιρεῖ τι, τῷ δ' ἔργον
γεννῆσαί τιν' οὐσίαν οὐκ οὖσαν. ἀλλ' οὐδετέρῳ γε
τοῦτ' αὐτὸ πρόκειται καθ' αὑτό, τῷ μὲν ἐκκόψαι τι, τῷ
δὲ γεννῆσαι, ὥσπερ οὐδὲ καῦσαί τι καὶ τεμεῖν οὐδ'
ἄλλο τῶν πάντων οὐδὲν αὐτὸ δι' ἑαυτὸ μεταχειρίζεταί
τις, ἀλλ' ὡς ἄνευ τούτου τυχεῖν ὑγιείας οὐ δυνάμενος,
ἐφ' ἣν ἐπείγονται μὲν ἅπαντες, οὐ μὴν ταῖς αὐταῖς
ὁδοῖς χρώμενοι.

διὰ μὲν δὴ τὸ κοινὸν, τοῦ σκοποῦ πάντες ἰατροὶ
καλοῦνται, διὰ δὲ τὸ τῆς ἐνεργείας ἢ τῆς ὕλης ἢ τοῦ
μορίου διάφορον ἤτοι χειρουργὸς ἢ φαρμακευτὴς ἢ
ὀφθαλμικός, χειρουργὸς μὲν ἀπὸ τῆς ἐνεργείας, φαρμακευτὴς δ' ἀπὸ τῆς ὕλης, ὀφθαλμικὸς δ' ἀπὸ τοῦ
μορίου. τῷ γὰρ καὶ τὰ μόρια θεραπευόμενα διαφέρειν
ἀλλήλων οὐκ ὀλίγον αὐτούς τε τοὺς θεραπεύοντας
ἐνεργείαις τε καὶ ὕλαις χρῆσθαι διαφόροις οἱ μὲν ἀπὸ
τῶν ἐνεργειῶν, οἱ δ' ἀπὸ τῶν μορίων, οἱ δ' ἀπὸ τῶν
ὑλῶν ὠνομάσθησαν ὀφθαλμικοί τε καὶ χειρουργικοὶ
καὶ φαρμακευταί, κοινῇ δ' ἅπαντες ἀπὸ τοῦ τέλους

850K

---

44 ἐναντία μὲν οὕτως καὶ πάντη H; ἐναντίον μὲν οὕτως καὶ
πάντα Ku

art. Anyway, this seems to bring together these dissimilar things, and to bind and compel them all to contribute to one art. Someone may cut out or cut off what is putrefied, while another may cause the cavity to grow up and become enfleshed. These will seem to be opposites and the matters to be different in every way, for the activities differ greatly, and the outcome from them is opposite. For one takes away what exists, whereas the task of the other is to generate some substance that does not exist. In fact, neither of the two—neither the excision nor the generation—is proposed for its own sake, just as one does not cauterize and cut, or undertake any other of all these procedures for its own sake, but because without it, it is not possible to achieve health. Due to this they are all pressing matters, although they do not use the same paths.

Certainly, all doctors call themselves doctors because of the commonality of the objective (*scopus*), whereas they are either a surgeon, an apothecary, or an ophthalmologist due to the difference of the action, or the material, or the part—a surgeon from the action, an apothecary from the material, and an ophthalmologist from the part. And because the parts being treated differ from one another to no small extent, and those treating use different actions and materials, some are named from the actions, some from the parts, and some from the materials, so there are ophthalmologists, surgeons and apothecaries, while by common consent all are doctors determined by the end

850K

297

ἰατροί. δέδεικται γὰρ ἔμπροσθεν, ὡς ἅπασα τέχνη
περὶ τὸ τῆς ὑποβεβλημένης οὐσίας ἀγαθὸν ἔσπευκεν,
ἔνθ᾽ ἔν[45] τι καθ᾽ ἑκάστην αὐτῶν ἐστι τὸ πρῶτον ἀγαθόν. εἰ δ᾽, ὅτι καλῶς ἀναρράπτων ὅδε τις τὰ βλέφαρα
κακῶς, εἰ τύχοι, φαρμακεύει καί τις ἄριστος ὢν ἐν
φαρμακείᾳ διαιτᾶν ὀρθῶς οὐκ ἐπίσταται καὶ τοῦτ᾽ ἄλλος ἠσκηκὼς οὐκ ἐγυμνάσατο τὰς χεῖρας, ἑτέρας ἀλλήλων πάντῃ τὰς τέχνας ὑποληψόμεθα, πρῶτον μὲν
οὐ τρεῖς μόνον ἀλλὰ καὶ τριακοσίας οὕτω γε ποιήσο-
μεν· ὁ μὲν γάρ τις καθετῆρι χρῆται καλῶς, ὁ δὲ κλυστῆρι, φλεβοτομεῖ δ᾽ ἄλλος, ἀρτηριοτομεῖ δ᾽ ἕτερος·
εἶθ᾽ ὅταν εὑρεθῇ τις ἅπαντα καλῶς ποιῶν, οὕτως αὖ
851K   πάλιν ἅπασαι μία γενήσονται. καὶ μὴν ἑκατέρως μο-
χθηρόν, ἢ διὰ τὴν ἀφυΐαν τῶν τεχνιτῶν εἰς πολλὰς
κατακερματίζειν τὴν μίαν ἢ διὰ τὴν εὐφυΐαν εἰς μίαν
ἀνάγειν τὰς πολλάς. ὁ μὲν γὰρ πρότερος λόγος οὔτε
τὴν ῥητορικὴν ἐάσει μίαν εἶναι τέχνην οὔτε τὴν ἀριθ-
μητικὴν ἢ γεωμετρίαν ἢ μουσικὴν οὐδ᾽ ἄλλην τινὰ
τῶν ἀξιολόγων, ἃς διὰ τὸ μέγεθος οἱ πολλοὶ τῶν με-
ταχειριζομένων ἀδυνατοῦσιν ὅλας ἐκμανθάνειν· ὁ δὲ
δεύτερος εἰς μίαν ἀνάξει τέχνην ἐνίοτε τὰς μηδαμῶς
κοινωνούσας· εἰ γὰρ ὁ αὐτὸς ἄνθρωπος ἀριθμητικός
θ᾽ ἅμα καὶ γραμματικὸς εἴη καὶ φιλόσοφος, ἐξέσται
τινὶ μιᾶς εἶναι νομίζειν ἁπάσας μόρια.

25. Μὴ τοίνυν μήτε τῷ πλήθει τῶν ἐργαζομένων
τὰς τέχνας ἀλλὰ τοῖς προκειμένοις σκοποῖς διαιρώ-

---

[45] See H's note on p. 64 regarding the alternative ἔνθέν τοι.

(*telos*). For it has been shown before that every art has striven for the good of the underlying substance, hence there is one primary good in relation to each of these. If, however, because someone is good at lifting the eyelids with sutures, but is bad at prescribing medications, as may happen, or if someone is excellent at pharmaceutics but doesn't understand regimen correctly, or another is well-versed in this but is not practiced in the use of his hands, we shall assume the arts are different from each other in every way. In this way we shall first make not only three arts but three hundred. It may be that someone uses a catheter well, and someone else a clyster, while another is good at phlebotomy, and another at arteriotomy. Then, when someone is found who does all these things well, all the arts will, in this way, become one again. Either way it is bad—to divide up the one art into many due to the lack of natural skill of the practitioners, or to bring the many arts into one due to the natural skill. For the first line of reasoning would not allow rhetoric to be one art, or arithmetic, geometry or music, or any other of the worthwhile arts, which, because of their magnitude, the majority who practice them are not able to learn in their entirety. The second line of reasoning would bring together those having nothing at all in common into one art, for if a particular man is an arithmetician, and at the same time also a grammarian and a philosopher, it will be possible for someone to think these are all parts of one art. 851K

25. Accordingly, let us not divide the arts on the basis of the number of things done but on the proposed objec-

μεθα μήτ᾽ εἰς τὰς κατὰ μέρος ἐνεργείας ἀποβλέπωμεν ἀλλ᾽ εἰς τὰς καθόλου. κοινὸν γὰρ ἐν ἐπὶ πάσαις ταῖς κατὰ μέρος ἐνεργείαις ἑκάστῃ τῶν τεχνῶν εὑρήσεις τι· διὸ καίτοι πάμπολυ δοκοῦσαι διαφέρειν ὅμως οὐ κεκώλυνται μιᾶς εἶναι τέχνης μόρια. τὸ γοῦν ὑφαίνειν

852K ἱμάτιον οὐδὲν ἄλλ᾽ ἐστὶν ἢ κρόκας στήμοσι διαπλέκειν. ἆρ᾽ οὖν ἕτερόν τι τοῦδε τὸ ῥάπτειν ἐστίν; οὐδαμῶς, ἀλλὰ κἂν τούτῳ διαπλέκονται κρόκαι στήμοσιν. ὡς εἴ γε καὶ κατ᾽ ἀρχὰς εὐθὺς ἐβούλετό τις ἄνευ τῆς ὑφαντικῆς ἐνεργείας ἑτέρῳ τρόπῳ διαπλέξαι τὰς κρόκας τοῖς στήμοσιν, ἤτοι καθ᾽ ὃν νῦν ἠπῶνται τὰ ῥαγέντα τῶν ἱματίων ἢ καθ᾽ ὃν τοὺς ταλάρους ἢ τὰς σπυρίδας ἢ τὰ δίκτυα πλέκουσιν, ἐποίησε μὲν ἂν οὐδὲν ἧττον ἱμάτιον, ἀλλ᾽ ἐν χρόνῳ παμπόλλῳ. στοχαζόμενοι τοίνυν οὐχ ἁπλῶς τοῦ τέλους, ἀλλὰ τοῦ θᾶττον ἐξικέσθαι πρὸς αὐτὸ τὴν διὰ τῶν ἱστῶν ἐπενόησαν ὑφαντικὴν οὐκ ἐν τῷ καθόλου τῶν εἰρημένων διαλλάττουσαν ἀλλ᾽ ἐν τῷ κατὰ μέρος.

οὕτω δὲ κἀπὶ τῆς ἰατρικῆς ἀνάριθμα μὲν οὖν ὡς οὕτω φάναι τὰ κατὰ μέρος, ἀλλὰ τό γε καθόλου πᾶσιν ἔργον κοινόν· ὅ τε γὰρ ἀφαιρῶν τι τοῦ σώματος ὡς περιττὸν ὅ τε προστιθεὶς ὡς λεῖπον ἐν ἄμφω ποιοῦσι καθόλου τὴν κατὰ φύσιν ἐκπορίζοντες τῷ σώματι συμμετρίαν, ἥτις ἦν ὑγίεια· οὕτω δὲ καὶ ὁ θερμαίνων καὶ ὁ ψύχων καὶ ὁ ξηραίνων καὶ ὁ ὑγραίνων.

853K 26. Ὡς γάρ, εἰ καὶ κατ᾽ ἀρχὰς εὐθὺς οἷόν τ᾽ ἦν ἡμῖν γεννῆσαι ζῷον, οὐκ ἄλλῳ μὲν οὖν ἄν τινι λόγῳ τὴν γένεσιν αὐτοῦ τὴν πρώτην, ἄλλῳ δ᾽ ἂν ἐποιούμεθα

tives (*scopos*). And let us not look at the actions individually, but at these in general. For you will discover what is common and singular in each of the arts from all the actions individually. On which account, even though they seem to differ very greatly, nonetheless they are not prevented from being parts of one art. Anyway, to weave a cloak is nothing other than to interlace the horizontal threads (woof) with the vertical threads (warp). Is sewing, then, something other than this? Not at all; in this also, horizontal threads are interwoven with vertical threads. So if someone, right from the start, apart from the weaving action, wished to interweave the horizontal threads with the vertical threads in another way, either in the way torn cloaks are now mended, or in the way people weave wicker baskets, creels or hunting nets, he would have made a cloak no less, but in a much longer time. Therefore, they are not simply aiming at the end (*telos*), but trying to reach this quicker by thinking of weaving with looms the things mentioned, differing not in general but individually.

852K

In this way too, in the case of the medical art, one might then say the individual actions are innumerable, but the task is in fact common to all of them. Thus, someone who removes something from the body as superfluous and someone who adds something that is deficient are both doing the one thing in general, which is providing for the body the balance that accords with nature, and this is health. The same applies to those who heat, cool, dry and moisten.

26. Thus, if it were possible for us to create an animal right from the start, it would not be by any other method than its first generation, nor would it be different if we

853K

τὴν ἐπανόρθωσιν, οὕτως, ἐπεὶ μηδέτερον τούτων αὐτοὶ
πρώτως[46] ποιοῦμεν, ἀλλ᾽ ἡ φύσις ἐστὶν ἡ καὶ δημι-
ουργοῦσα πρώτη τὸ ζῷον καὶ νῦν ἰωμένη νοσοῦν, οὐκ
ἄλλῳ μὲν τρόπῳ[47] σάρκα πρότερον ἐδημιούργησεν,
ἄλλῳ δὲ νῦν, οὐδ᾽ ἄλλῳ μὲν τρόπῳ τὴν τροφὴν ἐπὶ
τῶν κυουμένων, ἄλλῳ δὲ νῦν ἐπισπᾶται,[48] ἀλλὰ καὶ
ταῦθ᾽ ὁμοίως ἐργάζεται καὶ τρέφει καὶ διακρίνει καὶ
ἀποκρίνει καὶ πάνθ᾽ ἁπλῶς εἰπεῖν ὡσαύτως ἐργάζεται
νῦν τε καὶ τότε.

ὅτι δ᾽ ὁ ἰατρὸς τῆς φύσεώς ἐστιν ὑπηρέτης [οὐδὲ
πρώτη τέχνη τις ἐκείνης][49] καὶ ὅτι καλῶς εἴρηται "φύσις ἐξαρκεῖ πάντασιν," ἀλλὰ καὶ ὡς τὰς νόσους
αὐτὴ[50] κρίνει καὶ ὡς εἰσὶν αἱ φύσεις τῶν νούσων ἰατροί, παλαιοῖς ἀνδράσιν αὐτάρκως εἰρημένα, τί ἂν ἔτι
δεοίμην ἐγὼ διέρχεσθαι; τὸ γὰρ εἰς τὰ παρόντα χρήσιμον ἀναμιμνήσκω μόνον, ὡς οὐκ ἄλλης μὲν τέχνης
854K   τὸ ποιεῖν, ἄλλης δ᾽ ἐπανορθοῦσθαι τὸ πεπονθὸς ἀλλὰ
τῆς αὐτῆς· εἰ δὲ καὶ μὴ τῆς αὐτῆς, ἀλλὰ τὴν ἰατρικὴν
ἐδείξαμεν ἔμπροσθεν οὐχ ὥσπερ τὴν ἱματιουργικὴν
εἰς οὐσίαν ἄγουσαν ὃ πρότερον οὐκ ἦν, ἀλλὰ τῇ τὰ
πεπονηκότα τῶν ἱματίων ἐπανορθουμένῃ προσεοι-
κυῖαν. ἐδείξαμεν δὲ καί, ὡς τὸ φυλάττειν ὁτιοῦν διττόν

---

[46] αὐτοὶ πρώτως H; αὐτῶν πρῶτοι Ku

[47] post οὐκ ἄλλῳ μὲν τρόπῳ add. σάρκα πρότερον ἐδημιούργησεν, ἄλλῳ δὲ νῦν, οὐδ᾽ ἄλλῳ μὲν τρόπῳ (H, note,
p. 66—e cod. L restitui)   [48] ἐπισπᾶται H; ἐπίσταται Ku

[49] H has this phrase in parentheses; Ku has ἡ δὲ instead of
οὐδὲ—see H's note, p. 57.   [50] αὐτὴ H; αὐτῶν Ku

made some restoration. So, since we ourselves do neither of these things primarily, but Nature is both the primary creator of the animal and what now cures it if it is diseased, flesh is not created by any other method than that by which it was first created. Nor is the nourishment of the fetus done by any other method than that by which [the animal] is now cared for; rather, these things are done similarly—that is to say, nutrition, separation and elimination. In short everything is done now as it was done then [in the first formation].

That the doctor is a servant of Nature and not some primary deviser of that,[18] and that "Nature is quite sufficient all in all"[19] is well said. But also, that Nature brings the diseases to a crisis, and that our individual natures are the doctors of the diseases have been adequately stated by the ancients, so what do I still need to go over! Let me call to mind only what is useful for present purposes, which is that there is not one art for making and another for restoring what has been adversely affected—it is the same art. If it is not the same, it is as we have shown before that the medical art is not like the art of cloak making, which brings into existence what did not previously exist, but resembles the art which restores cloaks that have been damaged. We have also shown however, that the preservation of anything whatsoever is twofold, and that there is one class

854K

18 There is some doubt about this phrase—see note to Greek text on the facing page.

19 Hippocrates, *Nutriment* 15, LCL 147 (W. H. S. Jones), 346–47.

GALEN

ἔστι καὶ ὡς θάτερον αὐτοῦ γένος, ἐξ οὗπέρ ἐστι καὶ
τὸ προκείμενον ἡμῖν νῦν τὸ ὑγιεινόν, ἐπανορθοῦται
κατὰ βραχὺ τὰ πονοῦντα καὶ ταύτῃ λανθάνει τοὺς
πολλοὺς ὡς ἕτερον ὂν τῷ γένει τοῦ θεραπευτικοῦ.

27. Τὸ δ' οἴεσθαι τέλος εἶναί τινι τοῦτ' αὐτὸ τὸ
ποιεῖν οἰκίαν ἢ ἱμάτιον ἢ σκεῦος ἢ ὑγίειαν, ἀνθρώπων
ἐστὶ μὴ δυναμένων ἀπὸ τοῦ τέλους αὐτοῦ διακρῖναι
τὴν πρὸ τοῦ τέλους ἐνέργειαν· οὐ γὰρ οἰκοδομεῖν οἰκίαν αὐτὸ δὴ τοῦτο τέλος ἐστὶ τῆς οἰκοδομικῆς, ἀλλ'
οἰκία, καθάπερ οὐδ' ὑφαίνειν ἐσθῆτα καὶ ναῦν συμ-
πήττειν καὶ σκίμποδα καὶ τῶν ἄλλων ἕκαστον, ἀλλ'
αὐτὸ τὸ δημιουργηθέν, ὃ δὴ καὶ παυσαμένων τῆς
ἐνεργείας τῶν τεχνιτῶν ἔτι διαμένει. καὶ ταύτῃ γε δι-
ήνεγκαν αἱ ποιητικαὶ τέχναι τῶν μόνον πρακτικῶν,
ὅτι τῶν μὲν πρακτικῶν, ὅταν ἐνεργοῦσαι παύσωνται,
πέπαυται καὶ τὸ τέλος. οὐδὲν γοῦν ἔστι δεῖξαι τῆς
ὀρχηστικῆς παρ' αὐτὴν τὴν ἐνέργειαν, οἷον τῆς τεκτονικῆς τὸν σκίμποδα καὶ τῆς οἰκοδομικῆς τὴν οἰκίαν
καὶ τῆς ἰατρικῆς τὴν ὑγίειαν.

οὐκ οἰκοδόμησις οὖν, ὥσπερ ὄρχησις, ἐστὶ τὸ τέλος τῆς οἰκοδομικῆς, ἀλλ' οἰκία, τῆς ἐνεργείας ἕτερόν
τι· κατὰ ταὐτὰ δὲ καὶ τῆς ὑφαντικῆς οὔθ' ὕφανσις
οὔθ' ὑφαίνειν οὔτ' ἐσθῆτος ποίησις ἢ γένεσις, ἀλλ'
ἐσθής. ὡσαύτως δ' οὐδὲ ποίησις ἢ γένεσις ἢ ἐπανόρθωσις ὑγιείας ἐστὶ τὸ τέλος τῆς ἰατρικῆς ἀλλὰ ταῦτα
μὲν ἐνέργειά τις ἥ γε καθόλου τοῦ τεχνίτου,[51] καθάπερ αἱ κατὰ μέρος ἦσαν ἐν τῷ τέμνειν καὶ καίειν καὶ
κατατείνειν ἄρθρα καὶ κῶλα καὶ διαπλάττειν καὶ ἐπι-

855K

304

which includes the matter now before us—to wit, hygiene. This restores those things that have suffered gradually and by this escapes the notice of the majority. The other exists in the class of the therapeutic.

27. If it is thought that the end (*telos*) for something is the making of this actual thing, whether it be a house, a cloak, an implement or health, this is because people are unable to separate the action preceding the end from the end itself. Thus, in building a house, the actual end is not the building of the house but the house itself, just as it is not weaving garments or constructing a boat or a bed, and each of the other things; it is what is actually created, which in fact is also what still remains when the craftsmen end the action. And the productive arts differ from the 855K practical arts only in this—that when the actions of the practical arts cease, the end also has ceased. Anyway, dancing has nothing like the bed of the carpenter, the house of the builder and the health of the doctor to show—there is nothing apart from the action itself.

Thus, house building is not the end (*telos*) for the housebuilder, like dancing [is for the dancer]; a house, which is something other than the action, is. In the same way too, neither weaving nor to weave, neither the making nor the creation of the garment is the end for the weaver; the garment is. Similarly, neither the making, nor the creation, nor the restoration of health is the end for the doctor; these are the actions of the practitioner in general, just as the individual actions are in cutting and cauter-

---

51 τοῦ τεχνίτου H; τῆς τέχνης Ku

δεῖν καὶ τούτων ἔτ᾽ ἀνωτέρω καὶ γενικώτεραι χειρουρ-
γεῖν καὶ φαρμακεύειν καὶ διαιτᾶν, ἀλλ᾽ ὑγίεια τὸ
τέλος ἐστίν, ὃ καὶ παυσαμένου τῆς ἐνεργείας τοῦ
τεχνίτου δεικνύειν ἔχομεν. οὐδ᾽ οὖν οὐδὲ τὸ φυλάττειν
ὑγίειαν ἢ ἐπανορθοῦσθαι τέλος οὐδ᾽ ὅλως τὸ ὑγιάζειν,
856K ἀλλὰ ταῦτα μὲν ἐνέργεια, τέλος δ᾽ ἡ ὑγίεια.

τὴν δὲ κοινὴν καὶ γενικὴν ἐπὶ πάσαις ταῖς κατὰ
μέρος ἐνεργείαις ἄλλος μὲν ἄλλως ὀνομάζει, δηλοῦσι
δ᾽ ἅπαντες ἓν καὶ ταὐτὸν ὅ θ᾽ ὑγιάζειν λέγων ὅ θ᾽
ὑγίειαν ποιεῖν ὅ τε ποίησιν ὑγιείας ὅ θ᾽ ὑγίανσιν.
ἑτέρῳ δὲ τρόπῳ καὶ οἵδε ταὐτὸν τούτοις λέγουσιν ὅ τε
τὴν ἴασιν τῶν νοσημάτων εἰπὼν ἢ τὴν θεραπείαν ἢ
τὴν θεράπευσιν τῶν νοσούντων ἢ τὸ ἰᾶσθαι νὴ Δί᾽
τὸ τὰς νόσους ἐκκόπτειν ἢ τὸ τὴν ὑγίειαν ἀντεισάγειν
ὅ τε πάντα τὰ δέοντα πράττειν ὅ τε τὰς νοσοποιοὺς
αἰτίας ἐξάγειν. οὐδεὶς γὰρ τούτων τὸ τέλος ἀλλὰ τὴν
πρὸ τοῦ τέλους ἐνέργειαν λέγει, καθάπερ εἰ καὶ πρα-
κτική τις ἦν ἡ τέχνη περὶ τὸ σῶμα παραπλησίως
ὀρχηστική τε καὶ ὑποκριτικῇ.

28. Ἐπεὶ τοίνυν πολυειδῶς ἀποδέδεικται τὸ μὴ δεῖν
κρίνειν τὰς τέχνας ταῖς κατὰ μέρος ἐνεργείαις ἀλλ᾽
εἰς τὰς καθόλου τε καὶ πλησίον τοῦ τέλους ἀνάγειν,
εἶτ᾽ αὐτῶν ἐκείνων πειρᾶσθαι διορίζειν τὸ τέλος, ἑξῆς
857K ἴδωμεν, εἰ καὶ περὶ τὰς ὕλας αὐτῶν ὁμοίως ἔχει.
φαίνεται γοῦν ὕλη τε μία πολλαῖς ὑποβεβλημένη
τέχναις καὶ τέχνη μία παμπόλλαις ὕλαις χρωμένη.
ξύλον μὲν γὰρ ὕλη κοινὴ καὶ ναυπηγῷ καὶ τέκτονι
καὶ μηχανοποιῷ καὶ οἰκοδόμῳ καὶ ἄλλοις μυρίοις καὶ

izing, reducing dislocated joints, and molding and binding limbs. Higher generically than these are surgery, pharmaceutics and regimen. But health is the end. When the action of the practitioner ceases, this is what we have to show. Therefore, the end is not at all to preserve or restore health, or to make healthy—these are actions; the end is 856K health itself.

Different people name differently the common and generic which follows all these actions individually, although they all indicate one and the same thing, saying to make healthy, to create health, the creating of health, and the restoration of health. In another way too, they say the same as these; the cure of diseases, or the therapy or treatment of those who are diseased, or the curing, or by Zeus, or to eradicate the diseases, or to reintroduce health, or doing all that is needed, or removing disease-producing causes. None of these refers to the end but to the action that precedes the end, as if the art were a practical one concerning the body, similar to dancing and acting.

28. Accordingly, since it has been shown in various ways that it is necessary not to distinguish the arts by their individual actions, but to make reference to the actions in general and in close association with the end (*telos*), and then to attempt to distinguish the end from those actions themselves, next, let us look at whether the same applies 857K regarding the materials of the arts. Anyway, it seems that one material may underlie many arts and one art may use very many materials. Thus, wood is a material common to the shipbuilder, carpenter, engineer and housebuilder, and to countless others, while clay is a material of many

πηλὸς δὲ πολλῶν τεχνῶν ἐστιν ὕλη καὶ λίθοι καὶ τἆλλα σύμπαντα. ἰατρικῆς δ᾽ αὖ τέχνης μιᾶς οὔσης ὗλαι μυρίαι, τό τε σῶμα αὐτὸ τὸ τὴν ὑγίειαν δεχόμενον ἐδέσματά τε καὶ πόματα καὶ φάρμακα σύμπαντα καὶ διαιτήματα. ταυτὶ μὲν οὖν ἰατρῶν ὗλαι, τὸ σῶμα δ᾽ ὡς ἐξ οὗ τὸ τέλος ἢ ἐν ᾧ.

δῆλον οὖν, ὡς οὐδ᾽ ἐκ τῶν ὑλῶν χρὴ κρίνειν τὰς τέχνας. οὐδὲ γὰρ εἴ τις ἑτέρας μὲν τὰς κοινὰς ὕλας καὶ ὡς ἂν εἴποι τις κατὰ συμβεβηκός, ἑτέρας δὲ τὰς οἰκείας τε καὶ προσεχεῖς ὑποθέμενος ἑκάστῃ τῶν τεχνῶν, ᾗ μέν ἐστι φυσικὸν σῶμα τὸ ἡμέτερον, ὕλην αὐτὸ φαίη τῆς φυσικῆς ἐπιστήμης ὑπάρχειν, ᾗ δ᾽ ὑγιαστόν[52] ἐστι, τῆς ἰατρικῆς, ᾗ δ᾽ εὐεκτικὸν ἢ εὐεξίας δεκτικόν, τῆς γυμναστικῆς, οὐδ᾽ οὕτως ἑτέρῳ τινὶ

858K   πλὴν τῷ τέλει διακρίνει τὰς τέχνας· τὸ γὰρ ὑγιαστὸν[53] τόδε τι λέγειν σῶμα γιγνώσκοντός ἐστιν ἤδη τὸ τέλος. οὐ μὴν οὐδ᾽ εἰ τοῖς θεωρήμασι διακρίνει τις τὰς τέχνας, ἑτέρωθέν ποθεν ἄρχεται, τῇ δυνάμει τοῦ τέλους· τὰ γὰρ οἰκεῖα τῶν τεχνῶν ἑκάστης θεωρήματα τὸ τέλος ὁρίζει τε καὶ κρίνει. ταῦτα γοῦν οἰκεῖα τῆς τέχνης ἐστίν, ἃ γιγνώσκων τις εἰς τὸ τέλος ὠφελεῖται· καὶ γραμματικῆς δὲ καὶ μουσικῆς καὶ τεκτονικῆς καὶ τῶν ἄλλων ἑκάστης ὅσα μὲν ἂν ᾖ αὐτὸ τὸ τέλος ἄντικρυς ἢ αὐτὸ τὸ βέλτιον ἢ τὸ θᾶττον ἐν αὐτῷ δύνηται παρέχειν, οἰκεῖα τῆς τέχνης ἐστίν, ὅσα δ᾽ οὐδὲν ὠφελεῖ πρὸς τὴν τοῦ τέλους ποίησιν, οὐκ οἰκεῖα.

29. Πανταχόθεν οὖν ὁ λόγος ἐπὶ τὸ τέλος ἔρχεσθαι κελεύει καὶ τούτῳ κρίνειν τὰς τέχνας· διὰ τοῦτο γὰρ

arts, as are stones and all the other things. On the other hand, there are countless materials of the medical art, which is one art: the body itself which receives health, foods, drinks, and all medications and regimens. These very things, then, are materials for doctors, while the body is that from which and in which the end [is realized].

It is clear, then, that we must not distinguish the arts on the basis of the materials—not if someone proposes some materials that are common, and as one might say, contingent, and others that are specific and are attached to each of the arts, and might say our body is a material of the physical science, inasmuch as it is physical; or of the medical art, inasmuch as it is capable of health; or of the gymnastic art, inasmuch as it is receptive of *euektikos* and *euexia*. One does not distinguish the arts in any other way   858K
apart from by the end. For to say this body is something capable of receiving health is to already recognize health as the end. Nor does someone distinguish the arts from the theories, but begins from some other point, taking the potency of the end, for the theories characteristic of each of the arts define and determine the end. At all events, it is these characteristic theories of the art which, when someone knows them, he is benefitted regarding the end. And of grammar, music, carpentry and each of the other arts, those things that are able to provide the end itself directly, or are able to do something better or quicker, are characteristic to the art, whereas those that are of no benefit toward producing the end are not characteristic.

29. In every way, then, the argument directs us to come to the end (*telos*), and uses this to distinguish the arts. It

---

52 ὑγιαστόν H; ὑγιενόν Ku   53 ὑγιαστόν H; ὑγιενόν Ku

ἐν ταῖς ἐνεργείαις ταῖς καθόλου καὶ τοῖς οἰκείοις θεωρήμασι καὶ ταῖς ἀρχαῖς διαλλάττουσιν.⁵⁴ ἐξαλλάτ-
τονται γὰρ οὖν καὶ αἵδε κατὰ τὰ τέλη τῶν τεχνῶν. ἡ
γὰρ ὑγίεια τὸ τέλος ἐστίν, ἐν θερμοῖς δ' αὕτη καὶ
859K ψυχροῖς καὶ ξηροῖς καὶ ὑγροῖς. ἐν τούτοις δηλονότι
καὶ ἡ τῆς ὕλης ἀρχὴ καὶ ἡ τῆς θεωρίας ἀρχὴ περὶ
τὴν τούτου γνῶσιν, ἀλλ' οὐχ ὡσαύτως ἰατρῷ τε καὶ
φυσικῷ. καὶ τοῦτο δ' αὐτὸ τὸ οὐχ ὡσαύτως ἀπὸ τοῦ
τέλους τῶν ἐπιστημῶν ἐγνώσθη τε καὶ διεκρίθη καὶ
προσηκόντως ἀεὶ τὸ τέλος ὁ λόγος ἐξευρίσκει κανόνα
τε καὶ κριτήριον ἁπάντων τῶν κατὰ τὰς τέχνας.

ἡ γάρ τοι σύστασις ἁπάσης τέχνης τὴν ἀρχὴν
ἐντεῦθεν εἴληφεν. οὔτε γὰρ ἰατρικὴν ἄν τις ἔσπευσε
συστήσασθαι, μὴ προποθέσας ὑγιείας, οὔτ' οἰκοδομι-
κήν, εἰ μὴ κἀνταῦθα οἰκίας ὠρέχθη, οὔθ' ὑφαντικήν,
εἰ μὴ πρότερον ἐσθῆτος. ἀλλὰ μὴν καὶ ἡ τὰς τέχνας
ἁπάσας συνιστᾶσα τὴν ἀρχὴν ἐντεῦθεν εἴληφεν. ἐπιδέδεικται δ' ἑτέρωθι, πῶς ἄν τις ὑποκειμένου τοῦ τέλους, τὴν τέχνην αὐτοῦ κατὰ μέθοδον ἐξευρίσκοι, καὶ
χρὴ πρότερον ἐν ἐκείνῳ γυμνάσασθαι τῷ λόγῳ τὸν
ἀκριβῶς θέλοντα τοῖς ἐνεστῶσιν ἀκολουθεῖν. εἴσεται
γὰρ ἐναργῶς ὑπὸ μίαν τε καὶ τὴν αὐτὴν θεωρίαν
ἄμφω πίπτοντα τὰ μόρια τὸ ὑγιεινὸν καὶ τὸ θεραπευτικόν. ἐγὼ δ' ἂν εἰς τόδε τὸ γράμμα πάντα μεταφέρων
860K τὰ δι' ἑτέρων ἐπὶ πλεῖστον ἠκριβωμένα λάθοιμ' ἂν

⁵⁴ διαλλάττουσιν add. H

is because of this, they differ in the actions in general, in the specific theories and in the starting points (principles). Thus, these also change in relation to the ends of the arts. Health is the end and is itself in things that are hot, cold, dry and moist. In these, clearly, are both the principle of 859K the material and the principle of the theory concerning the knowledge of this, but not similarly for the doctor and the natural scientist. However, this itself was not similarly recognized and distinguished from the end (*telos*) of the sciences. Fittingly reason always discovers the end, rule and criterion of all the matters relating to the arts.

Certainly, the constitution of every art has taken the starting point (principle) from that source. Someone would not have striven to construct the medical art, if he didn't have a prior desire for health; nor for housebuilding, if there were not here the desire for a house; nor for weaving, if there were not a preexisting desire for clothes. But also the constructing of all the arts has taken the starting point (principle) from this source. It has been shown elsewhere how someone, having assumed the end, might discover the art of this [end] according to method.[20] And it is necessary for someone who wishes to follow accurately the present matters to first become practiced in that argument. For it will be clear that hygiene and therapeutics are both parts falling under one and the same theory. However, if I were to transfer to this work all things thor- 860K oughly investigated by others to the fullest extent, I would

[20] Presumably *Const. Art. Med.*, I.224–304K; Johnston, *On the Constitution*.

ἐμαυτὸν ὑπὲρ τὰς Μενεμάχου καὶ Μηνοδότου βί-
βλους ἀποτείνας αὐτό.

30. Ἀλλ᾽ ὅτι μὲν ἡ περὶ τὸ τοῦ σώματος ἀγαθὸν
τέχνη μία πάντως ἐστίν, ἔκ τε τῶν εἰρημένων ἔμπροσθεν αὐτάρκως οἶμαι δεδεῖχθαι κἀκ τῶν ἑξῆς ῥηθησομένων οὐδὲν ἧττον δειχθήσεται. οὐ μὴν ὅ τί γε
προσῆκεν ὀνομάζειν αὐτήν, ἤδη πω πέφανται· τάχα
γὰρ οὔτ᾽ ἰατρικὴν οὔτε γυμναστικὴν ἀλλ᾽ ἕτερόν τι.
τάχα δ᾽ οὐκ ἔστιν ὅλως ὄνομα τῆς τέχνης ἐκείνης
ὥσπερ οὐδ᾽ ἄλλων πολλῶν. ἀλλὰ τοῦτο μὲν ὀλίγον
ὕστερον ἐπισκεψόμεθα.

τὸ δ᾽ οὖν ἓν μὲν εἶναι τὸ τοῦ σώματος ἡμῶν ἀγαθόν,
ὅ τί περ ἂν ᾖ, καθάπερ καὶ τῶν ἄλλων ἁπασῶν οὐ-
σιῶν, εἶναι δὲ καὶ τέχνην αὐτοῦ μίαν, εἴπερ τι καὶ
ἄλλο τῶν πάντων, ἀληθὲς εἶναί φημι καὶ τὴν αὐτήν
γε μέθοδον ἀμφοῖν ὑπάρχειν. μιᾶς οὖν οὔσης περὶ τὸ
τοῦ σώματος ἀγαθὸν τέχνης ἤδη σκοπώμεθα σύμ-
861K παντ᾽ αὐτῆς τὰ μόρια· πάντως γάρ που κατοψόμεθα
καὶ τὸ καλούμενον ὑγιεινὸν ἥντιν᾽ ἔχει τὴν δύναμιν.
ἵνα δὲ κἀνταῦθα μεθόδῳ τινὶ ποιώμεθα τὴν τομήν,
ἐπισκεψώμεθα τὸ γένος τῆς περὶ τὸ σῶμα τέχνης. ὅτι
μὲν οὖν οὐκ ἐν αὐτῷ μόνῳ τῷ θεωρῆσαι τὸ τέλος αὐτῆς ἐστιν ὥσπερ τῆς ἀριθμητικῆς τε καὶ ἀστρονομι-

---

21 Menemachus of Aphrodisias (1st–2nd c. AD) was a Meth-
odist doctor mentioned a number of times by Galen. In his *De
methodo medendi,* the latter writes: "Once, when I questioned
him, he went over things at such length and so obscurely that I

find myself unwittingly extending it beyond the books of
Menemachus and Menodotus.[21]

30. But that there is, in all respects, one art concerning
the good of the body has, I think, been shown adequately
from those things said previously,[22] and will no less be
shown from those things that will be said to follow. What,
in fact, it is appropriate to name this art has not up to this
point already been made known. Perhaps it is neither
medical nor gymnastic but something else. Perhaps there
is no name for that art at all, just as there is not for many
others. But we shall consider this a little later.

Thus, there is one good for our body, whatever it may
be, just as there also is for all other existing things, and
there is one art for this, as there also is for everything else.
I say this is true and there is the same method for both.
Therefore, if there is one art concerned with the good of
the body, let us now consider all the parts of this, for at all
events, perhaps we shall see what potency the so-called   861K
[art of] hygiene has. However, so that here too we might
make the division by some method, let us consider the
class of the art concerning the body. It is clear to everyone
that the end (*telos*) of this art is not in theorizing itself
alone, as with arithmetic, astronomy and natural science;

understood nothing of what he said . . ." (X.54K, Johnston and
Horsley, *Galen: Method of Medicine*, 3.84–85). See also EANS,
546–47. Menodotus of Nicomedia (1st c. AD) was an Empiric
doctor also frequently mentioned by Galen—see his *Subfiguratio
Empirica*, and EANS, 549–50.

[22] For a similar discussion see the opening sections to Galen's
Const. Art. Med.; Johnston, *On the Constitution*, sections 1–2.

κῆς καὶ φυσικῆς, ἀλλά τι καὶ πράττει περὶ τὸ σῶμα,
παντὶ δῆλον· ὅτι δ' οὐδ' ἐν αὐτῷ τῷ πράττειν τελευτᾷ
καθάπερ ἡ ὀρχηστική, μηδέν, ὅταν ἐνεργοῦσα παύ-
σηται, δεῖξαι δυναμένη, καὶ τοῦτ' εὔδηλον εἶναι νο-
μίζω. δῆλον οὖν, ὡς ἤτοι τῶν ποιητικῶν ἢ τῶν κτητικῶν ἐστι τεχνῶν, ἐπειδὴ μήτε τῶν θεωρητικῶν εὑρέθη
μήτε τῶν πρακτικῶν. ἀλλ' οὐκ ἔστι τῶν κτητικῶν
ὥσπερ ἡ ἀσπαλιευτική τε καὶ ἡ ἀγκιστρευτικὴ καὶ τὸ
σύμπαν εἰπεῖν ἡ θηρευτική· χειροῦνται γὰρ αὗται τῶν
ὄντων τι καὶ κτῶνται, ποιοῦσι δ' οὐδὲν αὗται πρότερον οὐκ ὄν. ἐκ λοιπῶν οὖν τῶν ποιητικῶν ἐστιν ἡ περὶ
τὸ σῶμα τἀνθρώπου τέχνη.

διττὸν δὲ καὶ τούτων ἐστὶ τὸ ἔργον· ἢ γὰρ ὅλον τι
862K ποιοῦσι πρότερον οὐκ ὂν ἢ κατὰ μέρος ἐπανορθοῦνται
πεπονηκός. ἀλλ' ἡ ζητουμένη νῦν τέχνη ποιεῖν μὲν
ἀδύνατος ὅλον ἀνθρώπου σῶμα, κατὰ μέρη δ' ἐπανορθοῦσθαι δύναται παραπλησίως τῇ τῶν ἱματίων
ἀκεστικῇ καὶ οὐδ' ἐνταῦθα πάντα τὸν αὐτὸν ἐκείνῃ
τρόπον. ἡ γὰρ φύσις οὕτω γε καὶ ποιεῖ τὸ σῶμα καὶ
αὖθις ἐπανορθοῦται κάμνον ὡς ἡ περὶ τὴν ἐσθῆτα
τέχνη. ταύτης δ' ὑπηρετική τίς ἐστιν ἡ νῦν ζητουμένη.
συγχωρείσθω οὖν, ἵν' ὁ λόγος προΐῃ, καλεῖν ἡμᾶς
αὐτὴν ἐπανορθωτικήν. ἀλλ' ἤτοι κατὰ μεγάλα τὴν
ἐπανόρθωσιν ἢ κατὰ σμικρὰ ποιεῖται. τὸ μὲν δὴ κατὰ
μεγάλα μόρια αὐτῆς ἰατρικόν τε καὶ θεραπευτικὸν
ὀνομαζέσθω, τὸ δὲ κατὰ σμικρὰ φυλακτικόν. ἔστι μὲν
οὖν καὶ αὐτοῦ τοῦ κατὰ μεγάλα παμπόλλη τις ἡ δια-
φορά· πᾶν γὰρ ποσὸν εἰς ἀνάριθμόν τι πλῆθος ἐγχω-

314

there is also something to do concerning the body. On the other hand, I think it is also quite clear that the end is not in the doing itself, as it is with dancing. In this, when the activity stops, nothing can be shown. It is obvious, then, that it is either of the productive or acquisitive arts, since it is not to be found among either the theoretical or practical. But it is not among the acquisitive arts, like angling and fishing, and to speak generally, hunting, for these arts capture and acquire some existing thing, whereas they make nothing that did not exist before. From what remains, then, the art concerning the human body is of the productive arts.

The action of these is twofold: they either make some whole thing that did not exist before, or they restore a part   862K that has suffered damage. But the art we are seeking now is not able to make a whole human body. It is, however, able to restore it in parts, similar to a cloak that is fit to be mended, and not here everything in the same manner to that. For Nature in this way both makes the whole body and in turn restores a damaged body, as the art concerning garments does. The art we are now seeking is a servant of Nature. So the discussion may proceed, let us agree to call this a restorative art. But the restoration it makes is either in relation to large or small things. The component of this in relation to large things, let us call medical and therapeutic; the component of this in relation to small things, let us call preservative. The variance of the former, which relates to large things, is very great; it is possible to divide the whole amount into a countless number. But since it is

ρεῖ τέμνειν. ἀλλ' ἐπειδὴ μὴ πρόκειται νῦν ὑπὲρ αὐτῶν
λέγειν, ἀφείσθω μὲν τοῦτο, τεμνέσθω δὲ τὸ κατὰ σμικρά.

δοκεῖ δὲ καὶ τοῦτο τέμνεσθαι διαφοραῖς ἐπισήμοις
τριχῶς. ἢ γὰρ τὸν ἄκρως τε καὶ τελέως ὑγιαίνοντα
863K παραλαβὼν ἐν τούτῳ φυλάττει καὶ τουτὶ μὲν αὐτὸ τὸ
τμῆμα προσαγορεύουσιν εὐεκτικόν· ἕτερον δ' ἀναθρε-
πτικὸν τῶν νενοσηκότων, ὃ δὴ καὶ καλοῦσιν ἔνιοι τῶν
νεωτέρων ἰατρῶν ἀναληπτικόν· ἐν τῷ μέσῳ δ' ἀμφοῖν
ἐστι τὸ καλούμενον ὑπ' αὐτῶν ἰδίως ὑγιεινόν· κοινῶς
γὰρ καὶ τοῦτο φυλακτικὸν ἅμα καὶ ὑγιεινὸν ὀνομάζεται. ταῦτ' οὖν τὰ τρία μόρια τοῦ φυλακτικοῦ τὰ νῦν
εἰρημένα τὸ ἀναληπτικόν, τὸ ὑγιεινόν, τὸ εὐεκτικὸν
ἐστι μὲν τοῦ κατὰ σμικρὰ ἐπανορθωτικοῦ μέρους τῆς
τέχνης, ἀλλήλων δὲ διαφέρει τῷ μᾶλλόν τε καὶ ἧττον·
μᾶλλον μὲν γὰρ κατὰ σμικρὰ τὸ εὐεκτικόν, ἧττον δ'
αὐτοῦ τὸ ὑγιεινόν, ἐκείνου δ' ἧττον κατὰ σμικρὰ τὴν
ἐπανόρθωσιν ποιεῖται τὸ ἀναληπτικόν. ἔνιοι δὲ καὶ
τέταρτον ἐν τούτῳ τῷ φυλακτικῷ τμῆμα προστιθέασι,
τὸ καλούμενον ὑπ' αὐτῶν ἰδίως προφυλακτικόν, ἀντί-
στροφον ἔχον δύναμιν τῷ τοὺς ἐκ νοσημάτων ἄρτι
πεπαυμένους ἀνακομίζοντι. δῆλον δ', ὡς ἄμφω ταῦτ'
ἐπαμφοτερίζει πως τοῖς ἐναντίοις μορίοις ὅλης τῆς
τέχνης, λέγω δὲ τὸ κατὰ μεγάλα καὶ τὸ κατὰ σμικρά.
τὸ μὲν γὰρ θεραπευτικὸν ὅτι κατὰ μεγάλα, πρόδηλον

---

23 The term *analeptic* has transferred to English and is found

not proposed to speak about these now, let this be set aside for the present, and let that relating to small things be divided.

It seems this is divided in a threefold way by distinct divisions. Taking someone at the peak of perfect health and preserving him in this is also the actual division they call *euektic;* another is the nurturing of those who have been diseased, which some of the more recent doctors also call "analeptic" (restorative).[23] Between these two is what they call specifically hygiene. The terms, "preservation" and "hygiene" are commonly jointly applied to this at the same time. Thus, these three parts of the preservative art now spoken of—the analeptic, the hygienic and the *euektic*—are of the part of the art of restoration relating to small things. They differ from each other in terms of more and less. Thus, more in relation to small is the *euektic;* less than this is the hygienic, less than that in relation to small makes the restorative—i.e., the analeptic. Some also add a fourth division in this preservative part; they call this specifically "prophylactic," which is the converse in potency to that which restores those who have just now ceased from being diseased. It is clear, however, that in some way both have an ambiguous status in relation to the opposite parts of the whole art.[24] I speak of that in relation to large things and that in relation to small things, for it is clear to everyone that the therapeutic relates to the large,

863K

---

in a modern English dictionary such as Chambers (11th ed., 2008). The OED (1930) quotes J. Quincy: "Analeptical medicines cherish the nerves, and renew the spirit and strength."

[24] See Galen, *Const. Art. Med.*; Johnston, *On the Constitution*, section 18.

864K παντί, τὸ δ' εὐεκτικόν τε καὶ ὑγιεινὸν ὅτι κατὰ σμικρά, καὶ τοῦτ' εὔδηλον. ἐν τῷ μέσῳ δ' ἀμφοῖν ἐστι τὸ
ἀναληπτικὸν ὀνομαζόμενον καὶ τὸ προφυλακτικόν, ὡς
μὲν πρὸς τὸ ὑγιεινὸν οὐ κατὰ σμικρὰ τὴν ἐπανόρθωσιν ποιούμενον, ὡς δὲ πρὸς τὸ θεραπευτικὸν οὐ κατὰ
μεγάλα. καί μοι δοκοῦσιν οὐδ' ὅσοι ταῦτ' ἄμφω τὰ
μόρια προσαγορεύουσιν οὐδέτερα, κακῶς δοξάζειν.
ἀλλὰ περὶ μὲν τούτων ἕτερος λόγος.

31. Ἐν δὲ τῷ νῦν προκειμένῳ σκέμματι τοῦ φυλακτικοῦ μέρους τῆς περὶ τὸ σῶμα τέχνης ὑποκείσθω
τριττὰ τμήματα, τὸ μὲν ἐπὶ τῶν κατὰ σχέσιν ὑγιαινόντων σωμάτων τεταγμένον, τὸ δ' ἐπὶ τῶν καθ' ἕξιν,
τὸ δ' ἐπὶ τῶν εὐεκτούντων· καὶ τὸ μὲν ἀναληπτικὸν
ὀνομαζέσθω, μηδὲν γὰρ ὑπὲρ ὀνόματος αὐτοῖς νῦν
διαφερώμεθα, τὸ δ' ὑγιεινόν, τὸ δ' εὐεκτικόν, ἐπὶ μὲν
τὸ βέλτιον ἄγοντα τὰ πρότερα δύο, φυλάττον δ' ἐν
αὐτῷ τὸ εὐεκτικόν· καὶ τούτων οὕτως ἐχόντων ἀναμνη-
σθῶμεν τοῦ προκειμένου σκέμματος.

865K ἦν δ', ὡς οἶμαι, τοιόνδε, πότερον ἰατρικῆς ἐστιν ἢ
γυμναστικῆς μόριον τὸ καλούμενον ὑγιεινόν. εὐλόγως
οὖν εὐθὺς ἀπ' ἀρχῆς ἔφαμεν, ὡς εἰς ὁρισμὸν ἐμπίπτει
τὸ σύμπαν ζήτημα. τί ποτε γάρ ἐστιν ἰατρικὴ καὶ τί
ποτε γυμναστικὴ γνόντες οὐκ ἂν ἔτι χαλεπῶς εὕροι-
μεν, ὁποτέρας αὐτῶν ἐστι τὸ ὑγιεινόν. αὐτὸ μὲν γὰρ
ὅ τί ποτ' ἐστὶ τὸ ὑγιεινόν, ἤδη μοι δοκῶ πεφάνθαι
σαφῶς· ἢ γὰρ τὸ φυλακτικὸν ἅπαν ἢ τῶν τούτου μορίων τριῶν ὑπαρχόντων τὸ ἐν τῷ μέσῳ, ὃ περὶ τὴν

while that the *euektic* and hygienic relate to the small is   864K
also quite clear. In between both are the so-called analeptic and prophylactic; as regards the hygienic, the restoration they make does not relate to the small, while as regards the therapeutic, the restoration they make does not
relate to large things. And those who term both these parts
"neither" seem to me not to have conceived of the matter
badly.[25] But there is a different discussion about these.

31. In the subject now before us, which is that of the
preservative part of the art concerning the body, let us
assume three divisions: one relates to healthy bodies disposed according to *schesis* (a temporary or unstable state);
one relates to those disposed according to *hexis* (a permanent or stable state); and one relates to those that are
euektic (in a good state). And let the first be termed *analeptic* (for let us not now differ about names for these); the
second, hygienic; and the third, *euektic*. The first two lead
to what is better; the other preserves the *euektic* state as
it is. And these being as they are, let us remind ourselves
of the subject before us.

This was, as I think, as follows: whether what is called   865K
hygiene (healthfulness) is part of medicine or gymnastics.
We said, with good reason, right at the start that the whole
inquiry comes down to definition. For if we know what
medicine is and what gymnastics is, we should not still
have any difficulty discovering to which of these hygiene
belongs. What hygiene itself is seems to me to have been
clearly explained. It is either preservative as a whole, or
the one in the middle of the three parts existing in this,

[25] This is discussed in Galen's *Ars M.*, sections 1–2 particularly—see Johnston, *On the Constitution*, 160–69.

319

καθ᾽ ἕξιν ὑγίειαν ἀναστρέφεται. ἐμοὶ μὲν οὖν δοκεῖ
λοιπὸν ὑπὲρ ὀνόματος ἡ ζήτησις εἶναι, τὸ δὲ πρᾶγμ᾽
αὐτὸ μηκέτ᾽ ἀμφισβήτησιν ἔχειν, ἔστ᾽ ἂν ὁμολογῆταί
τε καὶ μένῃ τὰ προαποδεδειγμένα·[55] καὶ δὴ διελθὼν
αὐτὰ διὰ κεφαλαίων ἐπὶ τὴν τῶν ὀνομάτων ἐξήγησιν
οὕτω τρέψομαι.

φημὶ δή, καθάπερ ἀμφιεσμάτων ἀγαθῶν καὶ οἰκίας
τε καὶ ὑποδημάτων, οὕτω καὶ τοῦ σώματος εἶναί τινα
τέχνην, ἧς μιᾶς οὔσης δύο ταῦθ᾽ ὑπάρχειν μόρια, δη-
μιουργοῦν μὲν τὸ ἕτερον, ἐπανορθούμενον δὲ θάτερον,
καὶ ταῦθ᾽ ὑπάρχειν ἄμφω τῇ φύσει πρώτως,[56] συνερ-
γάζεσθαι δ᾽ εἰς τὸ ἕτερον αὐτῶν, τὸ ἐπανορθωτικόν,
ἀνθρωπίνην[57] τέχνην, ἧς εἶναι διττὰ μόρια, θεραπευτικὸν μὲν καὶ ἰατρικὸν τὸ ἕτερον, φυλακτικὸν δὲ θάτερον. ὀνομάζεσθαι δὲ τὸ φυλακτικὸν τοῦτο καὶ ὑγιει-
νόν· εἶναι δ᾽ αὐτοῦ τρία τμήματα, τὸ μὲν ἀναληπτικόν,
τὸ δ᾽ εὐεκτικόν, τὸ δ᾽ ὁμωνύμως τῷ παντὶ προσαγο-
ρευόμενον ὑγιεινόν. εἴτ᾽ οὖν τὸ σύμπαν φυλακτικὸν
εἴτε τὸ μόριον αὐτοῦ ζητεῖ τις ὁποτέρας ἐστὶ τέχνης,
ἆρ᾽ ἰατρικῆς ἢ γυμναστικῆς, ὅλον αὐτὸν χρὴ τὸ
πρᾶγμα διελθόντα, καθάπερ ἐγὼ νῦν ἐποίησα, λέγειν
οὕτως ἐφεξῆς, ὡς μιᾶς οὔσης τῆς περὶ τὸ σῶμα τέχνης ἔξεστιν, ὦ τάν,[58] εἰ βούλει, καλεῖν αὐτὴν ἰατρικήν, εἰ δ᾽ οὐκ ἐθέλεις οὕτω, γυμναστικήν· εἰ δ᾽ οὐδὲ
τοῦτο, τὴν ὅλην ἀνώνυμον εἰπὼν εἶναι τὸ μὲν ἕτερον
αὐτῶν τμῆμα κάλει ἰατρικήν, τὸ δ᾽ ἕτερον γυμναστι-

---

[55] προαποδεδειγμένα H; προδεδειγμένα Ku

which comes down to what concerns health in relation to *hexis*. Thus it seems to me that what remains is the inquiry about a name, while the matter itself is no longer in dispute, if those things that have been proven besides are agreed upon and remain so. And now, having gone through these under the chief points, I shall turn to the interpretation of the names.

So then I say, just as there is a certain art of excellent clothes, houses and shoes, in the same way too, there is a certain art of the body, which although it is the one art, exists in these two parts: one is the productive and the 866K other is the restorative. These are both primarily [carried out] by Nature, although a human art contributes to one of them, which is the restorative. There are two parts of this: one is the therapeutic and medical while the other is the preservative. This preservative part is also called hygiene, and of this there are three divisions: the *analeptic,* the *euektic* and what is called homonymously with the whole, the hygienic. If, therefore, one enquires of either the preservative as a whole or a part of this, to which art it belongs, whether to the medical or the gymnastic, one must go through the whole matter, just as I now did, speaking as follows: As there is one art concerning the body, it is possible, my friend, to call this medical if you wish, whereas, if you do not wish to do this, it is possible to call it gymnastic. On the other hand, if this is not the case, and you say the whole art is without a name, you may call one division of these medical and the other gymnastic.

---

56 πρώτως H; πρότερον Ku
57 ἀνθρωπίνην H; ἀνθρωποτέραν Ku
58 *H's note (p. 76) reads:* ὦ τάν L, ὅταν P, *om.* A

κήν· εἰ δὲ μηδ᾽ οὕτως ἐθέλεις, ὑγιεινήν τινα τέχνην,
τὴν ἀντιδιαιρουμένην τῇ θεραπευτικῇ, καὶ ταύτης, εἰ
867K βούλει, μέρη διαιτητικὴν εἰπὲ καὶ γυμναστικήν. εἰ
γὰρ καθ᾽ ἕκαστον ἐξηγήσῃ τῶν ὀνομάτων, οὕτως ἀκο-
λουθήσει σοι τὸ σύμπαν.

32. Ἰώ, φασί τινες ἐπὶ τούτοις, ἀλλ᾽ οὐχ, ὡς ἂν ἐγὼ
φάναι βουληθῶ, διαιρεῖσθαι χρὴ περὶ τῶν ὀνομάτων,
ἀλλ᾽ ὡς ὀρθῶς ἔχει διελθεῖν. αὖθις οὖν ἡμῖν ἐστι καὶ
τούτοις ἀποκριτέον ὡδί πως· εἰ μέλλεις μνημονεύειν,
ὦ οὗτος, ὡς οὐκέτι περὶ πραγμάτων ἡ σκέψις, ἀλλ᾽
ἐξήγησιν ὀνομάτων μοι προβέβληκας,[59] οὐδὲ τοῦτ᾽
ἀναδύομαι. φημὶ δή σοι καθόλου[60] περὶ τῶν ὀνομάτων
οὐ μόνον τούτων ἀλλὰ καὶ τῶν ἄλλων ἁπάντων οὐδὲν
ἔχειν εἰπεῖν σοφόν, ἀλλ᾽ εἴτ᾽ ἐκ τῆς τῶν Ἀσσυρίων εἴη
φωνῆς τοὔνομα, παρὰ τῶν Ἀσσυρίων αὐτῶν μανθάνειν χρῆναι τὸ πρᾶγμα, καθ᾽ οὗ τὸ ὄνομα λέγουσιν,
εἴτ᾽ ἐκ τῆς τῶν Περσῶν ἢ Ἰνδῶν ἢ Ἀράβων ἢ Αἰθιό-
πων ἢ ὅλως ὡντινωνοῦν, ἐκείνων πυνθάνεσθαι. τὸ γὰρ
ὄνομα ῥηθὲν αὐτὸ καθ᾽ αὑτὸ μόνον οὐδὲν ἐνδείκνυται.
τὸ μὲν οὖν ἐμὸν ἀκήκοας.

εἰσὶ δ᾽ οἳ λέγουσιν ἐνδείκνυσθαι σφίσι τοὔνομα καὶ
868K τούτους ἐγὼ πάμπολλα συναθροίσας ὀνόματα Κελ-
τῶν καὶ Θρᾳκῶν καὶ Μυσῶν καὶ Φρυγῶν ἐκέλευον ἐφ᾽
ἑκάστου λέγειν τὸ δηλούμενον πρᾶγμα· τῶν δ᾽ ἐν τῇ
τῶν Ἑλλήνων φωνῇ μόνῃ δύνασθαι τοῦτο ποιεῖν εἰπόντων, ἑξῆς προὔτεινα λιμένα· τῶν δ᾽ εἰπόντων, ἵνα-

But if you do not wish to do this, you may call it a certain
healthy art, distinguished from the therapeutic, saying, if
you wish, a part of this is dietetic and a part gymnastic. If   867K
you were to give an interpretation in relation to each of
the names, the whole matter would fall into place for you.

32. Alas! Some say in response to these things, that it
is not, as I would wish to say, necessary to distinguish
names, but to go over what there is correctly. In turn, then,
it behooves me to answer them as follows: if you are going
to call to mind, my good fellow, that the speculation is no
longer about matters, but you have presented me with an
explanation of names, I do not resile from this. In fact, I
say to you in general, that I not only have nothing wise to
say about these names but also about all the others, apart
from this. If the name is from the language of the Assyrians, it is necessary to learn from the Assyrians themselves
what the matter is in relation to which they say the name.
If it is from the language of the Persians, Indians, Arabs
or Ethiopians, or of any other people whatsoever, it is to
be learned from those people. For the actual name that is
spoken relates only to itself and indicates nothing. And so,
you have heard my opinion.

There are, however, those who say the name does indicate something to them. I have collected together from   868K
these people a large number of names—from the Celts,
Thracians, Mysians and Phrygians—and exhorted each to
state the matter signified. When they said they were only
able to do this in the language of the Greeks, I next presented the word *limēn* (harbor). When they said it was the

---

59 προβέβληκας H; προβέβληκεν Ku
60 καθόλου *add.* H

περ αἱ νῆες ὁρμοῦσιν, ἀλλὰ Θετταλούς γ᾽ ἔφην τὴν
ὑφ᾽ ἡμῶν προσαγορευομένην ἀγορὰν οὕτως ὀνομάζειν. οἱ δ᾽ ἠρνοῦντο τὴν τῶν Θετταλῶν ἐπίστασθαι
διάλεκτον, ὥσπερ οὐκ αὐτὸ δὴ τοῦθ᾽ ὁμολογοῦντες,
ὅπερ ἐξ ἀρχῆς ἐλέγετο, μηδὲν τῶν ὀνομάτων, ἐφ᾽ οὗ
κεῖται πράγματος, ἄλλως δύνασθαι μαθεῖν ἢ παρ᾽ αὐτῶν τῶν θεμένων διδαχθέντα.

πρὸς μὲν δὴ τοὺς οὕτως ἐμπλήκτους ἕτερός μοι
γέγραπται λόγος ἰδίᾳ, τῷ δὲ τὰ πράγματ᾽ αὐτὰ μαθεῖν ἐθέλοντι μάλιστά τε καὶ πρῶτα καὶ τοῦθ᾽ ὡς ἔργον χρηστὸν σπουδάζοντι,[61] τὰ δ᾽ ἐπ᾽ αὐτοῖς ὀνόματα
τῆς πρὸς ἀλλήλους ἕνεκα διαλέκτου μαθεῖν ἐπιπο-
θοῦντι τὴν τῶν Ἑλλήνων ἐξηγήσομαι χρῆσιν οὐδ᾽
οὖν τούτων ἁπάντων ἐπὶ πᾶσιν ὀνόμασιν, τουτὶ μὲν
869K γὰρ ἑρμηνευτικῆς τινός ἐστιν ἢ γραμματικῆς ἐμπει-
ρίας, ἀλλὰ τῶν Ἀττικῶν μὲν μάλιστα, δεύτερον δ᾽ ἤδη
καὶ τῶν Ἰωνικῶν ἔχειν ἐμπειρίαν ὁμολογήσω τινὰ καὶ
τῶν Δωρικῶν ὀνομάτων ὥσπερ γε καὶ τῶν Αἰολικῶν.
ἀλλ᾽ ἐν ταύταις μὲν ταῖς διαλέκτοις ἀγνοεῖν μᾶλλον
ἢ γιγνώσκειν τὰ πλεῖστα, τῆς Ἀτθίδος δ᾽ αὖ γιγνώσκειν τὰ πλείω ἢ ἀγνοεῖν ὁμολογήσαιμ᾽ ἄν. εἰ δὲ βούλει σὺ καὶ περὶ τῶν ὀνομάτων εἰπεῖν, Ὅμηρος μὲν
οὕτω φησίν·

ἰητρὸς γὰρ ἀνὴρ πολλῶν ἀντάξιος ἄλλων ἰούς
τ᾽ ἐκτάμνειν ἐπί τ᾽ ἤπια φάρμακα πάσσειν.

---

[61] post σπουδάζοντι: τὰ δ᾽ ἐπ᾽ H; τῷ τε τὰ ἐπ᾽ Ku—see H's
note, p. 77.

name of the place where ships set out, I said, "but the Thessalians use the term for what we call the *agora.*" They, however, denied any knowledge of the Thessalian dialect, as if not agreeing with what I said at the beginning—that there is nothing from which the matter is established, and it is not possible otherwise to learn what they are applied to apart from being taught by them.

I have written another work specifically for those who are stupid in this way.[26] However, for someone who wishes to learn the matters themselves particularly and primarily, is diligent in his study of this as a useful task, and is desirous as well of learning the names for those things for the sake of discourse with others, I shall relate in detail the usage among the Greeks, although not for all the terms among all the Greeks, for this would be a kind of hermeneutical or grammatical exercise. But I will acknowledge 869K that I have experience of the Attic names particularly, and secondly also of the Ionic, and also some of the Doric names, as also of the Aeolian, but in these dialects, I am more ignorant than knowledgeable regarding the majority of terms. Conversely, I would acknowledge that of the Attic, I know more than I am ignorant of. If you wish also to speak about names, Homer has this to say:

A man who is a healer is equivalent to many others,
cutting out arrows then sprinkling soothing
    remedies.[27]

[26] I am not clear which work Galen is referring to here. It is presumably one of the several works on names listed in sections 11 and 17 of *Libr. Propr.* (XIX.39–45 and 48K).

[27] Homer, *Iliad* 11.514–15.

GALEN

ἑτέρωθι δέ φησι·[62]

φάρμακα, πολλὰ μὲν ἐσθλὰ μεμιγμένα, πολλὰ δὲ
λυγρά·
ἰητρὸς δὲ ἕκαστος ἐπιστάμενος περὶ πάντων

ὡς τῆς ἰατρικῆς τέχνης ἰωμένης τὰ κάμνοντα σώματα
διά τε φαρμάκων καὶ χειρουργίας.

33. Εἰ δ' ἔτι καὶ τρίτον ἄλλο μόριον ἰάσεως ὑπῆρχε
τὸ διαιτητικὸν ἐν τοῖς καθ' Ὅμηρον χρόνοις, ἐγὼ μὲν
οὐκ ἔχω συμβαλεῖν, ὁ δ' ἐμοῦ πρεσβύτερος θ' ἅμα
καὶ τὰ τῶν Ἑλλήνων πράγματα πιθανώτερος ἐπίστασθαι Πλάτων ὁ φιλόσοφος οὐ πάνυ τι χρῆσθαί φησι
τοὺς παλαιοὺς Ἀσκληπιάδας τούτῳ τῷ μέρει τῆς τέχνης. ἀλλ' ὅτι γε τῆς ἰατρικῆς ἐστι μέρη ταῦτα τὰ
τρία καὶ ὡς ἡ τὰ παρὰ φύσιν ἔχοντα σώματα θερα-
πεύουσα τέχνη πρὸς ἁπάντων Ἑλλήνων ἰατρικὴ καλεῖται, σχεδὸν οὐδεὶς ἀντιλέγει. γυμναστικῆς δὲ
τέχνης οὔπω μὲν ἦν τοὔνομα καθ' Ὅμηρον οὐδὲ κα-
λεῖταί τις ὅλως γυμναστής, ὥσπερ ἰατρός, ὅπου γε
καὶ παρὰ Πλάτωνι τὸ μὲν τῆς γυμναστικῆς ὄνομα οὐ
πολλάκις εὑρεῖν ἔστι, παιδοτρίβην μέντοι καλεῖ μᾶλλον ἢ γυμναστὴν τὸν τεχνίτην αὐτῆς. ἤρξατο γὰρ
ὀλίγον ἔμπροσθεν τῶν Πλάτωνος χρόνων ἡ τέχνη
τῶν γυμναστῶν, ὅτε περ καὶ τὸ τῶν ἀθλητῶν ἐπιτήδευμα συνέστη. πάλαι μὲν γὰρ εἷς ἀνὴρ ἐργάτης τῶν
κατὰ φύσιν ἔργων ἀληθῶς εὐεκτικὸς εἰς ἀγῶνα κατα-

870K

---

[62] ἑτέρωθι δέ φησι add. H

326

And elsewhere, he says:

> With medicines many are excellent mixed but many
>   are baneful.
> Each healer knows all about this.[28]

Which is that it is of the medical art to heal damaged bodies through medications and operations.

33. I have nothing to contribute on the question of
whether there was also another, third part of healing, the
dietetic, in Homeric times. However, my elder, Plato the
philosopher who is at the same time quite credible when
it comes to knowing the matters of the Greeks, says the
ancient Asclepiadians did not use this part of the art at all.   870K
But that there are these three parts of the medical art, and
that the art that treats the bodies in a state contrary to
nature is called medical among all Greeks, almost no one
would gainsay. However, in Homer's time, there was not
yet the name of a gymnastic art, nor was anyone at all
called a "gymnastic trainer" as they were a "doctor," and
in fact the term "gymnastics" is not often found anywhere
in Plato; he calls the practitioner of this a "physical trainer"
rather than a "gymnastic trainer."[29] The art of the gymnastic trainers began a little earlier than Plato's times when
they were also involved in the practice of the athletes. For
in ancient times, one man, who was truly in excellent condition (*euektikos*) practiced the activities in accord with

---

[28] Homer, *Odyssey* 4.230–31.

[29] The two terms are παιδοτρίβης (physical trainer, gymnastic trainer—Plato, *Protagoras* 312B) and γυμναστικός (gymnastic practitioner, trainer of athletes—Plato, *Protagoras* 313D).

βαίνων οὐ πάλην μόνον ἀλλὰ καὶ δρόμον ἠγωνίζετο
καί τις ἐνίκα πολλάκις εἰς ἄμφω τε ταῦτα καὶ ἀκοντί-
ζων καὶ τοξεύων καὶ δισκοβολῶν καὶ ἅρματος ἐπι-
ιστατῶν, ὕστερον δὲ διεκρίθη καὶ οἱόνπερ ἕνα πεποίη-
κεν Ὅμηρος Ἐπειὸν εἰς μὲν πάντα τὰ κατὰ φύσιν
ἔργα πάντων ὕστατον, ἀγαθὸν δὲ πυγμήν, ἧς ἐν
871K   ἄθλοις μόνοις ἡ χρεία, τοιοῦτοι πάντες ἐγένοντο, μήτ'
ἀρόσαι μήτε σκάψαι μήθ' ὁδὸν ἀνύσαι μηδ' ἄλλο μηδὲν εἰρηναῖον ἔργον, ἔτι δὲ μᾶλλον πολέμιον ἐργάσασθαι καλῶς δυνάμενοι.

34. Τὴν τούτων τῶν ἀνθρώπων εὐεξίαν ἔμπροσθεν
ἐμεμψάμεθα καὶ τὴν ἐπιστατοῦσαν αὐτοῖς ὁμοίως
γυμναστικήν, ἄλλην δ' εὐεξίαν ἐπηνέσαμεν ἀσφαλῆ
θ' ἅμα καὶ πρὸς τὰς κατὰ φύσιν ἁπάσας ἐνεργείας
χρηστήν, ἧς ὗλαι ποιητικαί θ' ἅμα καὶ φυλακτικαὶ
δύο ἐστόν, ἥ θ' ὑγιεινὴ προσαγορευομένη δίαιτα καὶ
γυμνάσια. τεττάρων γὰρ οὐσῶν τῶν πασῶν ὑλῶν
κατὰ γένος, ὑφ' ὧν ἀλλοιοῦται τὸ σῶμα, χειρουργίας
καὶ φαρμακείας καὶ διαίτης καὶ γυμνασίων, ἡ μία μὲν
ἄχρηστος τοῖς νοσοῦσιν, αἱ δύο δὲ τοῖς κατὰ φύσιν
ἔχουσιν. ὁ μὲν γὰρ νοσῶν εἰς πολλὰ καὶ φαρμάκων
καὶ χειρουργίας καὶ διαίτης δεόμενος εἰς οὐδὲν δεῖται
γυμνασίων, ὁ δ' ὑγιαίνων ἀκριβῶς γυμνασίων μὲν
χρῄζει καὶ διαίτης τινός, οὔτε δὲ φαρμάκων οὔτε χειρουργίας προσδεῖται.

872K   35. Τῆς οὖν ὑγιεινῆς τέχνης μέρος ἐστὶν ἡ γυμναστικὴ καὶ αὐτάρκως ὑπὲρ ἀμφοῖν ἡμᾶς ἐδίδαξεν Ἱπ-

nature, coming down to the arena and contending, not only in wrestling but also in running. And he often triumphed in both these, and also in javelin throwing, archery, hurling the discus and racing war chariots. Later, there was separation and one like Homer's Epeius[30] was created. He was last in all of the activities in accord with nature, but he excelled at boxing, which is only of use in contests. All such people who arose were not able to plow, dig or complete a journey, nor carry out any other peaceful task, much less perform well in battle.

871K

34. I found fault before with the *euexia* of these men and similarly with the gymnastic master in charge of them, although I approved of another, safe *euexia* which is at the same time useful for all actions in accord with nature. There are two materials which are productive and preservative of this—the so-called healthy regimen and exercises. Thus, there are four materials in all according to class by which the body is changed; surgery, medications, regimen and exercise. One of these is useless for those who are sick while two are useless for those who are in accord with nature. Someone who is sick needs, in many instances, medications, surgery and regimen, but has no need at all of exercises. On the other hand, one who is perfectly healthy also needs in addition exercises and some regimen, but neither medications nor surgery.

35. Therefore, the gymnastic art is a part of the hygienic art, and Hippocrates has taught us enough about

872K

---

[30] Epeius was the son of Panopeus—see Homer, *Odyssey* 8.493. For his prowess as a boxer, see Homer, *Iliad* 23.664–99 and 839–40; for his incompetence in other matters—see Homer, *Iliad* 23.839–40.

ποκράτης, ὅσα τε χρὴ γιγνώσκειν ἀέρων τε πέρι καὶ
χωρίων καὶ ὑδάτων καὶ ἀνέμων καὶ ὡρῶν, ὡσαύτως
ἐδεσμάτων τε πέρι καὶ πομάτων καὶ ἐπιτηδευμάτων
ἀκριβέστατα γράψας ἅπαντα· συμπληροῦται γὰρ ἐκ
τούτων ἡ δίαιτα. κατὰ δὲ τὸν αὐτὸν τρόπον ὑπέρ τε
καιροῦ καὶ ποσότητος καὶ ποιότητος οὐ γυμνασίων
μόνον ἀλλὰ καὶ τρίψεως αὐτάρκως διεξῆλθεν. ἔοικε δὲ
Πλάτων ἀπὸ μέρους ὀνομάζειν τὸ σύμπαν μόριον οὐχ
ὑγιεινὴν τέχνην ἀλλὰ γυμναστικὴν προσαγορεύων,
ὅτι τ᾽ ἐξαίρετον ἴδιον ὑπάρχει τοῖς ὑγιαίνουσιν, ὡς ἂν
μηδ᾽ ὅλως αὐτῷ χρωμένων τῶν νοσούντων, ὅτι τε
τοῦτο μόνον ἐνόμιζεν ἐπιστάτου δεῖσθαι. τὸ γὰρ ὑγιαῖνον ἀκριβῶς σῶμα ταῖς κατὰ φύσιν ὀρέξεσι χρώμενον οὔτ᾽ ἐν ταῖς ποιότησιν οὔτ᾽ ἐν τῷ καιρῷ τῆς
χρήσεως ἐξαμάρτοι ἄν τι περὶ τὰ σιτία. περὶ μὲν δὴ
τούτου λόγος ἕτερος. ἀλλ᾽ ὅτι γε καὶ Πλάτων ἑτέραν
μὲν οἶδε τὴν ὡς μέρος τῆς περὶ τὸ σῶμα τέχνης γυμ-
873K ναστικήν, ἑτέραν δὲ τὴν νῦν εὐδοκιμοῦσαν, ἐκ τῶνδ᾽
ἂν μάλιστα μάθοις· ὑπογράψω γάρ σοι τὰς ῥήσεις
αὐτοῦ, πρώτην μὲν τὴν ἐκ Γοργίου τόνδε τὸν τρόπον
ἔχουσαν·

> δυοῖν ὄντοιν τοῖν[63] πραγμάτοιν δύο λέγω τέ-
> χνας· τὴν μὲν ἐπὶ τῇ ψυχῇ πολιτικὴν καλῶ, τὴν
> δ᾽ ἐπὶ τῷ σώματι μίαν μὲν οὕτως ὀνομάσαι οὐκ
> ἔχω σοι, μιᾶς δ᾽ οὔσης τῆς τοῦ σώματος θερα-
> πείας δύο μόρια λέγω, τὴν μὲν γυμναστικήν,
> τὴν δ᾽ ἰατρικήν.

both. He wrote with great accuracy about all those things it is necessary to know of airs, and about places, waters, winds and seasons, and similarly about foods, drinks and practices. Regimen is made up entirely from these. In the same way, he went over sufficiently the time, amount and kind, not only of exercises but also of massage. Plato seems to name the whole from a part, terming the part not the hygienic art but the gymnastic, because this is chosen specifically by those who are healthy, since those who are sick do not use it at all, and because he thought this alone needs an overseer. For the perfectly healthy body, using the appetites in accord with nature is not to be mistaken in the qualities or the time of use of something involving the foods. Of course, there is another discussion about this. But that Plato also thought there was one art, the gymnastic, as part of the art concerning the body, different, however, from that now highly regarded, you may 873K learn particularly from the following. I shall write down for you his statements; first there is that in the following manner from the *Gorgias.*

Since there are two matters, I say there are two arts. The one to do with the soul, I shall call "political," while the one to do with the body, as I do not have a name for you like this, is one for the treatment of the body, which I say has two parts: the gymnastic and the medical.[31]

31 Plato, *Gorgias* 464B, LCL 166 (W. R. M. Lamb), 316–17.

---

63 τοῖν *add.* H

ἐνταῦθα μὲν οὖν ὅτι τε μία τοῦ σώματός ἐστιν ἡ
θεραπευτικὴ τέχνη δύο ἔχουσα τὰ πρῶτα μόρια, σαφῶς ἐδήλωσεν ὁ Πλάτων· ὅτι δὲ τῶν μὲν μορίων
αὐτῆς ἐστιν ὀνόματα, τῆς δ' ὅλης οὐκ ἔστιν ὅτι τε καὶ
πρὸς τὸ βέλτιστον ἀποβλέπει καὶ ὡς ἡ εὐεξία τοῦτ'
ἔστιν, ἐκ τῶν τοιῶνδε φανερόν, ὧν ἕκαστα ἐν ταὐτῷ
βιβλίῳ[64] διῆλθε·

Σῶμά που καλεῖς τι καὶ ψυχήν;

Πῶς γὰρ οὔ;

Οὐκοῦν καὶ τούτων οἴει τινὰ εἶναι ἑκατέρου εὐεξίαν;

Ἔγωγε.

Τί δέ; δοκοῦσαν μὲν εὐεξίαν, οὖσαν δ' οὔ; οἷον
τοιόνδε λέγω· πολλοὶ δοκοῦσιν εὖ ἔχειν τὰ σώματα, οὓς οὐκ ἂν ῥᾳδίως αἴσθοιτό τις, ὅτι οὐκ
874K   εὖ ἔχουσιν, ἄλλος ἢ ἰατρός τε καὶ τῶν γυμναστικῶν τις.

ἔν τε τοῖς ἑξῆς ὅτι τε πρὸς τὸ βέλτιστον μὲν ἀπο-
βλέπουσιν οἶδε, πρὸς δ' αὖ τὸ ἥδιστον ὀψοποιός τε
καὶ κομμωτής, αὐτός γε διεξέρχεται.

36. Τὴν μὲν τῶν τοὺς ἀθλητὰς γυμναζόντων γυμναστικήν, ὑποδυομένην μὲν ὀνόματι σεμνῷ, κακοτεχνίαν δ' οὖσαν, οὔπω μὲν εἰς τοσοῦτον ἀποκεχωρη-

---

[64] βιβλίῳ add. H

Here, then, that Plato clearly showed that there is one therapeutic art for the body which has two primary parts, that there are names for the parts of this art, but there is not for the whole art, that the art directs attention toward what is best, and that this is *euexia* is clear from these things, each of which he went through in the same book.

Do you call something body and something soul?

How could I not?

Therefore, do you also think for each of these is a certain *euexia* (good state)?

I certainly do.

What then? May these seem to be *euexia* when they really are not? For example, I say the following, Many people seem to be in a good state in their bodies, although it would not be easy for someone to perceive that they are not good, other than the doctor and one of the gymnastic trainers.[32]   874K

In what follows, Plato goes on [to show] that the former (i.e., doctors and gymnastic trainers) devote their attention to what is best, while conversely, a cook and a hairdresser go to what is most pleasant.

36. Plato found fault with the gymnastic art of those training athletes, [seeing it as] a base art hiding under a pompous name, which had not yet departed from an ac-

---

[32] Plato, *Gorgias* 463E–464A, LCL 166 (W. R. M. Lamb), 314–17.

κυῖαν τοῦ κατὰ φύσιν, εἰς ὅσον αὐτὴν νῦν προήχασιν,[65] ἤδη δ᾽ ἀρχομένην οὐ πρὸς τὸ βέλτιον ἀποβλέπειν, ἀλλ᾽ ἁπλῶς ὡς ἰσχὺν τῶν ἀντιπάλων καταβλητικὴν ἐν τῷ τρίτῳ τῆς Πολιτείας ἐμέμψατο λέγων ὡδί·

ναὶ μὰ τὸν Δία, ἦ δ᾽ ὅς, σχεδόν γέ τι πάντων μάλιστα ἤ γε περαιτέρω γυμναστικῆς ἡ περιττὴ αὕτη ἐπιμέλεια τοῦ σώματος· καὶ γὰρ πρὸς οἰκονομίας καὶ πρὸς στρατείας καὶ πρὸς ἑδραίους ἐν πόλει ἀρχὰς δύσκολος. τὸ δὲ δὴ μέγιστον, ὅτι καὶ πρὸς μαθήσεις ἀστιναοῦν καὶ ἐννοήσεις τε καὶ μελέτας πρὸς ἑαυτὸν χαλεπή, κεφαλῆς τινας ἀεὶ διατάσεις καὶ ἰλίγγους ὑποπτεύουσα καὶ αἰτιωμένη ἐκ φιλοσοφίας ἐγγίγνεσθαι, ὥστε,[66] ὅπῃ ταύτῃ ἀρετὴ ἀσκεῖται καὶ δοκιμάζεται, πάντῃ ἐμπόδιος· κάμνειν γὰρ οἴεσθαι ποιεῖ ἀεὶ καὶ ὠδίνοντα μήποτε λήγειν περὶ τοῦ σώματος.

875K

ἔτι δὲ σαφέστερον ἐν τοῖς ἑξῆς, ὡς οὐκ ἰσχὺν καταβλητικὴν οἴεται τέλος εἶναι τῆς γυμναστικῆς, ἀλλὰ τὴν πρὸς τὰς κατὰ φύσιν ἐνεργείας χρείαν, ἐνεδείξατο λέγων ὡδί.

αὐτὰ μὴν τὰ γυμνάσια καὶ τοὺς πόνους πρὸς τὸ θυμοειδὲς τῆς φύσεως βλέπων κἀκεῖν᾽ ἐπεγείρων πονήσει μᾶλλον ἢ πρὸς ἰσχύν, οὐχ ὥσπερ οἱ ἄλλοι ἀθληταὶ ῥώμης ἕνεκα σιτία καὶ πόνους μεταχειρίζονται.

[65] προήχασιν H; προσαγηόχασιν Ku

334

cord with nature to the degree it now has, although it was already starting not to look at what is better but simply at the strength to throw opponents in wrestling. In the third book of *The Republic* he writes as follows:

> Yes, indeed, he said, this excessive care for the body that goes beyond simple gymnastics is almost the greatest of all obstacles. For it is troublesome in household affairs, military service and sedentary occupations in a city. And chief of all, it puts difficulties in the way of any kind of learning, thinking and private contemplation of oneself, forever imagining headaches and dizziness and attributing their origin to philosophy. So that, wherever this kind of virtue is practiced and sanctioned, it is in every way a hindrance. For it makes a man always think he is sick and he never stops anguishing about his body.[33]

875K

That Plato does not think strength in wrestling contests is the end (*telos*) of the gymnastic art, but rather the use toward the functions in accord with nature was shown even more clearly in what follows, when he speaks thus:

> And even the actual exercises and toils of gymnastics he will undertake with a view to the spirited part of his nature to arouse that rather than for mere strength, unlike other athletes, who treat food and exertion only as a means to strength.[34]

[33] Plato, *Republic* 407B–C, LCL (Paul Shorey, 1930), 276–79.
[34] Plato, *Republic* 410B, LCL (Paul Shorey, 1930), 286–87.

---

66 *post* ὥστε: ὅπῃ ταύτῃ ἀρετὴ ἀσκεῖται καὶ δοκιμάζεται
H; ὅπῃ αὕτη, ἀρετῇ ἀσκεῖσθαι καὶ δοκιμάζεσθαι Ku

δῆλος οὖν ἐξ ἁπάντων ὁ Πλάτων ἐστὶ τὴν Ἱπποκράτους ἀκριβῶς φυλάττων γνώμην ὑπὲρ τῆς γυμναστικῆς τέχνης ταύτης, ἧς τὸ τέλος ἐστὶν ἡ τῶν ἀθλητῶν εὐεξία· μέμφεται γὰρ αὐτὴν ὡς ἄχρηστον εἰς τὰς
πολιτικὰς πράξεις ἁπάσας, ὅπερ ἑνὶ λόγῳ περιλαβὼν
ἐκεῖνος ὡδί πως ἀπεφήνατο· "διάθεσις ἀθλητικὴ οὐ
φύσει, ἕξις ὑγιεινὴ κρείσσων." ὅτι δὲ καὶ σφαλερὰ
876K πρὸς ὑγίειαν ἡ μάλιστα κατορθωμένη διάθεσις αὐτῶν
ἐστιν, ἐφ᾽ ἣν σπεύδουσι, καὶ ὡς καὶ τοῦτ᾽ ἐγίγνωσκον
Ἱπποκράτης τε καὶ Πλάτων, εἴρηται μὲν καὶ δι᾽ ἄλλων.

37. Οὐ μὴν ἀλλ᾽ ἐπειδήπερ ἅπαξ κατέστην εἰπεῖν
τι περὶ τῆς μοχθηρᾶς εὐεξίας τε καὶ γυμναστικῆς,
ὑπομνησθήσομαι καὶ τούτων ὡς ἕνι μάλιστα διὰ
βραχυτάτων. τῆς ὑγιείας ἐν συμμετρίᾳ τινὶ τεταγμέ-
νης ἀμετρίας ἐστὶ δημιουργὸς ἡ τοιαύτη γυμναστικὴ
πολλὴν καὶ πυκνὴν αὔξουσα σάρκα καὶ πλῆθος αἵματος ὡς ἕνι μάλιστα γλισχροτάτου παρασκευάζουσα.
βούλεται γὰρ οὐ τὴν ἰσχὺν αὐξῆσαι μόνον ἀλλὰ καὶ
τὸν ὄγκον τε καὶ τὸ βάρος τοῦ σώματος, ὥστε καὶ
ταύτῃ χειροῦσθαι τὸν ἀνταγωνιστήν. οὔκουν ἔτι χαλεπὸν οὐδὲν ἐξευρεῖν, ὡς διὰ τοῦτ᾽ ἄχρηστός τε πρὸς
τὰς κατὰ φύσιν ἐνεργείας ἐστὶ καὶ ἄλλως σφαλερά.
ὅπερ γὰρ ἐν ἁπάσαις ταῖς ὄντως τέχναις ἀγώνισμά
ἐστι μέγιστον εἰς ἄκρον ἀφικέσθαι τοῦ σκοποῦ, τοῦτ᾽
ἐν ταύτῃ μοχθηρότατόν ἐστιν, ὡς οὐ φυσικήν τινα
877K διάθεσιν, ἀλλ᾽, ὡς Ἱπποκράτης ἔλεγεν, οὐ φύσει
κατασκευαζούσῃ.

336

It is clear, then, from all these things, that Plato holds exactly the same opinion as Hippocrates on this gymnastic art, of which the end is the *euexia* of the athletes. For he finds fault with this as being useless for all political activities. This is understood from the brief statement Hippocrates made, to wit: "An athletic condition is not natural; a healthy state (*hexis*) is better."[35] That the successfully attained condition of these, which is what they strive for, is particularly dangerous to health was also known to both Hippocrates and Plato, as stated in other writings.

876K

37. Since I have come to speak definitively about nothing other than bad *euexia* and the gymnastic art, I shall mention these things as briefly as possible. Although health is established in a certain balance, such a gymnastic art is the producer of an imbalance, increasing flesh in amount and thickness, and providing an abundance of blood which is, in a word, very viscid. For it would fain increase not only the strength, but also the bulk and weight of the body, and as a means of subduing the opponent. It is not, therefore, difficult to discover that, because of this, it is useless for the functions that accord with nature and dangerous in other ways. What one strives for in all the true arts is to reach the highest peak of the objective (*scopos*); in this art this is very bad, as it does not provide some natural condition but, as Hippocrates said, one that is unnatural.

877K

[35] Hippocrates, *Nutriment* 34, LCL 147 (W. H. S. Jones), 352–53.

τὰ μὲν γὰρ τῆς φύσεως ἀγαθὰ προϊόντα τε καὶ
ἐπιδιδόντα καὶ αὐξανόμενα γίγνεται βελτίω, τὰ δ' οὐ
φύσει πάντα τοσούτῳ χαλεπώτερα ὅσῳ καὶ μείζω.
ὅθεν ἄφωνοί τινες αὐτῶν ἐξαίφνης, ἕτεροι δ' ἀναίσθη-
τοι καὶ ἀκίνητοι καὶ τελέως ἀπόπληκτοι γίγνονται τοῦ
παρὰ φύσιν ὄγκου τοῦδε καὶ τοῦ πλήθους ἀποσβέσαν-
τός τε τὴν ἔμφυτον θερμασίαν ἐμφράξαντός τε τὰς
διεξόδους τοῦ πνεύματος. ὅσοι δ' ἂν αὐτῶν τὰ πρᾷό-
τατα πάθωσιν, ἀγγεῖον ῥήξαντες ἐμοῦσιν ἢ πτύουσιν
αἷμα. τοὺς μὲν δὴ τῆς τοιαύτης εὐεξίας δημιουργούς,
ὧν ἐστι τὰ θαυμαστὰ ταυτὶ συγγράμματα νῦν ὑπὸ
τῶν τὰ ὦτα κατεαγότων περιφερόμενα, τελέως ἤδη
τοῦδε τοῦ γράμματος ἀποδιοπομπησόμεθα. γιγνώσκεις γὰρ δήπου καὶ σύ, Θρασύβουλε φίλτατε, μηδ'
ἀποκρίσεως αὐτοὺς ἀξιουμένους ὑπ' ἐμοῦ. τί γὰρ ἂν
καὶ πλέον εἴη τοῖς χθὲς μὲν καὶ πρώην πεπαυμένοις
τοῦ παρὰ φύσιν ἐμπίπλασθαί τε καὶ κοιμᾶσθαι, τόλ-
μης δ' εἰς τοσοῦτον ἥκουσιν, ὥσθ' ὑπὲρ ὧν[67] οὐδ' οἱ
878K ἱκανῶς ἠσκηκότες τὴν ἀκολούθων τε καὶ μαχομένων
διάγνωσιν ἔχουσιν εὐπετῶς ἀποφήνασθαι, περὶ τούτων ἀναισχύντως διατείνεσθαι; τί μάθοιεν ἂν οὗτοι
βαθὺ καὶ σοφὸν καὶ ἀκριβὲς ἀκούσαντες θεώρημα;
θαυμαστὸν μέντ' ἂν ἦν, εἰ τοῖς μὲν ἐκ παίδων ἀσκου-
μένοις ἐν τοῖς ἀρίστοις μαθήμασιν οὐχ ἅπασιν ὑπάρχει κριταῖς ἀγαθοῖς εἶναι τῆς τοιαύτης θεωρίας, ὅσοι
δ' ἀσκοῦνται μέν, ὥστ' ἐν ἄθλοις νικᾶν, ἀφυεῖς δ'
ὄντες κἀκεῖ στεφάνων μὲν ἠτύχησαν, ἐξαίφνης δ'

The good things of nature become better when they go
forward, advance and increase, whereas all those things
that are unnatural become more problematic the bigger
they are. From the latter, some people suddenly become
aphonic, while others become anesthetic and immobile,
and completely apoplectic due to this unnatural mass, the
excessive quenching of the innate heat and obstruction of
the outflow channels of the *pneuma*. The most mild of
these things they may suffer is that when a vessel is rup-
tured, they vomit and expectorate blood. We shall now, in
this account, completely set aside the producers of such
euexia, whose remarkable books are presently carried
around by those with damaged hearing. For you too, my
dearest Thrasybulus, surely know that it is not worth my
while to answer them. What more would there be for those
who just yesterday or the day before have left off stuffing
themselves with food contrary to nature and sleeping, but
come to such a degree of arrogance that about these things
in which they are insufficiently practiced, they are ready 878K
to state the recognition of consequences and conflicts?
What would these men learn, if they heard a theory that
was profound, wise and accurate? Indeed, it would be
remarkable if those trained from childhood in the best
disciplines were not all the best judges of such a theory,
whereas those trained so they triumph in contests, being
without natural talent, fail to gain crowns of victory, but

---

⁶⁷ *post* ὑπὲρ ὧν: οὐδ' οἱ ἱκανῶς ἠσκηκότες τὴν ἀκολούθων
H; οὐδὲ τοὺς ἱκανῶς ἠσκηκότας ἀκολούθιας Ku

ἀνεφάνησαν γυμνασταί, τούτοις ἄρα μόνοις ὑπάρξει
νοῦς περιττός.

καὶ μὴν ἐγρήγορσις μᾶλλον καὶ φροντὶς οὐκ ἀμα-
θὴς ἢ ὕπνος ὀξὺν τὸν νοῦν ἀπεργάζονται καὶ τοῦτο
πρὸς ἁπάντων σχεδὸν ἀνθρώπων ᾄδεται, διότι πάντων ἐστὶν ἀληθέστατον,[68] ὡς γαστὴρ ἡ παχεῖα τὸν
νοῦν οὐ τίκτει τὸν λεπτόν. ἴσως οὖν ἡ κόνις ἔτι μόνη
σοφίαν αὐτοῖς ἐδωρήσατο. τὸν μὲν γὰρ πηλόν, ἐν ᾧ
πολλάκις ἐκυλινδοῦντο, τίς ὑπολαμβάνει σοφίας εἶναι δημιουργὸν ὁρῶν γε καὶ τοὺς σῦς ἐν αὐτῷ διατρί-
βοντας; ἀλλ᾽ οὐδ᾽ ἐν τοῖς ἀποπάτοις εἰκός, ἐν οἷς δι-
ημέρευον, ἀγχίνοιαν φύεσθαι. καὶ μὴν παρὰ ταῦτ᾽
οὐδὲν ἄλλο πρότερον ἔπραττον· ὅλον γὰρ ἑωρῶμεν
αὐτῶν τὸν βίον ἐν ταύτῃ τῇ περιόδῳ συστρεφόμενον
ἢ ἐσθιόντων ἢ πινόντων ἢ κοιμωμένων ἢ ἀποπατούν-
των ἢ κυλινδουμένων ἐν κόνει τε καὶ πηλῷ.

38. Τούτους οὖν ἀποπέμψαντες, οὐ γὰρ κακοτεχνίας, ἀλλὰ τέχνας ἥκομεν ἐπισκεψόμενοι, τοὺς τῆς
ὄντως γυμναστικῆς ἐπιστήμονας ἤδη καλῶμεν, Ἱπποκράτην τε καὶ Διοκλέα καὶ Πραξαγόραν καὶ Φιλότι-
μον Ἐρασίστρατόν τε καὶ Ἡρόφιλον ὅσοι τ᾽ ἄλλοι
τὴν ὅλην περὶ τὸ σῶμα τέχνην ἐξέμαθον. ἤκουσας

879K

___

[68] post ἀληθέστατον: ὡς γαστὴρ ἡ παχεῖα τὸν νοῦν οὐ
τίκτει τὸν λεπτόν. H; Παχεῖα γαστὴρ λεπτὸν οὐ τίκτει νόον.
Ku—see H's note, p. 85.

suddenly emerge as gymnastic trainers, were the only ones in whom there was prodigious intelligence.

Truly, it is wakefulness and attention more than dullness and sleep that make the mind sharp, and this is pleasant to almost all people because it is truest of all that a full stomach does not beget a fine mind. Perhaps, then, it is only the dust [of the exercise arena] that presented them with wisdom. For who would assume the mud in which they often rolled around to be the producer of wisdom, seeing that pigs also spend their time in this? Nor is it likely that sagacity is generated in the lavatories in which 879K they spend much of the day. And apart from this, they do not customarily do anything else prior, for we are used to seeing them during this period of their lives gathering together, either eating, drinking, sleeping or excreting, or rolling around in dust and mud.

38. Therefore, dismissing these people, for it is not base arts but arts we come to consider, let us call on those who were already truly knowledgeable in the gymnastic art—Hippocrates, Diocles, Praxagoras, Philotimus, Erasistratus and Herophilus,[36] and those others who thoroughly learned the whole art concerning the body. Of

[36] These were all notable men in the history of medicine often referred to by Galen. Only Hippocrates has extant writings. Dates for them all, together with the main collection of fragments for those other than Hippocrates, are as follows: Hippocrates of Cos (ca. 440–370 BC) ; Diocles of Carystus (4th c. BC, van der Eijk, 2000–2001); Praxagoras of Cos (ca. 325–275 BC—Steckerl, 1958); Philotimus (ca. 330–270 BC—Steckerl, 1958); Erasistratus of Cos (fl. 260–240 BC—Garafolo, 1988); Herophilus of Chalcedon (fl. 280–260 BC—von Staden, 1989).

δήπου ἀρτίως Πλάτωνος λέγοντος, ὡς οὐδέν ἐστιν
ἴδιον ὄνομα. μὴ τοίνυν ζήτει ἓν ὄνομα κατὰ πάσης
τῆς περὶ τὸ σῶμα τέχνης, οὐ γὰρ εὑρήσεις· ἀλλ' εἴπερ ποτὲ κατασταίης ὑπὲρ αὐτῆς εἰπεῖν, ἀρκείτω σοι
Πλάτωνα μιμησαμένῳ διελθεῖν, ὡς μιᾶς οὔσης τῆς
τοῦ σώματος θεραπείας δύο μόρια λέγω, τὴν μὲν γυμναστικήν, τὴν δ' ἰατρικήν, ὑγιαινόντων μὲν δηλονότι
880K τὴν γυμναστικήν, νοσούντων δὲ τὴν ἰατρικήν. ἀλλ'
ἐκεῖνο μᾶλλον ἐπισκέψεως ἄξιον, ὅτι μὴ τὴν ὑγιεινὴν
τέχνην ὁ Πλάτων ἀντιδιεῖλε τῇ ἰατρικῇ, καθάπερ
ἐποίησαν οἱ προειρημένοι πάντες ἄνδρες, ὧν, εἰ βούλει, μνημονεύσωμεν ἑνός, ἐπειδὴ καὶ πρόχειρα πᾶσίν
ἐστιν αὐτοῦ τὰ συγγράμματα.

λέγει τοίνυν Ἐρασίστρατος ἐν τῷ προτέρῳ τῶν
Ὑγιεινῶν· "οὐδὲ γὰρ ἰατρὸν εὑρεῖν ἔστιν, ὅστις δέδω-
κεν αὑτὸν εἰς τὴν περὶ τῶν ὑγιεινῶν πραγματείαν."
εἶτ' ἐφεξῆς· "αἱ μὲν γὰρ παρά τι πάθος γιγνόμεναι
ἀπεψίαι καὶ αἱ τούτων θεραπεῖαι εἰς ἰατρικὴν καὶ οὐκ
εἰς τὴν τῶν ὑγιεινῶν πραγματείαν πίπτουσιν." εἶτ'
αὖθις ἔτι προελθών· "εἰ δέ τις εἴη φαυλότης περὶ τὸ
σῶμα, δι' ἣν τὰ προσφερόμενα αἰεὶ διαφθαρήσεται,
καὶ οὕτως εἰς τὴν αὐτὴν κακοχυμίαν τοῖς προϋπάρ-
χουσιν ἥξει, τὴν τοιαύτην διάθεσιν ἰατροῦ καὶ οὐχ
ὑγιεινοῦ λύειν." εἶτ' ἐφεξῆς· "τὸ λέγειν περὶ τούτων ἢ
λύειν ταύτας ἰατροῖς καὶ οὐχ ὑγιεινοῖς ἐπιβάλλει."

---

[37] See section 35.

course, we heard just now that Plato says there is no specific name. Accordingly, you should not seek one name relating to the whole art concerning the body, for you will not find it. But if at some time you are appointed to speak about this, let it suffice for you, imitating Plato, to recount [what he said]: "I say there is one treatment of the body which has two parts—the gymnastic and the medical."[37] Obviously the gymnastic is for those who are healthy, while the medical is for those who are sick. But what is more worthy of inquiry is that Plato did not distinguish on logical grounds the hygienic art from the medical, as all those men previously mentioned did, of whom, if you wish, let us call to mind one, since his writings are available to all.

880K

Accordingly, Erasistratus, in the first book of his *Hygiene*,[38] says: "It is impossible to find a doctor who has dedicated himself to the matter of hygiene." And subsequently, "The *apepsias* occurring along with some affection, and the treatments of these, fall to medicine and not to the matter of hygiene." Then, going on still further: "If there is some ill health involving the body, due to which those things administered will always be destroyed, and in this way will come to the same *kakochymia* as those previously existing, it is for a doctor to resolve such a condition and not a hygienist." Then, subsequently: "It is for the doctors to speak about these [conditions] and to resolve them; this should not fall to hygienists."

[38] This work is no longer extant. Scarborough (EANS, 296) writes: ". . . in Erasistratus' *Hygiene* (fragments 115–167, Garofolo) he uses a healthy life-style (regimen) to prevent *plethora,* and mild intervention to restore displaced matter."

φαίνεται γὰρ οὗτος οὐ μόνον ὑγιεινήν τιν᾽ ὀνομά-
881K ζων τέχνην ὁμοίως τοῖς ἄλλοις ἅπασιν, ἀλλὰ καὶ τὸν
τεχνίτην αὐτῆς ὑγιεινόν, ὥσπερ οἶμαι καὶ τὸν τῆς ἰατρικῆς ἰατρόν, ἵνα τῆς περὶ τὸ σῶμα θεραπευτικῆς
τέχνης, ἧς οὐδὲν ἦν ὄνομα τοῖς Ἕλλησιν ἴδιον, εἰς
δύο τὰ πρῶτα τμηθείσης, ὥσπερ αὐτὰς τὰς τέχνας
ἰατρικήν τε καὶ ὑγιεινήν, οὕτω καὶ τοὺς τεχνίτας αὐτῶν ὀνομάζωμεν ὑγιεινόν τε καὶ ἰατρόν· ὡσαύτως δὲ
καὶ ἄλλοι πολλοὶ τῶν ἰατρῶν ἐχρήσαντο τοῖς ὀνόμασιν.

39. Ἀλλ᾽, ὡς ἔοικεν, οὔπω κατὰ Πλάτωνα σύνηθες
ἦν τοῖς Ἕλλησιν οὔθ᾽ ὑγιεινὴν τὴν τέχνην οὔτε τὸν
ἐπιστήμονα καλεῖν ὑγιεινόν· οὐδὲ γὰρ οὐδὲ τὸ τοῦ
Ἱπποκράτους ὑγιεινὸν ὑγιεινόν, ἀλλὰ τὸ μέν τι περὶ
διαίτης ἐπιγέγραπται, τὸ δέ τι περὶ ὑδάτων τε καὶ
ἀέρων καὶ τόπων. τάχα δ᾽, ὡς ἔφην, οὐδὲ χρείαν εἶναι
διαίτης ὁ Πλάτων ἐνόμιζε τοῖς ὑγιαίνουσιν, ἀλλ᾽ ἀρ-
κεῖν μόνην γυμναστικήν. οὐδὲ γὰρ τὸ ἀναληπτικὸν
μόριον ἢ τὸ προφυλακτικὸν ἴσως ἡγεῖται προσήκειν
τῇ γυμναστικῇ· μέσα γάρ, ὡς ἔφην, αὐτὰ τεταγμένα
δυνατόν ἐστιν, ὁποτέρῳ τις βούλεται, προσνέμειν.
882K ἀλλ᾽ εἰ καὶ ταῦτά τις τοῖς ἰατροῖς ἐπιτρέψειεν, ὀλίγον
ἐστὶ τὸ ὑγιεινὸν ἀέρων τε πέρι καὶ ὑδάτων καὶ χωρίων
καὶ γυμνασίων καὶ σιτίων ἐπισκοπούμενον· καίτοι γ᾽
οὐδὲ περὶ σιτίων ἁπάντων, ἀλλ᾽ ὅσα τοῖς ὑγιαίνουσιν
ἁρμόττει μόνον. τάχα δ᾽, ἐπειδὴ καὶ τὰ τῶν ἀέρων τε
καὶ χωρίων καὶ ὑδάτων ἔφθανεν ἤδη γιγνώσκεσθαι
τοῖς ἰατροῖς, ὡς ἂν καὶ πρώτου τοῦ θεραπευτικοῦ

It appears that Erasistratus not only named something the art of hygiene similar to all the others, but also named 881K the practitioner of this art a hygienist, just as, I think, he named the practitioner of the medical art a doctor, so that, of the therapeutic art concerning the body, for which there was no specific name among the Greeks, there is a division into two primary parts. And just as we name these same arts medicine and hygiene, so too we name their practitioners hygienist and doctor. Many other doctors similarly used these names.

39. But, as it seems, it was not yet customary in Plato's time for the Greeks to call the art, "hygiene" or the one skilled in it, "hygienist." Nor was the hygiene of Hippocrates called hygiene. Rather, he wrote about regimen, and about waters, airs and places. But perhaps, as I said, Plato didn't think there was need of regimen for those who are healthy; the gymnastic art alone suffices. For he perhaps did not believe the *analeptic* (restorative) part or the prophylactic belong to the gymnastic. As I said, it is possible to place this in an intermediate position, assigning it to one or other, as one might wish. But if someone will hand these over to doctors, there is little for the hygienist apart from 882K the consideration of airs, waters, places, exercises and foods, and indeed, not of all foods but only those that are suitable for the healthy. Perhaps however, since the matters of airs, places and waters were already recognized beforehand by doctors as also comprising the primary

μορίου συστάντος, οὐδὲν ἔτι προσεπιμαθεῖν αὐτῶν
ἐνέλιπεν εἰς τὸ μὴ μόνον ἰατροῖς ἀλλὰ καὶ ὑγιεινοῖς
γενέσθαι πλὴν τῆς περὶ τὰ γυμνάσια τέχνης. οὔκουν
οὐδ' ἀπεικὸς οὐδ' ἀπὸ μορίου[69] τοῦ προσαγορευθέντος
ὀνομασθῆναι τὸ σύμπαν. ἀλλὰ τοῦτο μὲν ὑπὲρ ὀνόματός ἐστιν, οὐ περὶ πράγματος σκοπεῖσθαι.

40. Καὶ εἴ τις, ὅπερ καὶ πρόσθεν εἶπον, αὐτὸ δὴ
τοῦτο γιγνώσκων, ὅτι μὴ καὶ περὶ πράγματος ὁ λόγος ἀλλ' ὀνόματός ἐστιν, ἐμὲ κελεύσειεν ἀποφαίνε-
σθαί τι καὶ περὶ τοῦδε, βέλτιον ὀνομάζειν ἐρῶ τοὺς
δύο τὰ πρῶτα μόρια φάσκοντας τῆς ὅλης τέχνης ὑγιεινόν τε καὶ θεραπευτικόν, εἴ γ' ἐν ταῖς μεθόδοις ταῖς
883K διαιρετικαῖς ἐμάθομεν ἀντιδιαιρεῖν ἀλλήλοις ὁμογενῆ
πράγματα. τῷ γάρ, εἰ τύχοι, τῶν ζῴων εἰπόντι τὰ μὲν
ἐναέρια, τὰ δ' ἔνυδρα προσθεῖναι τούτοις τὰ λογικὰ
προφανῶς ἄτοπον. ἡ μὲν γὰρ προτέρα διαίρεσις ἔμ-
πυρά τε καὶ ἔγγεια προσκεῖσθαι ποθεῖ· ταῦτα γὰρ
ἀνάλογόν ἐστι τοῖς ἐναερίοις τε καὶ ἐνύδροις· ἡ δ' αὖ
δευτέρα τοῖς λογικοῖς ἀντιδιαιρεῖσθαι βούλεται τὰ
ἄλογα, κατὰ ταὐτὰ δὲ καὶ τῷ μὲν θνητῷ τὸ ἀθάνατον,
ἀγρίῳ δὲ τὸ ἥμερον καὶ πεζῷ τὸ πτηνὸν καὶ τὸ νη-
κτόν· ἐσχάτη γὰρ ἀτοπία τῶν ζῴων, εἰ τύχοι, τὰ μὲν
ἀθάνατα λέγειν εἶναι, τὰ δὲ πεζά, τὰ δὲ δίποδα.

κατὰ δὲ τὸν αὐτὸν τρόπον ἐπὶ τῆς περὶ τὸ σῶμα
τέχνης εἴ τις ἐθέλοι τέμνων ἐξευρίσκειν τὰ μόρια, τὸ
μὲν τοὺς νοσοῦντας ἰώμενον ἰατρικῆς τέχνης λεχθή-

---

[69] μορίου H; μόνου Ku

therapeutic part, there is nothing of these left still to learn, not only for doctors but also for hygienists, apart from the art concerning exercises. It is not, therefore, unlikely for the whole to be named from what the part is called. But this is to inquire about a name and not about a matter.

40. And, as I also said before, if someone who realized this itself—that the discussion is not also about a matter but is about a name—were to direct me to say something about this too, I shall say it is better to name the two primary parts of the whole art, calling them hygienic and therapeutic, if in fact we learned in the diairetic methods[39] to make a logical division of matters of the same class from each other. For to someone speaking of "air-dwelling" and "water-dwelling" animals, should it so happen, it would clearly be absurd to add to these "rational," for the former division requires the addition of "fire-dwelling" and "earth-dwelling," these being analogous to "air-dwelling" and "water-dwelling," while the second division in turn requires the "irrational" to be divided off from the "rational," and in the same way too, the "immortal" from the "mortal," the "tame" from the "wild" and the "flying" and "swimming" from the "walking." It would be absurd in the extreme, should it so happen, to say of animals, some are "immortal" while some are "terrestrial" and some are "two-footed."

In the same way, in the case of the art concerning the body, if someone should wish to discover the parts by division, it will be said that the healing of those who are dis-

883K

---

[39] *Diairetic* is taken here to be a term of Logic—see Galen, *MM*, X.115K; Johnston and Horsley, *Galen: Method of Medicine*, 1.178–79, and Ammon. in APr. 7.31.

σεται, τὸ δὲ τῶν ὑγιαινόντων προνοούμενον ὑγιεινῆς.
καὶ τούτου γ᾽ ἔτι κάλλιον, ὡς ἡμεῖς ἔμπροσθεν ἐλέγομεν, τῆς περὶ τὸ σῶμα τέχνης ἐπανορθωτικῆς τινος
οὔσης τὸ μὲν κατὰ μεγάλα τὴν ἐπανόρθωσιν ποιούμενον ἰατικόν τε καὶ ἰατρικὸν[70] ὀνομασθήσεται, τὸ δὲ

884K κατὰ σμικρὰ καὶ διὰ τοῦτο λανθάνον, εἰ τὴν ἀρχὴν
ἐπανορθοῦται, φυλακτικόν. ἔτι δὲ κάλλιον ὀνομάσεις
ἀπὸ τῶν ὑλῶν αὐτά, νοσερὸν μὲν τὸ ἕτερον, ὑγιεινὸν
δὲ θάτερον. εἰ δ᾽ αὖ πάλιν ἐθέλοις τέμνειν τὸ ὑγιεινὸν
ἢ φυλακτικόν, εἴρηται γάρ, ὡς ἑκατέρως ἐγχωρεῖ[71]
προσαγορεύειν, ἀπὸ μὲν τῆς ὑποκειμένης ὕλης εἰς
τρία τεμεῖς, ὡς ἔμπροσθεν εἴρηται, τό τ᾽ ἀναληπτικὸν
ὀνομαζόμενον καὶ τὸ κατ᾽ εἶδος ὑγιεινὸν καὶ πρὸς
τούτοις τὸ εὐεκτικόν. ὑποβέβληται γὰρ ἡ ἴδιος ἑκάστῳ τῶν εἰρημένων ὕλη, τῷ μὲν τὸ κατὰ σχέσιν
ὑγιαῖνον σῶμα, τῷ δὲ τὸ καθ᾽ ἕξιν, τῷ τρίτῳ δὲ τὸ
καλούμενον εὐεκτικόν· συμβέβηκε γὰρ ὁμωνύμως λέγεσθαι τοῖς μορίοις τῆς τέχνης αὐτὰ τὰ σώματα.

κατὰ δὲ τὰς ὕλας τῶν βοηθημάτων εἴπερ ἡ τομή
σοι γίγνοιτο, τετραχῇ διαιρήσεις τὴν φυλακτικήν τε
καὶ ὑγιεινὴν ταύτην τέχνην. ἔσται γὰρ αὐτῆς ἡ μὲν
ἐν προσφερομένοις, ἡ δ᾽ ἐν ποιουμένοις, ἡ δ᾽ ἐν κε-
νουμένοις, ἡ δ᾽ ἐν τοῖς ἔξωθεν προσπίπτουσιν ἀνα-
στρεφομένη· διὰ τούτων γὰρ ἡ ὑγίεια φυλάττεται. τὸ
μὲν οὖν ἐν τοῖς προσφερομένοις τῆς ὑγιεινῆς τέχνης

---

[70] ἰατικόν τε καὶ ἰατρικὸν H; ἰατρικόν τε καὶ ἰατρικὴ Ku
[71] ἐγχωρεῖ H; ἐγχωρήσει Ku

eased belongs to the medical art, while the providential
care of those who are healthy belongs to the hygienic art.
And even better than this, as we said before, of the restorative art that exists concerning the body, that which makes
the restoration in relation to large matters will be termed
"healing" and "medical," while that which makes the restoration in relation to small matters that are, because of
this, hidden, if restored from the beginning, will be called   884K
"preservative." Better still will be for you to apply terms
to these from the materials—the diseased in one case and
the healthy in the other. If, on the other hand, you wish to
divide the hygienic or preservative for as we said it is possible to name this either way, dividing it into three parts
from the underlying material, as was said before, they are
termed "analeptic," "health according to kind" and in addition to these, *euektic*. The material underlying each of
those mentioned is specific: the first is the healthy body in
relation to *schesis;* the second is the healthy body in relation to *hexis;* and the third is what is called *euektic,* for the
bodies themselves happen to be named homonymously to
the parts of the art.

If your division occurs in relation to the materials of
the remedies, you will divide this preservative and hygienic art into four parts. The *anastrophe* (reversal)[40] will
be in things administered, things done, things evacuated,
and things befalling externally, for through these things
health is preserved. Thus, the part of the hygienic art that

---

[40] The verb here is ἀναστρέφω—somewhat unusual in a
medical context.

885K μέρος ἐδεσμάτων τ᾽ ἐστὶ καὶ πομάτων ἐπιστήμη τῶν
εἰς ὑγιείας φυλακὴν ἀνηκόντων, τὸ δ᾽ ἐν τοῖς κενουμέ-
νοις ἱδρώτων τε καὶ διαχωρημάτων οὔρων τε καὶ ὅλως
ἁπάντων τῶν ἐκ τοῦ σώματος ἡμῶν ἐκκενοῦσθαι δεομένων, τὸ δὲ τῶν ἔξωθεν προσπιπτόντων, ἀέρος ὕδατος, ἅλμης θαλάττης, ἐλαίου τῶν τ᾽ ἄλλων ἁπάντων
τῶν τοιούτων. ἔτι δὲ τὸ τέταρτόν τε καὶ λοιπὸν ὅπερ
ἐν τοῖς ποιουμένοις ἐτέτακτο μέρος τῆς τέχνης ἐν
γυμνασίοις ἐστὶ καὶ τοῖς καλουμένοις ἐπιτηδεύμασι.
καὶ γὰρ ἐγρήγορσις καὶ ἀγρυπνία καὶ ὕπνος καὶ
ἀφροδίσιά τε καὶ θυμὸς καὶ φροντὶς καὶ λουτρὸν ἐκ
τούτου τοῦ γένους ἐστὶ καὶ χρὴ διαγιγνώσκειν τὸν
ὑγιεινὸν ἑκάστου τῶν εἰρημένων ποσότητά τε καὶ ποιότητα καὶ καιρόν.

41. Πολλοστὸν οὖν μέρος γίγνεται τῆς ὑγιεινῆς
τέχνης ἡ περὶ τῶν γυμνασίων ἐπιστήμη· δικαιότερον
γὰρ οἶμαι καλεῖν ἐστι γυμναστικὴν τέχνην, ἥτις ἂν
ἐπιστήμη τῆς ἐν ἅπασι τοῖς γυμνασίοις ᾖ δυνάμεως,
ὡς τό γε περὶ τῶν κατὰ παλαίστραν ἐπίστασθαι μόνον ἐσχάτως τε[72] μικρόν ἐστι καὶ πᾶσιν ἥκιστα χρή-
886K σιμον. ἐρέσσειν γοῦν ἄμεινόν ἐστι καὶ σκάπτειν καὶ
θερίζειν καὶ ἀκοντίζειν καὶ τρέχειν καὶ πηδᾶν καὶ
ἱππάζεσθαι καὶ κυνηγετεῖν ὁπλομαχεῖν τε καὶ σχίζειν
ξύλα καὶ βαστάζειν καὶ γεωργεῖν ἅπαντά τε τἆλλα
κατὰ φύσιν ἐργάζεσθαι κρεῖττον ἢ κατὰ παλαίστραν
γυμνάζεσθαι. καί σοι πάρεστιν ἤδη σκοπεῖν, οὐ μόνον ὅτι πολλοστόν ἐστι μόριον τῆς ὑγιεινῆς τέχνης ἡ
γυμναστική, ἐλάχιστον δ᾽ αὖ πάλιν ἐστὶ τῆς γυμνα-

350

lies in the things administered is a knowledge of those   885K
foods and drinks connected to the preservation of health.
In the evacuations, on the other hand, it is a knowledge of
sweats, feces, urine, and in general of all those things we
need to evacuate from our bodies. That of the things befalling externally involves air, water, seawater and olive oil,
and all other such things. The fourth, and still remaining
part of the art, which had been placed in the things done,
is in exercises and so-called pursuits. And wakefulness,
insomnia, sleep, sexual activity, anger, anxiety and bathing
are from this class. Also, the quantity, quality and timing
of each of the things mentioned must be recognized in
respect of health.

41. The knowledge of exercises is, then, one small part
of the hygienic art. What I think is more fitting is to call it
the "gymnastic art," which would be knowledge of the
potency in all the exercises, as in fact only a very small part
of these are set up in a wrestling school, and these are
the least useful for all. Anyway, better are rowing, digging,   886K
reaping, javelin throwing, running, leaping, horse riding,
hunting, fighting with heavy arms, chopping wood, lifting
and plowing, and all the other things done naturally; these
are better than exercising in a wrestling school. And now
it is in your power to consider that not only is the gymnastic art a very small part of the hygienic art, but in turn that
which takes place in a wrestling school is the least part of

---

[72] τε H; πω Ku

στικῆς τὸ περὶ τῶν κατὰ παλαίστραν μόριον ὅτι τε
τούτῳ παράκειται τὸ τῶν ἀθλητῶν ἐπιτήδευμα, καὶ
διαθέσεώς τινος ἀπεργαστικὸν οὐ φύσει καὶ κακοτεχνίαν ἐπιστατοῦσαν ἔχον, ὑποδυόμενον μὲν τέλει
χρηστῷ τῆς εὐεξίας, ἄλλο δέ τι[73] μᾶλλον ἢ εὐεξίαν
ἐργαζόμενον.

ἀλλὰ τῆς μὲν τοιαύτης κακοτεχνίας οἱ δυστυχῶς
ἀθλήσαντες ἐξαίφνης ἐπιστήμονες ἀναφαίνονται, τῆς
δ' ὄντως γυμναστικῆς Ἱπποκράτης καὶ ὁ προειρημένος ἅμ' αὐτῷ χορὸς ἐπιστήμων ἐστίν. ὥσπερ δ' ἐν
τοῖς κατ' ἰατρικὴν οὐχ ἅπασιν ὀρθῶς ἅμα εἴρηται
πάντα, κατὰ τὸν αὐτὸν τρόπον οὐδὲ περὶ γυμνασίων.
ἀλλὰ νῦν οὐ πρόκειται διελέγχειν τοὺς κακῶς ὑπὲρ
887K   αὐτῶν ἐγνωκότας, ἀλλ' ἐξηγεῖσθαι τοὔνομα τῆς γυμναστικῆς, ὡς ἔστιν ἐπιστήμη τις ἥδε τῶν ἐν γυμνασίοις δυνάμεων, ὥσπερ ἡ φαρμακευτικὴ τῶν ἐν φαρ-
μάκοις· ἄμφω γὰρ ταῦτα τὰ μόρια τὴν κλῆσιν ἀπὸ
τῶν ὑλῶν ἔσχηκεν ὁμοίως τοῖς ἄλλοις, οἷς ὀλίγον
ἔμπροσθεν εἶπον.

42. Ὧι καὶ δῆλον, ὡς[74] ὅσοι τὴν γυμναστικὴν ἀντι-
διαιροῦσι τῇ ἰατρικῇ, κακῶς γιγνώσκουσιν· ἡ μὲν
γὰρ ἀπὸ τῆς ὕλης, ἡ δ' ἀπὸ τῆς καθόλου τε καὶ γε-
νικῆς ἐνεργείας ὠνόμασται. πρῶται μὲν γάρ εἰσιν αἱ
κατὰ μέρος ἐνέργειαι, καθῆραι δι' ἐλλεβόρου καὶ
σκαμμωνίας, εἰ τύχοι, καὶ τεμεῖν φλέβα καὶ ὀστοῦν
ἐκκόψαι καὶ ἀσιτῆσαι προστάξαι καὶ δοῦναι τροφήν·
ἐπ' αὐταῖς δ' ἕτεραι γενικώτεραί τε καὶ ἤδη καθόλου,
φαρμακεῦσαι καὶ χειρουργῆσαι καὶ διαιτῆσαι καὶ

the gymnastic art, and closely connected to this is the practice of athletes, which is the cause of an unnatural condition and has as a guide a bad art, assuming a final good of *euexia,* although it produces anything but *euexia.*

But those unfortunates who practice athletics suddenly proclaim themselves knowledgeable of such a bad art, whereas Hippocrates and the group previously mentioned along with him[41] are experts in what is truly gymnastics. However, in all the things in relation to medicine, not everything said [by them] is correct, and the same applies to gymnastics. But it is not now proposed to go over those things poorly understood by them, but to interpret the 887K name of the gymnastic art as what is knowledge of the powers in exercises, just as pharmaceutics is knowledge of powers in medications. For both these parts take the name from the materials, similar to the others I spoke of a little earlier.

42. It is, alas, also clear that those who make a logical distinction of the gymnastic art from the medical have a wrong understanding, for one is named from the material and the other from the general and generic action. First, there are the individual actions—for example, purging with hellebore and scammony, venesection, cutting away bone, prescribing fasting and giving nourishment. In addition to these, there are other more generic and general activities—pharmaceutics, surgery and dietetics. What is

41 See section 38 and note 36.

---

73 δέ τι H; δ' ἔτι Ku
74 Ὧι καὶ δῆλον, ὡς H; Καὶ δῆλον ὡς, Ku

τούτων ἁπασῶν κοινή, τὸ ἰάσασθαι, καθάπερ οἶμαι
πάλιν ἐπὶ ταῖς περὶ τὰ ὑγιαίνοντα γιγνομέναις ἐνεργείαις ἁπάσαις κοινὴ ἡ φυλακή. καὶ ὅστις ἀντιδιαι-
ρεῖται τῷ ἰατρικῷ τὸ φυλακτικόν, ὁμογενῶν ποιεῖται

888K τὴν ἀντίθεσιν. ἄμφω γὰρ ἀπὸ τῆς ἐνεργείας ὠνόμασται, καθάπερ αὖ πάλιν ἀπὸ τῆς ὑποκειμένης ὕλης τό
τε νοσερὸν καὶ ὑγιεινόν, ἀπὸ δὲ τῆς τῶν βοηθημάτων
ὕλης τό τε φαρμακευτικὸν ὠνόμασται καὶ τὸ γυμνα-
στικόν· τὸ μὲν γὰρ τοῦ νοσεροῦ τε καὶ ἰατρικοῦ μέρος
ὑπάρχει, τὸ δὲ τοῦ ὑγιεινοῦ τε καὶ φυλακτικοῦ.

43. Προσέχειν δ᾽ ἐνταῦθα τὸν νοῦν χρή, μή πη
λάθωμεν τὸν αὐτὸν ἀποδείξαντες τῷ γυμναστικῷ τὸν
ἐπιστάτην τῶν παλαισμάτων καὶ μόριόν τι ποιήσαν-
τες ὅλης τῆς γυμναστικῆς τὴν ὡς ἂν εἴποι τις παλαι-
στρικήν· ὁ γὰρ ἐπιστάμενος τά τε παλαίσματα σύμπαντα καὶ τῶν τρίψεων ἁπάσας καλῶς ἐργάζεσθαι
τὰς κατὰ μέρος ἐνεργείας ἀνάλογόν ἐστι σιτοποιῷ
καὶ μαγείρῳ καὶ οἰκοδόμῳ, δημιουργεῖν μὲν ἐπιστα-
μένοις ἄρτους τε καὶ ὄψα καὶ οἰκίας, οὐ μὴν ἐπαΐουσί
γ᾽ οὐδὲν οὐδὲ γιγνώσκουσιν, ὅ τί τε χρηστὸν ἐν αὐτοῖς καὶ μὴ χρηστὸν ἥντινά τε δύναμιν ἔχον ἕκαστον
αὐτῶν ἐστι πρὸς ὑγίειαν. ἡ μὲν οὖν προνοουμένη τοῦ
σώματος ἡμῶν τέχνη μία, καθάπερ εἴρηται πολλάκις,

889K αἱ δ᾽ ἄλλαι τὰς ὕλας ταύτῃ παρασκευάζουσιν. οὔτε
γὰρ οἷς ἄμεινόν ἐστιν ὑποδεδέσθαι μὲν καὶ ἀνυπο-
δέτοις εἶναι σκυτοτόμος οἶδεν οὔθ᾽ οἷς ὑποδεδέσθαι
μέν, ἀλλὰ τοῖον ἢ τοῖον ὑπόδημα.

τοῦτο μὲν γὰρ Ἱπποκράτης γιγνώσκει καὶ κελεύει

common to all these is the curing, just as, in turn, I think, what is common to all those actions occurring in regard to healthy bodies is the preserving. And whoever logically distinguishes the preservative from the medical makes the antithesis between things of the same class. For both are named from the activity, just as, on the other hand, diseased and healthy are named from the underlying material, and pharmaceutics is named from the material of the remedies, as is gymnastics; the former is part of the morbid and medical while the latter part of the hygienic and preserving.

43. We must direct our attention here to the question of whether we have not somehow unwittingly shown the overseer of wrestling contests to be the same person as the gymnastic trainer, and made some part of the whole gymnastic art, which someone might call "of the wrestling school." For one who knows all the wrestling holds and does all the individual actions of the massages well is analogous to a baker, cook and housebuilder. These are people who know how to produce bread, food and houses, but neither understand nor know what is useful or useless in these, and what kind of potency each of them has toward health. Therefore, the art that provides for our body is one, as I have often said, while the other arts prepare the materials for this. For a shoemaker is not the person to know for whom it is better to wear shoes or go barefoot, or for those who wear shoes, which shoes to wear.

Hippocrates, however, knows this, and in fact directs a

γε τῷδέ τινι φορεῖν ἀρβύλας τὰς πηλοπάτιδας, οὐ
μὴν αὐτός γε δημιουργήσει τὰς ἀρβύλας, οὐ μᾶλλον
ἢ ὁ στρατηγὸς τὸ κράνος καὶ τὸν θώρακα καὶ τὸ δόρυ
καὶ τὴν ἀσπίδα καὶ τὴν μάχαιραν καὶ τὰς κνημῖδας.
οὐδὲ γὰρ ὁ οἰκοδόμος οἶδεν ὑψηλῆς πέρι καὶ ταπεινῆς
οἰκήσεως οὐδὲ τῆς πρὸς ἀνατολὴν ἢ δύσιν ἐστραμ-
μένης οὐδέ γε τῆς πρὸς βορρᾶν τε καὶ ψυχρᾶς ἢ
νοτίου καὶ θερμῆς οὐδὲ σκοτεινῆς ἢ φωτεινῆς οὐδὲ
καταγαίου τε καὶ ὑπερῴου καὶ νοτερᾶς καὶ ξηρᾶς,
ἀλλ᾽ ὅλως ἀγνοεῖ, τίνα μὲν ὠφέλειαν ἕκαστον τούτων
ἔχει, τίνα δὲ βλάβην, ὥσπερ γε καὶ ὁ σιτοποιὸς οὐκ
οἶδεν, ὅτῳ χρὴ καθαρὸν ἄρτον διδόναι καὶ ὅτῳ συγ-
κομιστὸν οὐδ᾽ ὅτῳ πολὺν ἢ ὀλίγον οὐδ᾽ ἐν ὅτῳ καιρῷ,
κατασκευάζει μέντοι καλῶς αὐτούς. ὡσαύτως δὲ καὶ
ὁ μάγειρος ἢ φακῆν ἢ πτισάνην ἢ χόνδρον ἢ τεῦτλον
ἤ τι τῶν ἄλλων ἐπίσταται σκευάζειν οὐκ εἰδὼς οὐδενὸς αὐτῶν τὴν δύναμιν. αἱ μὲν δὴ τοιαῦται τέχναι
τὰς ἁπάσας ὕλας παρασκευάζουσι τῇ θεραπευτικῇ
τοῦ σώματος, ὥσπερ αὐταῖς ταύταις ἕτεραι, τῷ μὲν
σιτοποιῷ τοὺς πυροὺς ὁ γεωργὸς καὶ κρίβανον ὁ
ἰπνοπλάστης καὶ ὁ ὑλοτόμος τὰ ξύλα καὶ ὁ τέκτων
τὸν ἄβακα, τῷ τέκτονι δ᾽ αὐτῷ πάλιν ἄλλος μὲν
ἀξίνην, ἄλλος δὲ στάθμην, ἄλλος δ᾽ αὖ τὰ ξύλα, καὶ
σκυτοτόμῳ δὲ κατὰ ταὐτὰ τὴν μὲν σμίλην ἡ χαλκευ-
τική, τὰ σκύτη δ᾽ ἡ βυρσοδεψική, τὸν καλόποδα δ᾽ ἡ
τεκτονική· καὶ οἰκοδόμῳ δὲ λατύποι τε καὶ λιθοτόμοι
καὶ πλινθουργοὶ καὶ τέκτονες, οἱ μὲν λίθους, οἱ δὲ
πλίνθους, οἱ δ᾽ ἐπιτήδεια ξύλα προπαρασκευάζουσιν.

890K

particular person to wear half-boots for walking in mud, but will not himself make the half-boots, any more than the general will make the helmet, breastplate, spear, shield, dagger and greaves. The housebuilder does not know whether the house should be on high ground or low, or turned toward the east or west, or toward the north and cold, or toward the south and heat, or dark or light, or underground or above ground, and damp or dry; he is altogether ignorant as to whether each of these has a certain benefit or harm. In the same way, the baker does not know to whom he must give pure bread and to whom a mixed loaf, and to whom a large loaf or a small, or at what time. Nevertheless, he prepares these well. Like- 890K wise, the cook, who knows how to prepare lentils, barley water (ptisan), wheat gruel, beet, or one of the others, knows nothing of their potency. Certainly, such arts prepare all the materials for the therapeutic art of the body, just as other arts do from these same materials: the farmer provides the wheat for the breadmaker; the oven maker, the oven; the woodcutter, the wood; and the carpenter, the wooden board. However, someone else in turn supplies the carpenter himself with an ax, someone else with a line, and someone else with the wood. In the same way, the smith provides the knife for the shoemaker; the tanner's art, the skins; the carpenter's art, the last. And the stonemasons, stonecutters, brickmakers and carpenters provide the housebuilder with stones, bricks and wooden fittings.

αἱ μὲν οὖν τοιαῦται τέχναι σύμπασαι προσδέονταί τ᾽
ἀλλήλων καὶ παρασκευάζουσι ταῖς τελεωτέραις τάς
θ᾽ ὕλας αὐτάς, ἐξ ὧν ἐργάζονται τὸ τέλος, ὄργανά τε
χρηστά, δι᾽ ὧν ἐκείνας κοσμοῦσιν.

44. Ἁπάσαις δ᾽ αὐταῖς ἐφέστηκεν οἷον ἀρχιτεκτονική τις ἡ θεραπευτικὴ τοῦ σώματος καὶ κελεύει τῇ
891K μὲν οἰκοδομικῇ τοιάνδε τινὰ τὴν οἰκίαν ποιῆσαι,[75] τῇ
δὲ σκυτοτομικῇ τὸ ὑπόδημα καὶ τῇ μὲν σιτοποιητικῇ
τὸν ἄρτον, τῇ δ᾽ αὖ μαγειρικῇ τοὔψον ἑκάστῃ τε τῶν
ἄλλων, ὡς ἑκάστη πέφυκε. ἀλλὰ ταύτης τῆς περὶ τὸ
σῶμα θεραπευτικῆς ἡ ὑγιεινὴ μόριον ἦν καὶ ταύτης
τετραχῇ τεμνομένης ἑνὸς τῶν μορίων ἡ γυμναστικὴ[76]
μόριον. αὕτη τοίνυν οἷον ἀρχιτεκτονική τίς ἐστι το-
σῶνδε τεχνῶν, ἱππευτικῆς μέν, ὅταν ἱππάζεσθαι δέῃ,
διδάσκουσα μέτρον καὶ καιρὸν καὶ ποιότητα, κυνηγε-
τικῆς δὲ καὶ συμπάσης θηρευτικῆς, ὅταν αὖ τούτων
ἡ χρεία καλῇ, κατὰ ταὐτὰ δὲ[77] καὶ σκαπτόντων καὶ
θεριζόντων καὶ ὑλοτομούντων καὶ ἐρεσσόντων καὶ ὀρ-
χουμένων καὶ πᾶν ὁτιοῦν ἐνεργούντων ἐπιστατεῖ τε
καὶ ἄρχει καὶ προστάττει.

45. Μία τῶν τοιούτων ἐστὶ τεχνῶν καὶ ἡ περὶ τὰ
παλαίσματα, καλῶμεν δ᾽ αὐτήν, εἰ βούλει, παιδοτρι-
βικήν, αὕτη μὲν οὐδ᾽ ἐπαΐουσα τῶν πρὸς ὠφέλειάν τε
καὶ βλάβην τοῦ σώματος, εὐσχήμονας δὲ καὶ πολυ-
ειδεῖς καὶ καταβλητικὰς καὶ ἀπόνους ἐξευρίσκουσα
892K λαβάς θ᾽ ἅμα καὶ κινήσεις οὐδὲν ἧττον τῆς ὀρχηστικῆς. ἀλλ᾽ ἐκείνη μὲν οὐκ ἀντιποιεῖται τῆς περὶ τὸ
σῶμα θεραπείας· ἡ παιδοτριβικὴ δ᾽ οὐκ οἶδ᾽ ὅπως ὑπ᾽

Thus all such arts have a prior need for others and prepare the materials for the more complete arts, from which they make the end (*telos*), and useful instruments with which they bring order to these.

44. The therapeutic art of the body is set over all the others like a master art which directs the housebuilder to 891K make a particular kind of house, the shoemaker the shoes, the baker the loaf, the cook in turn the food, and each of the others as befits his nature. But the hygienic art is a part of the therapeutic art concerning the body, and when the former is divided four ways, the gymnastic part is one of the parts. Moreover, this is like a master art of many arts: of horsemanship when horse riding is required, teaching the measure, time and quality; and of hunting with hounds, or all forms of hunting, when in turn the need of these calls; and in the same way also of digging, reaping, wood-cutting, rowing and dancing, and every activity whatsoever it is in charge of, rules over and prescribes.

45. One of such arts is also that pertaining to wrestling bouts. Let us call this, if you will, gymnastic training; this does not claim any knowledge of things bringing benefit and harm to the body. No less than dancing, it will discover 892K holds and, at the same time, movements that are graceful, varied, capable of throwing and effortless. But that art does not lay claim to belong to the treatment concerning the body. Gymnastic training (I do not know how—per-

---

75 ποιῆσαι *add.* H
76 ἡ γυμναστικὴ H; ἦν γυμναστικὸν Ku
77 *post* ταὐτὰ δὲ *add.* ταῦτα Ku

ἀναισθησίας ἐσχάτης οἷον ἔμπληκτός τις οἰκέτης
ἐπανίσταται δεσπότῃ χρηστῷ, τῇ γυμναστικῇ, καθάπερ εἰ καί τις ὁπλίτης ἢ ἱππεὺς ἢ τοξότης ἢ σφεν-
δονήτης ἢ ἀκοντιστὴς ἀντιλέγοι τῷ στρατηγῷ παρα-
τάττοντί τε καὶ ὁπλίζοντι καὶ πρὸς τὸν πόλεμον
ἐξάγοντι καὶ αὖθις ἡσυχάζειν κελεύοντι· μαίνοιτο
γὰρ ἄν που κἀκεῖνος, εἰ τὸν στρατηγὸν εἰς τὰς κατὰ
μέρος ἐνεργείας προκαλούμενος, εἶτα βελτίων εὑρι-
σκόμενος ἢ κοινωνεῖν ἀξιοῖ τῆς ἀρχῆς ἢ μέρος ἀπο-
φαίνοι τὴν αὑτοῦ τέχνην τῆς στρατηγικῆς. μαίνοιτο
δ᾽ ἂν οὐδὲν ἧττον, οἶμαι, καὶ ὁ παιδοτρίβης ἢ κοινωνεῖν γυμναστικῆς ἢ μέρος ἔχειν αὐτῆς οἰόμενος· ὑπηρέτης γάρ ἐστι μόνον, ὥσπερ γ᾽ ὁ στρατιώτης τῆς
στρατηγικῆς, οὕτω καὶ ὁ παιδοτρίβης αὐτὸς γάρ, ᾗ
μὲν γυμνάζει, τῆς γυμναστικῆς, ᾗ δ᾽ ἀσκεῖ τέχνην
παλαισμάτων, ἑτέρας αὖ τινός ἐστιν ἐπιτηδεύσεως
ὑπηρέτης, ἣν ἐγὼ μὲν ὀνομάζω καταβλητικήν, οὐ μήν
γ᾽[78] αὐτοὺς οὕτω καλοῦσιν οἱ τοὺς ἀθλητὰς ἀσκοῦν-
τες, ἀλλὰ γυμναστὰς ὀνομάζουσι.

καίτοι κατά γε τὴν ἀλήθειαν ἕτερον μέν ἐστι παιδοτριβικὴ καθάπερ ἡ τοξευτική, καταβλητικὴ δ᾽ οἷόνπερ ἡ στρατηγική, γυμναστικὴ δ᾽ οἷον ἡ ἰατρική, καὶ
διτταῖς ὑπηρετεῖ τέχναις ἐφεστώσαις ἑκάστη τῶν τοιούτων. οὕτω γὰρ καὶ ἡ σκυτοτομικὴ τὸ μὲν ὡς στρα-
τιώτῃ τῷδὲ[79] τινι χρηστὸν ὑπόδημα παρὰ τῆς στρατηγικῆς ἐκδιδάσκεται, τὸ δ᾽ ὡς εἰς ὑγίειαν ἐπιτήδειον
ὑπὸ τῆς περὶ τὸ σῶμα τέχνης ἐργάζεσθαι κελεύεται.
ὡσαύτως δὲ καὶ ἡ μαγειρικὴ τὸ μὲν εἰς ὑγίειαν ὄψον

haps through extreme senselessness) is like some stupid
household slave rising up against a master who is good,
i.e., gymnastic training, or also like some heavily-armed
foot soldier, cavalryman, archer, slinger or javelin thrower
speaking out against the general, arranging and arming his
men, leading them against the enemy, and in turn ordering
them to halt. That man would somehow be mad if, calling
upon the general to perform the actions individually, then
finding himself better, either thought himself worthy to
share the command or declared his art was part of the art
of generalship. It would be no less mad, I think, for the
gymnastic trainer to claim commonality with a gymnastic
art, or think to possess part of this art, for he is only a
servant, just as the soldier is of the general, and in this way
too, the gymnastic trainer himself, if he trains in gymnastic exercises, is a servant of the gymnastic art, or practices
the art of wrestling holds, or is the servant of some other
refinement, which I call throwing, although those who   893K
train athletes do not refer to themselves in this way; they
call themselves gymnastic trainers.

Nevertheless, in truth, the art of gymnastic training is
different, just as that of archery is; the art of wrestling
throws is like that of generalship, while that of gymnastic
training is like the medical art; each of these serves two
arts set above them. In this way too, the shoemaker is
taught by the general on the matter of a useful shoe for a
certain soldier, while he is directed to make one that is
suitable for health by the art concerning the body. In like
manner also, the cook prepares a dish conducive to health

---

78 μήν γ´ H; μόνον Ku   79 τῷδέ add. H

ἰατρῷ τε καὶ ὑγιεινῷ παρασκευάζει, τὸ δ᾽ εἰς ἡδονὴν
τέχνῃ μὲν οὐκέτι τοῦτό γ᾽ οὐδεμιᾷ, κολακείᾳ δέ τινι
τὸ τέλος οὐ τὴν ὑγίειαν, ἀλλὰ τὴν ἡδονὴν πεποιη-
μένῃ. καὶ δὴ οὖν καὶ ἡ παλαιστικὴ τὸ μὲν εἰς ὑγίειαν
χρηστὸν ὑγιεινοῖς τε καὶ γυμνασταῖς παρασκευάζει,
τὸ δ᾽ εἰς διάθεσιν ἀθλητικὴν ἐκείνῃ τῇ κακοτεχνίᾳ τῇ
πολλάκις εἰρημένῃ, τῇ καλούσῃ μὲν ἑαυτὴν γυμναστικήν, κάλλιον δ᾽ ἂν ὀνομασθείη καταβλητική. καὶ
γὰρ οὖν καὶ ὀνομάζουσιν αὐτὴν οἱ Λάκωνες καββα-
894K λικὴν καὶ καββαλικώτερον ἑαυτῶν γέ φασιν εἶναι τὸν
ἐν ταύτῃ τεθραμμένον, οὐκ ἰσχυρότερον.

46. Ὅπου γε καὶ ἡ ὑγιαίνουσα πολιτεία μισεῖ τοῦτο
τὸ ἐπιτήδευμα καὶ βδελύττεται, πάσης μὲν τῆς εἰς τὸν
βίον ἰσχύος ἀνατρεπτικὸν ὑπάρχον, εἰς οὐκ ἀγαθὴν
δὲ τοῦ σώματος ἄγον διάθεσιν. ἐγὼ γοῦν ἐπειράθην
ἐμαυτοῦ πολλάκις ἰσχυροτέρου τῶν ἀρίστων εἶναι
δοκούντων καὶ πολλοὺς στεφανίτας ἀγῶνας ἀνηρη-
μένων ἀθλητῶν. ἔν τε γὰρ ὁδοιπορίαις ἁπάσαις
ἄχρηστοι τελέως ἦσαν ἔν τε ταῖς πολεμικαῖς πράξεσιν, ἔτι δὲ μᾶλλον ἐν πολιτικαῖς τε καὶ γεωργικαῖς, εἰ
δέ που καὶ φίλῳ νοσοῦντι παραμεῖναι δέοι, πάντων
ἀχρηστότατοι συμβουλεῦσαί τε καὶ συσκέψασθαι καὶ
συμπρᾶξαι, ταύτῃ μέν, ἥπέρ γε καὶ οἱ σύες. ἀλλ᾽
ὅμως οἱ τούτων ἀτυχέστατοι καὶ μηδεπώποτε νικήσα-
ντες ἐξαίφνης ἑαυτοὺς ὀνομάζουσι γυμναστάς, εἶτ᾽
οἶμαι καὶ κεκράγασιν οὐδὲν ἧττον τῶν συῶν ἐκμελεῖ
καὶ βαρβάρῳ φωνῇ. τινὲς δ᾽ αὐτῶν καὶ γράφειν ἐπι-

[instructed by] a doctor or hygienist, whereas he might have made what is pleasant by what is not an art at all, but is a kind of flattery whose end is not health but pleasure. And indeed also, the art of wrestling provides something useful to health for hygienists and gymnastic trainers, whereas something is provided to the athletic condition by that oft-mentioned base art, which calls itself "gymnastic," but would be better named the art of wrestling throws. For truly too, the Spartans speak of themselves as being "good at throwing" and "better at throwing," and in fact say 894K someone is reared in this and not "stronger."

46. But actually, the healthy polity hates and loathes this practice as being upsetting of every strength pertaining to life, and as leading to a condition of the body that is not good. Anyway, I have often proved myself stronger than athletes with the best reputations who have carried off many victors' crowns, for they were completely useless in all journeys and in military actions, and even more in political and agricultural matters. And if at any time they need to stand beside a sick friend, they are the most useless of all at advising, contemplating together and assisting—in this, they are about as much use as pigs. Nevertheless, the most unfortunate of them, who have never yet triumphed, suddenly call themselves gymnastic trainers, and then, I think, they have also squealed no less than pigs with a discordant and barbarous voice. Some of them even

895K χειροῦσιν ἢ περὶ τρίψεως ἢ εὐεξίας ἢ ὑγιείας ἢ γυμνασίων, εἶτα προσάπτεσθαι τολμῶσι καὶ ἀντιλέγειν
οἷς οὐδ᾽ ὅλως ἔμαθον, οἷος καὶ ὁ πρώην μὲν Ἱπποκράτει ἐγκαλῶν ὡς οὐκ ὀρθῶς ἀποφηναμένῳ περὶ τρίψεως. ἐπεὶ δ᾽ ἡμᾶς ἀφικομένους ἠξίωσάν τινες τῶν
παρόντων ἰατρῶν τε καὶ φιλοσόφων ἅπαντα διελθεῖν
αὐτοῖς τὸν λόγον, εἶτ᾽ ἐφαίνετο ἀπάντων πρῶτος ὑπὲρ
αὐτῆς Ἱπποκράτης ἀποφηνάμενος ἄριστα, παρελθὼν
εἰς τὸ μέσον ἐξαίφνης ὁ αὐτοδίδακτος ἐκεῖνος γυμνα-
στὴς ἐκδύσας παιδάριον ἐκέλευσεν ἡμᾶς τρίβειν τε
τοῦτο καὶ γυμνάζειν ἢ σιωπᾶν περὶ τρίψεως καὶ γυμνασίων, εἶτ᾽ ἐφεξῆς ἐβόα· ποῦ γὰρ Ἱπποκράτης εἰ-
σῆλθεν εἰς σκάμμα; ποῦ δ᾽ εἰς παλαίστραν; ἴσως οὐδ᾽
ἀναχέασθαι καλῶς ἠπίστατο. οὗτος μὲν οὖν ἐκεκρά-
γει τε καὶ ἄλλως οὐδὲ σιωπῶν ἀκούειν ἐδύνατο καὶ
μανθάνειν τὰ λεγόμενα, ἡμεῖς δὲ κατὰ σχολὴν τοῖς
παροῦσι διελέχθημεν, ὡς ὅμοιον ὁ κακοδαίμων ἐκεῖνος ἐργάζοιτο μαγείρῳ τε καὶ σιτοποιῷ περὶ πτι-
896K σάνης ἢ ἄρτου διαλέγεσθαι τολμῶντι κἄπειτα φά-
σκοντι· ποῦ γὰρ Ἱπποκράτης ἐν μαγειρείῳ διέτριψεν
ἢ ἐν μυλῶνι; σκευασάτω γοῦν μοι πλακοῦντα καὶ ἄρτον καὶ ζωμὸν καὶ λοπάδα, ἔπειθ᾽ οὕτως ὑπὲρ αὐτῶν
διαλεγέσθω.

47. Τί ποτ᾽ οὖν, φήσεις, ἰατροὺς ὀνομάζομεν Ἱπποκράτην τε καὶ τὸν ἀπ᾽ αὐτοῦ χορόν, εἴπερ οὐ κατὰ
πάσης τῆς περὶ τὸ σῶμα τέχνης τοὔνομα φέρομεν,

attempt to write, either about massage, or *euexia,* or   895K
health, or exercises, and then they have the temerity to
fasten onto and contradict those whom they have not stud-
ied at all—an example is a man who just a day or so ago
leveled a charge against Hippocrates of not speaking correctly about massage. When I arrived, some of the doctors
and philosophers present asked me to go over the whole
argument for them. It then became apparent that Hippocrates was the first of all to discuss this in the best way. Then
a fellow, self-taught in the gymnastic art, suddenly came
forward to center stage, stripped a young boy, and ordered
us to massage and exercise him, or remain silent on massage and exercises. And then he was shouting: "Where did
Hippocrates go to a wrestling arena?[42] Where did he go to
a wrestling school? Perhaps he didn't know how to pour
oil on properly." This man, then, kept up his clamor and
would not otherwise be silent so he could hear and learn
what was being said. However, at an opportune time, I
discussed with those present how that poor devil acted like
a cook or baker who had the audacity to discourse on
ptisane or bread, and then to say: "When did Hippocrates   896K
spend time in a kitchen or mill? Anyway, let him prepare
for me a flat cake, bread, sauce and a dish of food, and then
discourse about these things in this way."

47. Why in the world, you will say, do we call Hippocrates and the whole troop who followed him "doctors," if
we do not use the term to refer to the whole art concern-

---

[42] The term *skamma* basically refers to the action of digging.
However, LSJ also lists a meaning of "a place dug up and sanded
on which wrestlers practiced" (LSJ 1604). KLat has *caveam ludo-
rum,* which is essentially the same meaning.

ἀλλὰ κατὰ μόνου τοῦ μέρους αὐτῆς, ὃ τοὺς κάμνοντας
ἰᾶται· φαίνονται μὲν γὰρ ὅλην μεταχειρισάμενοι[80] τὴν
τέχνην, ὡς μηδὲ τὸ περὶ τὰ γυμνάσια μέρος αὐτῆς
ἀπολιπεῖν. ὅτι πρώτου συστάντος ὅλης τῆς τέχνης
μέρους τοῦ θεραπευτικοῦ, διότι καὶ μάλιστα κατήπει-
γεν, ὕστερον δὲ κατὰ πολλὴν σχολὴν αὐτῷ προστε-
θέντος τοῦ φυλακτικοῦ τε καὶ ὑγιεινοῦ προσαγορευο-
μένου, τὴν ὅλην τέχνην ἀπὸ τοῦ μέρους ὀνομασθῆναι
συνέβη, καθάπερ ἐπ' ἄλλων πολλῶν ἐγένετο· καὶ γὰρ
γεωμέτρας ὀνομάζομεν οὐ τοὺς περὶ τῶν ἐπιπέδων
σχημάτων μόνον, ἀλλὰ καὶ τοὺς περὶ τῶν στερεῶν
ἐπισταμένους οὐδ' ἔστιν οὐδεὶς νῦν ὥσπερ γεωμέτρης
οὕτω καὶ στερεομέτρης ὀνομαζόμενος, ἀλλ' ἐκλέλοιπε
897K τοὔνομα, καθάπερ καὶ τὸ τῶν ὑγιεινῶν, οὐ τῶν σωμάτων οὐδὲ τῶν διαιτημάτων λέγω τῶν ὑγιεινῶν, ἀλλὰ
τῶν ἐπισταμένων ἀνδρῶν ταῦτα, τῶν ὑπ' Ἐρασιστρά-
του τοῖς ἰατροῖς ἀντιδιηρημένων. ὡσαύτως δὲ καὶ τρι-
ηράρχας μὲν ὠνόμαζον οἱ παλαιοὶ τοὺς ἄρχοντας τῶν
τριήρων, νῦν δ' ἤδη πάντας οὕτω καλοῦσι τοὺς ὁπωσοῦν ἡγουμένους στόλου ναυτικοῦ, κἂν μὴ τριήρεις
ὦσιν αἱ νῆες.

ὅμοιόν τι τούτῳ περὶ τὴν ἰατρικὴν ἐγένετο καὶ τοὺς
ἰατρούς· ἀπὸ γὰρ τοῦ πρώτου συστάντος μέρους ὅλην
τε τὴν περὶ τὸ σῶμα τέχνην ὀνομασθῆναι συνέβη τοῦ
χρόνου προϊόντος ἰατρικὴν αὐτόν τε τὸν ἐπιστήμονα
τῆς τέχνης ἰατρόν. οὔκουν οὐδ' ἀπεικός ἐστιν ἐρωτη-

ing the body but only to the part of it which heals those
who are sick? For they appear to practice the whole art so
as not to leave aside the part of this to do with exercises.
It is that the therapeutic part of the whole art was the first
to be established because it was the most urgently required; the so-called preservative and hygienic parts were
added to it later in a much more leisurely fashion, and the
whole art came to be named from the part, just as occurred
in many other cases. Thus, we term geometers not only
those knowledgeable about plane (two-dimensional) figures, but also those knowledgeable about solid (three-
dimensional) figures; no one is now called in this way
a geometer or a stereometer, but [the latter] term is
omitted, just as I say "hygienists" and not "hygienists of   897K
bodies" or "hygienists of regimens." But men knowledgeable about these things were distinguished from doctors
by Erasistratus.[43] In like manner also, the ancients termed
trierarchs those in charge of triremes; now, however, everyone already calls those in charge of any naval activity
whatsoever *trierarchs,* even if the vessels are not triremes.

Something like this happened to the medical art and
doctors, for as time progressed, the whole art concerning
the body came to be named from the first part established,
and the one skilled in the art, a doctor. It is not, therefore,

---

[43] See section 38 above.

---

80 μεταχειρισάμενοι H; μεταχειρισάμενος Ku

θέντα τινὰ νῦν, ἧστινός ἐστι τέχνης μέρος τὸ ὑγιεινόν, ἀποκρίνασθαι τῆς ἰατρικῆς· ἐκταθείσης γὰρ ἐπὶ
πλέον τῆς προσηγορίας καὶ μηκέτι τὸ μέρος ἀλλ'
ὅλην τὴν περὶ τὸ σῶμα τέχνην σημαινούσης Ἱπποκράτης τε δικαίως καὶ οἱ νῦν ἅπαντες ἰατροὶ ὀνομά-
ζονται· ἴσασι γὰρ μόρια τῆς τέχνης αὐτῆς δύο τὰ
μέγιστα, θεραπευτικόν τε καὶ ὑγιεινόν. αὐτοῦ δ' αὖ

898K πάλιν τοῦ ὑγιεινοῦ μέρους ἴσασι τὸ γυμναστικόν, ὡς
καὶ πρόσθεν ἐπιδέδεικται. καθάπερ οὖν Ἱπποκράτης
καὶ Διοκλῆς καὶ Πραξαγόρας καὶ Φιλότιμος καὶ
Ἡρόφιλος ὅλης τῆς περὶ τὸ σῶμα τέχνης ἐπιστήμονες ἦσαν, ὡς δηλοῖ τὰ συγγράμματα[81] αὐτῶν, οὕτως
αὖ πάλιν οἱ περὶ Θέωνα καὶ Τρύφωνα τὴν περὶ τοὺς
ἀθλητὰς κακοτεχνίαν μετεχειρίσαντο, καθάπερ αὖ
καὶ τὰ τούτων δηλοῖ συγγράμματα, παρασκευήν τέ τι
γυμνάσιον ὀνομαζόντων καὶ αὖθις ἕτερόν τι μερι-
σμόν, ἔπειτ' ἄλλο τι τέλειον, ἀποθεραπείαν δ' ἄλλο,
καὶ ζητούντων, εἴτε κατὰ τὴν τοιαύτην περίοδον
ἀσκητέον ἐστὶ καὶ γυμναστέον τὸν ἀθλητὴν εἴτε[82]
κατ' ἄλλον τινὰ τρόπον. ᾗ καὶ θαυμάζειν ἐπέρχεταί
μοι τῶν νῦν τοὺς ἀθλητὰς γυμναζόντων, ὅταν ἀμφι-
σβητούντων ἀκούσω μέρος εἶναι τῆς ἑαυτῶν τέχνης

[81] συγγράμματα H; γράμματα Ku
[82] post τὸν ἀθλητὴν εἴτε: add. κατ' ἄλλον τινὰ τρόπον. ᾗ
καὶ θαυμάζειν ἐπέρχεταί μοι τῶν νῦν τοὺς ἀθλητὰς γυμναζόντων, ὅταν H (his note, p. 99)

unreasonable, when asked what art hygiene is part of, to answer "the medical art." Since the name has been extended further, and no longer signifies the part but the whole art concerning the body, Hippocrates and all the doctors of the present time are rightly so named, for they know the greatest parts of the art itself are two—therapeutic and hygienic. In turn, they know gymnastics is part 898K of hygiene itself, as has also been shown before. Therefore, just as Hippocrates, Diocles, Praxagoras, Philotimus and Herophilus[44] were knowledgeable in the whole art concerning the body, as their writings show, so conversely the followers of Theon and Tryphon practiced the base art concerning athletes,[45] as in turn the works of these men also reveal; they use the terms "preparatory" and "bodily exercise" and again speak of something "partial" and something "complete," and *apotherapy* (restoration), and inquire into whether the athlete must be trained and exercised according to such a course or in another way. And it comes as a surprise to me when I hear those training athletes now laying claim to hygiene as part of their own

[44] See section 38 and note 36.

[45] Theon of Alexandria (fl. AD 130–160) is described as follows by Keyser (EANS, 795): "Autodidact ex-athlete who wrote a work on exercise and a longer work, *Gymnastrion,* both known only from Galen, who praised him as wiser than other such writers, but chides him for thinking he knew better than Hippocrates [on several matters]." Tryphon is possibly the person who was sometimes called a surgeon and was renowned for his plasters. He is said to have worked with gladiators—see Scarborough, EANS, 817.

τὸ ὑγιεινόν. ὅπου γὰρ οὐδὲ τῆς ὄντως γυμναστικῆς μέρος ἐστὶν ἀλλ᾽ ἔμπαλιν ἐκείνη μέρος ὑγιεινοῦ, τί χρὴ περὶ τῆς τούτων κακοτεχνίας ἀμφιβάλλειν, ἢ μήτε μέρος ἐστὶν ὅλως τῆς περὶ τὸ σῶμα τέχνης ἐπιτηδεύματός τε προέστηκεν οὐχ ὑπὸ Πλάτωνος μόνον ἢ Ἱπποκράτους, ἀλλὰ καὶ τῶν ἄλλων ἁπάντων ἰατρῶν τε καὶ φιλοσόφων ἀτιμαζομένου;

art. For when hygiene is really not a part of gymnastics but conversely, gymnastics is part of hygiene, why must there be dispute about the base art of these men, which is altogether not a part of the art practiced regarding the body, and which is deemed unworthy not only by Plato and Hippocrates, but also by all other doctors and philosophers?

ΓΑΛΗΝΟΥ ΠΕΡΙ ΤΟΥ ΔΙΑ ΤΗΣ
ΣΜΙΚΡΑΣ ΣΦΑΙΡΑΣ
ΓΥΜΝΑΣΙΟΥ

ON EXERCISE WITH A
SMALL BALL

INTRODUCTION

Galen, in this short essay, makes and attempts to substantiate the claim that the group of exercises carried out with a small ball are the best forms of exercise across the board. He offers a number of cogent reasons for his conclusion and draws attention briefly to the negative aspects of some of the other common forms of exercise. However, his account is far from comprehensive. In particular, he does not systematically describe the exercises he is referring to, nor does he describe the ball or balls he classifies as small. Presumably, the exercises would be familiar to Galen's readers, while the balls would have been similar in size to modern tennis balls, cricket balls, and baseballs. Galen focuses on the following benefits:

1. Ease of finding time and a venue for the exercise.
2. No need for complicated equipment and associated expense.
3. The important attribute of bringing "delight to the soul."
4. The capacity to exercise all bodily regions equally or some preferentially as required.
5. Variations in form of the exercises allow their practice by one person alone, by two or more people, or by teams.

6. Variations in the level of intensity may make a particular exercise suitable for all age groups and different bodily conditions.

7. These exercises are essentially without danger to the participant(s).

Galen is unequivocal in his claim for the preeminence of exercise with the small ball over all other forms of exercise.

TEXTS AND TRANSLATIONS

The Greek text given is essentially that of Ernst Wenkebach (W) found in *Sudhoff's Archiv für Geschichte der Medizin und der Naturwissenschaften,* Bd. 31. H 4/5 (September, 1938), pp. 254–97. Comparison was made with the texts of Johannes Marquardt (M) in *Claudii Galeni Pergameni Scripta Minora* I (Leipzig: Teubner, 1884), pp. 93–102, and Carl-Gottlieb Kühn (Ku) in *Galeni Opera Omnia* V.899–910. Notes have been added to the Greek text when there are significant differences between the three texts listed, and particularly when these may bear on the translation. A German translation and commentary accompanies Wenkebach's text, and there are English translations by Waldo E. Sweet and P. N. Singer.[1] J. A. López Férez lists two other German translations (K. Schuetze [Berlin, 1936], and J. Marker, [Berlin, 1962]) as well as two studies, one in German (R. Heubaum [1939])

---

[1] Waldo E. Sweet, *Sport and Recreation in Ancient Greece: A Sourcebook and Translations* (Oxford: Oxford University Press, 1987); Singer, *Galen: Selected Works*.

and one in Italian (E. Valentini [1941]).[2] None of these four has been consulted. For an overview of games and sports in ancient Greece and Rome, the works by E. N. Gardiner and H. A. Harris, referred to in notes to the translation, are useful and interesting, as is the work by Zahra Newby.[3]

[2] J. A. López Férez, *Galeno: Obra, Pensamiento e Influencia* (Madrid: Universidad Nacional, 1991), 317.

[3] E. Norman Gardiner, *Athletics of the Ancient World* (Oxford: Oxford University Press, 1930); H. A. Harris, *Sport in Greece and Rome*; Zahra Newby, *Athletics in the Ancient World* (London: Bristol Classical Press, 2006).

ΓΑΛΗΝΟΥ ΠΕΡΙ ΤΟΥ ΔΙΑ ΤΗΣ ΣΜΙΚΡΑΣ ΣΦΑΙΡΑΣ ΓΥΜΝΑΣΙΟΥ

1. Πηλίκον μὲν ἀγαθόν ἐστιν, ὦ Ἐπίγενες,[1] εἰς ὑγίειαν γυμνάσια, καὶ ὡς χρὴ τῶν σιτίων ἡγεῖσθαι αὐτά, παλαιοῖς ἀνδράσιν αὐτάρκως εἴρηται, φιλοσόφων τε καὶ ἰατρῶν τοῖς ἀρίστοις· ὅσον δ' ὑπὲρ τἆλλα τὰ διὰ τῆς σμικρᾶς σφαίρας ἐστί, τοῦτ' οὐδέπω τῶν πρόσθεν ἱκανῶς οὐδεὶς ἐξηγήσατο. δίκαιον οὖν ἡμᾶς ἃ γιγνώσκομεν εἰπεῖν, ὑπὸ σοῦ μὲν κριθησόμενα[2] τοῦ πάντων ἠσκηκότος ἄριστα τὴν ἐν αὐτοῖς τέχνην, χρήσιμα δ',[3] εἴπερ ἱκανῶς εἰρῆσθαι δόξειε, καὶ τοῖς ἄλλοις, οἷς ἂν μεταδῷς τοῦ λόγου, γενησόμενα.

φημὶ γὰρ ἄριστα μὲν ἁπάντων γυμνασίων εἶναι τὰ μὴ μόνον τὸ σῶμα διαπονεῖν, ἀλλὰ καὶ τὴν ψυχὴν τέρπειν δυνάμενα. καὶ ὅσοι κυνηγέσια καὶ τὴν ἄλλην θήραν ἐξεῦρον, ἡδονῇ καὶ τέρψει καὶ φιλοτιμίᾳ τὸν ἐν αὐτοῖς πόνον κερασάμενοι, σοφοί τινες ἄνδρες ἦσαν

---

[1] See W's note, p. 258, on ὦ Ἐπίγενες, which is present in all three texts, but placed after ὑγίειαν in Ku and M.

[2] κριθησόμενα M, W; κριθησόμεθα Ku

[3] post χρήσιμα δ',: εἴπερ ἱκανῶς εἰρῆσθαι δόξειε, W; ἱκανῶς εἰρῆσθαι δόξειε, om. M; εἴπερ om. Ku

ON EXERCISE WITH A
SMALL BALL

1. How great a good for health exercises are, Epigenes,[1]   899K
and that they must precede food, was adequately stated by
men of earlier times—both the best philosophers and doctors. However, the extent to which exercises with the small
ball are superior to the others has never been set out in
sufficient detail by anyone previously. It is proper, then,
for me to state those things I know, so they will be judged
by you, a man best practiced of all in the art of these. If
stated adequately, they will seem useful to you and will be
also employed by others to whom you transmit this work.

    I say the best of all exercises are not only those which   900K
are able to exercise the body vigorously, but those which
are also able to delight the soul. Men who discovered
hunting with hounds and the other hunting, and combined
the labor in these activities with pleasure, delight and love

---

[1] Epigenes, who is also the dedicatee of Galen's work *De prae-
notione ad Epigenem* (Nutton, 1970, CMG, V.8.1) is simply described in the index to that work as, *amicus Galeni, cui hunc li-
brum dedicavit.* He is mentioned seven times in the work itself.
There is also an Epigenes mentioned in the pseudo-Galenic work
Hist. Phil., XIX.286, but this is probably the astrologer / astrono-
mer of the first century BC (see EANS, 290).

καὶ φύσιν ἀνθρωπίνην ἀκριβῶς καταμεμαθηκότες.
τοσοῦτον γὰρ ἐν αὐτῇ δύναται ψυχῆς κίνησις,[4] ὥστε
πολλοὶ μὲν ἀπηλλάγησαν νοσημάτων ἠσθέντες μόνον, πολλοὶ δ᾽ ἑάλωσαν ἀνιαθέντες. οὐδ᾽ ἔστιν οὐδὲν[5]
οὕτως ἰσχυρόν τι τῶν κατὰ τὸ σῶμα παθημάτων, ὡς
κρατεῖν τῶν περὶ τὴν ψυχήν. οὔκουν οὐδ᾽ ἀμελεῖν χρὴ
τῶν ταύτης κινήσεων ὁποῖαί τινες ἔσονται, πολὺ δὲ
μᾶλλον ἢ τῶν τοῦ σώματος ἐπιμελεῖσθαι τά τ᾽ ἄλλα
καὶ ὅσῳ[6] κυριώτεραι. τοῦτο μὲν δὴ κοινὸν ἁπάντων
γυμνασίων τῶν μετὰ τέρψεως, ἄλλα δ᾽ ἐξαίρετα τῶν
διὰ τῆς σμικρᾶς σφαίρας, ἃ ἐγὼ νῦν ἐξηγήσομαι.

2. Πρῶτον μὲν ἡ εὐπορία. εἰ γοῦν ἐννοήσειας, ὅσης
δεῖται παρασκευῆς θ᾽ ἅμα καὶ σχολῆς τά τ᾽ ἄλλα
πάντα τὰ περὶ θήραν ἐπιτηδεύματα καὶ τὰ κυνηγέσια,
901K   σαφῶς ἂν μάθοις, ὡς οὔτε τῶν τὰ πολιτικὰ πραττόντων οὐδεὶς[7] οὔτε τῶν τὰς τέχνας ἐργαζομένων δυνατὸς[8] μεταχειρίζεσθαι τὰ τοιαῦτα γυμνάσια. καὶ γὰρ[9]
πλουτοῦντος δεῖται πολλῶν καὶ σχολὴν ἄγοντος οὐκ
ὀλίγην ἀνθρώπου. τοῦτο δὲ μόνον οὕτω μὲν φιλάν-
θρωπον, ὡς μηδὲ τὸν πενέστατον ἀπορεῖν τῆς ἐπ᾽
αὐτὸ[10] παρασκευῆς οὐ γὰρ δικτύων οὐδ᾽ ὅπλων οὐδ᾽

[4] ψυχῆς κίνησις M, W; ψυχὴν κινῆσαι Ku
[5] οὐδ᾽ ἔστιν οὐδὲν M, W; οὐκ ἔστι δ᾽ Ku   [6] post ὅσῳ:
κυριώτεραι. W; κυριώτερον σώματος ἡ ψυχή. Ku; κυριωτερα
σώματος ἡ ψυχή. M   [7] οὐδεὶς M, W; οὐδενὶ Ku
[8] δυνατὸς W; δυνατὸν Ku; δύναται M
[9] post καὶ γὰρ: πλουτοῦντος δεῖται πολλῶν W; πολλῶν
ταῦτα δεῖται Ku; πλούτου ταῦτα δεῖται πολλοῦ M

of honor, were wise men with an accurate knowledge of human nature. Movement of the soul is possible to such a degree in this that many, by the pleasure alone, were freed from diseases, while many others, when distressed, were won over. None of the affections involving the body is so strong that it prevails over those involving the soul. One must not, therefore, neglect the movements of the latter, whatever kind they may be; one must take much more care of the movements of the soul than of those of the body by as much as the former are more important than the latter. This is certainly common to all exercises done with pleasure, but particularly notable among them are those with the small ball, which I shall now discuss in detail.

2. First, there is the ease of provision. At all events, if you were to consider how much preparation and leisure is required for all the other practices of hunting and hunting with hounds, you would clearly understand that none of 901K those who are involved in political activities, or of those carrying out the arts, would be able to practice such exercises. Such exercises need a man of considerable wealth and no little leisure time to carry them out. Exercise with the small ball is not, however, only for the *philanthropic*[2] man; even the poorest person is not without the means for it. There is no need for nets, implements, horses and hunt-

---

[2] The transliterated term here is taken to indicate someone with more than adequate means and the will to use them to benefit others, which of course accords with the modern sense, although this is not listed as such in LSJ.

---

10 ἐπ᾽ αὐτὸ M, W; ἑαυτοῦ Ku

ἵππων οὐδὲ κυνῶν θηρευτικῶν, ἀλλὰ σφαίρας μόνης
δεῖται καὶ ταύτης σμικρᾶς, οὕτω δ᾽ εὔγνωμον εἰς τὰς
ἄλλας πράξεις, ὥστ᾽ οὐδεμιᾶς αὐτῶν ὀλιγωρεῖν ἀναγ-
κάζει δι᾽ αὐτό. καίτοι τί ἂν εὐπορώτερον γένοιτο τοῦ
καὶ τύχην ἀνθρωπίνην ἅπασαν καὶ πρᾶξιν προσιεμέ-
νου; τῶν μὲν γὰρ ἀμφὶ τὰς θήρας γυμνασίων τῆς
χρήσεως οὐκ ἐφ᾽ ἡμῖν ἡ εὐπορία· πλούτου τε γὰρ δεῖται τὴν παρασκευὴν τῶν ὀργάνων ἐκπορίζοντος καὶ
ἀργίας σχολῇ[11] τὸν καιρὸν ἐπιτηρούσης. τούτου δ᾽ ἡ
τῶν ὀργάνων παρασκευὴ καὶ τοῖς πενεστάτοις[12] εὔ-
πορος, ὅ τε καιρὸς τῆς χρήσεως καὶ τοὺς ἱκανῶς
ἀσχόλους ἀναμένει. τὸ μὲν δὴ τῆς εὐπορίας αὐτοῦ
τηλικοῦτον ἀγαθόν.

902K   ὅτι δὲ καὶ πολυαρκέστατον τῶν ἄλλων γυμνασίων,
ὧδ᾽ ἂν μάλιστα μάθοις, εἰ σκέψαιο καθ᾽ ἕκαστον
αὐτῶν, ὅ τι τε δύναται καὶ οἷόν τι τὴν φύσιν ἐστίν.
εὑρήσεις γὰρ ἢ σφοδρὸν ἢ μαλακὸν ἢ τὰ κάτω μᾶλλον ἢ τὰ ἄνω κινοῦν ἢ μέρος τι πρὸ τῶν ἄλλων, οἷον
ὀσφὺν ἢ κεφαλὴν ἢ χεῖρας ἢ θώρακα, πάντα δ᾽ ἐξ
ἴσου τὰ μέρη τοῦ σώματος κινοῦν καὶ δυνάμενον ἐπί
τε τὸ σφοδρότατον ἀνάγεσθαι καὶ ἐπὶ τὸ μαλακώτα-
τον ὑφίεσθαι τῶν μὲν ἄλλων οὐδέν, τοῦτο δὲ μόνον τὸ
διὰ τῆς σμικρᾶς σφαίρας γυμνάσιον, ὀξύτατόν γ᾽ ἐν
μέρει καὶ βραδύτατον γενόμενον, σφοδρότατόν τε καὶ
πρᾳότατον, ὡς ἂν αὐτός τε βουληθῇς καὶ τὸ σῶμα
φαίνηται δεόμενον.[13] οὕτω δὲ καὶ τὰ μέρη κινεῖν ἔστι

ing dogs, but only for a ball, and this a small one. In this way it is considerate of other activities so that it does not force any of these to be neglected because of it. And indeed, what could be easier to do than something which allows of every human fortune and activity? For the ease of use of the exercises that are involved in hunting is not in our hands. It requires wealth to provide the preparation of the implements and the idleness of leisure to watch out for the appropriate time. On the other hand, for this exercise [with the small ball] the preparation of the implements is readily available to the poorest, as I said, and the time of use waits for those who are very busy. This is how great a good the ease of provision of this exercise is.

That it is also much more beneficial than the other exercises is something you would particularly learn, if you were to consider for each of the exercises with the small ball, what it is able to do and what the nature of it is. For you will find it is either vigorous or gentle, or moves the parts below more than those above, or some part rather than others—for example, the lower back, head, arms or chest. None of the other exercises moves all the parts of the body equally, and can be raised up to very vigorous and slackened off to very mild; this is a feature of the exercise with the small ball alone. At times, it is very rapid and at times very slow, at times very vigorous and at times very gentle, as the individual might wish and the body appears to need. In this way too, it moves all the parts of

902K

---

11 σχολῇ W; σχολῆς Ku; om. M
12 post πενεστάτοις add. ὡς εἴρηται Ku
13 δεόμενον W, M; ἐμμένον Ku

μὲν αὐτοῦ πάνθ᾽ ὁμοῦ, εἰ τοῦτο συμφέρειν δόξειεν, ἔστι[14] δὲ πρὸ ἄλλων ἄλλα, εἰ καὶ τοῦτό ποτε δόξειεν.

ὅταν μὲν γὰρ συνιστάμενοι πρὸς ἀλλήλους καὶ ἀποκωλύοντες ὑφαρπάσαι τὸν μεταξὺ διαπονῶσι, μέγιστον αὐτὸ καὶ σφοδρότατον καθίσταται πολλοῖς μὲν τραχηλισμοῖς, πολλαῖς δ᾽ ἀντιλήψεσι παλαιστικαῖς ἀναμεμιγμένον, ὥστε κεφαλὴν μὲν καὶ αὐχένα

903K διαπονεῖσθαι τοῖς τραχηλισμοῖς, πλευρὰς δὲ καὶ θώρακα καὶ γαστέρα ταῖς τε[15] τῶν ἁμμάτων περιθέσεσι καὶ ἀπώσεσι καὶ ἀποστηρίξεσι καὶ ταῖς ἄλλαις παλαιστικαῖς λαβαῖς. τούτῳ δὲ καὶ ὀσφὺς τείνεται σφοδρῶς καὶ σκέλη, καὶ δὴ ῥώννυται καὶ τὸ[16] ἑδραῖον τῆς βάσεως τῷ τοιούτῳ πόνῳ. τὸ δὲ καὶ προβαίνειν καὶ ὑποβαίνειν[17] καὶ εἰς τὰ πλάγια μεταπηδᾶν οὐ μικρὸν σκελῶν γυμνάσιον,[18] ἀλλ᾽, εἰ χρὴ τἀληθὲς εἰπεῖν, μόνον[19] δικαιότατα κινοῦν πάντ᾽ αὐτῶν τὰ μόρια. τοῖς μὲν γὰρ προϊοῦσιν ἕτερα νεῦρα καὶ μύες, τοῖς δ᾽ ὑποβαίνουσιν ἕτερα διαπονεῖται πλέον, οὕτω δὲ καὶ τοῖς εἰς τὰ πλάγια μεθισταμένοις ἄλλα. καὶ ὅστις καθ᾽ ἓν εἶδος κινήσεως κινεῖ τὰ σκέλη καθάπερ οἱ θέοντες, ἀνωμάλως οὗτος καὶ ἀνίσως τὰ μέρη γυμνάζει.

---

[14] ἔστι W; ἔτι Ku, M   [15] post ταῖς τε: τῶν ἁμμάτων περιθέσεσι W, M; τῶν ὀμμάτων μετάρσεσι, καὶ θέσεσι, Ku
[16] καὶ δὴ ῥώννυται καὶ τὸ add. W
[17] καὶ ὑποβαίνειν add. W   [18] From ἀλλ᾽ to γυμνάζει with the inclusion post γυμνάζει of τοῖς εἰς τὸ πλάγιον μεθισταμένους. ἀλλὰ καὶ constitutes the start of section 3 in Ku.
[19] post μόνον add. ἐστὶ Ku

the body equally, if this seems to be of benefit, or some rather than others, if on occasion this also seems to be of benefit.

Thus, when men place themselves across from each other and work hard at preventing the one in the middle intercepting [the ball], this sets up a very strong and vigorous exercise, with much seizing of the neck and many wrestling holds mixed together, so that head and neck are worked out strongly by the neck holds, while sides, chest 903K and abdomen are worked out hard by the encirclings of the arms, the thrustings away, establishing a firm position, and the other wrestling holds. By this also, the lower back and legs are strained strongly, and are certainly strengthened also by the effort to establish a firm base. Both moving forward and back, and leaping to the sides provide no little exercise for the legs.[3] But, if the truth must be told, this exercise alone is the most fitting to move all the parts of these. For in the advancing, some sinews and muscles are worked out, while in the withdrawing, others are worked out more, and this applies also to the changes of position to the sides. On the other hand, someone who moves the limbs in one kind of movement, as those who run do, exercises the parts unevenly and unequally.

[3] According to H. A. Harris, this is the game or exercise of *harpastum,* a name derived from the verb "to intercept"—see Harris, *Sport in Greece and Rome,* 92–93. See also Gardiner's comment on this in E. N. Gardiner, *Athletes of the Ancient World* (Oxford: Oxford University Press, 1930), 232–33.

3. Ὡς δὴ τοῖς σκέλεσιν, οὕτω καὶ ταῖς χερσὶ τὸ γυμνάσιον τοῦτο δικαιότατον ἐν παντὶ σχήματι λαμβάνειν ἐθιζομένων τὴν σφαῖραν. ἀνάγκη γὰρ κἀνταῦθα τὴν ποικιλίαν τῶν σχημάτων ἄλλοτ' ἄλλους τῶν μυῶν τείνειν σφοδρότερον, ὥστε πάντας ἐν μέρει

904K πονοῦντας ἴσον ἔχειν, ἀνάπαυλάν[20] τε τοῖς ἡσυχάζουσιν εἶναι τὸν χρόνον τῶν ἐνεργούντων, καὶ οὕτως ἐν μέρει πάντας ἐνεργοῦντάς τε καὶ ἀναπαυομένους οὔτ' ἀργοὺς μένειν τὸ πάμπαν οὔτε κόποις ἁλίσκεσθαι μόνους πονοῦντας. ὄψιν δ' ὅτι γυμνάζει,[21] μαθεῖν ἔνεστιν ὑπομνησθέντας, ὡς, εἰ μή τις ἀκριβῶς τὴν ῥοπὴν τῆς σφαίρας εἰς ὅ τι[22] φέροιτο προαισθάνοιτο, διαμαρτάνειν τῆς λαβῆς ἀναγκαῖόν ἐστιν αὐτόν.[23] ἐπὶ τούτῳ δὲ καὶ τὴν γνώμην θήγει τῇ φροντίδι τοῦ τε μὴ καταβαλεῖν καὶ τοῦ διακωλῦσαι[24] τὸν μέσον ἢ αὐτὸν ὑφαρπάσαι, εἴπερ ἐν τούτῳ κατασταίη. φροντὶς δὲ μόνη μὲν καταλεπτύνει, μιχθεῖσα δέ τινι γυμνασίῳ καὶ φιλοτιμίᾳ καὶ εἰς ἡδονὴν τελευτήσασα τὰ μέγιστα καὶ τὸ σῶμα πρὸς ὑγίειαν καὶ τὴν ψυχὴν εἰς σύνεσιν ὀνίνησιν.

οὐ σμικρὸν δὲ καὶ τοῦτ' ἀγαθόν, ὅταν ἄμφω τὸ γυμνάσιον ὠφελεῖν δύνηται, καὶ σῶμα καὶ ψυχήν, εἰς τὴν ἰδίαν ἑκάτερον ἀρετήν. ὅτι δ' ἀσκεῖν ἄμφω δύναται τὰς μεγίστας ἀσκήσεις, ἃς μάλιστα μετιέναι τοῖς

---

[20] post ἀνάπαυλάν: τε τοῖς ἡσυχάζουσιν εἶναι τὸν χρόνον τῶν ἐνεργούντων, W; τοῖς ἡσυχάζουσιν τὸν χρόνον τῶν ἐνεργούντων, Ku; τε . . . τῶν ἐνεργούντων, om. M

3. Certainly, as for the legs, so too for the arms, this exercise is the most well-balanced, since in every form [of the exercise] people are accustomed to catch the small ball. For of necessity here also, in the variety of forms, some of the muscles at one time and others at another time are strained quite strongly, so that all in turn are exercised equally, and rested during the time of the pauses that exist   904K in the activities. In this way, all function and all rest in turn; none remain idle during the whole time and none are seized by fatigues through working hard all the time. It is possible to learn that it also exercises vision, when you call to mind that, if someone does not perceive accurately the downward momentum of the ball, whither it may carry, it is inevitable that he fails to make the catch. In this also, it sharpens the mind by the anxiety of not letting the ball fall and of preventing the person in the middle snatching it away, if he should stop it. Anxiety alone causes thinning, while if it is mixed with some exercise and love of honor, and comes to end in pleasure, it is of the greatest benefit to the body in terms of health and to the soul in terms of sagacity.

And this good is by no means small, when the exercise is able to benefit both body and soul, each in regard to their specific excellence (*aretē*). That the greatest exercises were able to train both, and that these were exercises

---

21 *post* γυμνάζει *add.* δεῖ Ku

22 ὅ τι W, M; ὅποι Ku

23 ἐστιν αὐτόν W, M; αὐτῷ Ku

24 *post* διακωλῦσαι· τὸν μέσον ἢ αὐτὸν ὑφαρπάσαι, εἴπερ ἐν τούτῳ κατασταίη. W, Ku (*om.* εἴπερ); *om.* M

στρατηγικοῖς οἱ πόλεως βασιλεῖς νόμοι κελεύουσιν,
905K οὐ χαλεπὸν κατιδεῖν. ἐπιθέσθαι γὰρ ἐν καιρῷ καὶ
λαθεῖν ἐπιθέμενον καὶ ὀξυλαβῆσαι τὴν πρᾶξιν καὶ
σφετερίσασθαι τὰ τῶν ἐναντίων ἢ βιασάμενον ἢ καὶ
ἀδοκήτως ἐπιθέμενον καὶ φυλάξαι τὰ κτηθέντα τῶν
ἀγαθῶν στρατηγῶν ἔργα· καὶ τὸ σύμπαν φάναι, φύ-
λακά τε καὶ φῶρα δεινὸν εἶναι χρὴ τὸν στρατηγόν,[25]
καὶ ταῦτ᾽ αὐτοῦ τῆς ὅλης τέχνης τὸ κεφάλαιον. ἆρ᾽
οὖν ἄλλο τι γυμνάσιον οὕτω προεθίζειν ἱκανὸν ἢ φυλάττειν τὸ κτηθὲν ἢ ἀνασῴζειν τὸ μεθειμένον ἢ τῶν
ἐναντίων τὴν γνώμην προαισθάνεσθαι; θαυμάζοιμ᾽
ἄν, εἴ τις εἰπεῖν ἔχοι. τὰ πολλὰ γὰρ αὐτῶν αὐτὸ τοὐναντίον ἀργοὺς καὶ ὑπνηλοὺς καὶ βραδεῖς τὴν γνώμην
ἐργάζεται. καὶ γὰρ καὶ ὅσα κατὰ παλαίστραν πονοῦ-
σιν[26] εἰς πολλοὺς στεφανίτας ἀγῶνας, πολυσαρκίαν
μᾶλλον ἢ ἀρετῆς ἄσκησιν φέρει· πολλοὶ γοῦν οὕτως
ἐπαχύνθησαν, ὡς δυσχερῶς ἀναπνεῖν. ἀγαθοί γ᾽ οὐδ᾽
ἂν δύναινθ᾽[27] οἱ τοιοῦτοι πολέμου γενέσθαι στρατηγοὶ
ἢ βασιλικῶν ἢ πολιτικῶν πραγμάτων ἐπίτροποι· θᾶττον ἂν τοῖς ὑσὶν ἢ τούτοις τις ὁτιοῦν ἐπιτρέψειεν.

ἀλλ᾽ ἴσως οἰήσῃ με δρόμον ἐπαινεῖν καὶ τἄλλ᾽ ὅσα
906K λεπτύνει τὸ σῶμα γυμνάσια. τὸ δ᾽ οὐχ οὕτως ἔχει.
τὴν γὰρ ἀμετρίαν ἐγὼ πανταχοῦ ψέγω, καὶ πᾶσαν
τέχνην ἀσκεῖν φημι χρῆναι τὸ σύμμετρον, κἂν εἴ τι

[25] post τὸν στρατηγόν: καὶ ταῦτ᾽ αὐτοῦ τῆς ὅλης τέχνης
τὸ κεφάλαιον. W, M; καὶ ταῦτα τῆς τέχνης αὐτῆς ὅλης τὸ
κεφάλαιον. Ku

which the rulers assigned to the city ordered the generals
particularly to pursue is not difficult to see. For the tasks   905K
of good generals are to attack at the opportune time and
unseen, to quickly take control of the action, to appropriate the possessions of the enemy, attacking either by force
or unexpectedly, and to preserve what has been gained. In
short, the general needs to be a skilled guard and thief,
and this is the sum total of his whole art. Is there, then,
any other exercise which provides adequate prior training
in this way, in preserving what has been gained, in recovering what has been lost, or in becoming aware beforehand
of the mind of the enemy? I would be surprised if someone were to say there is, for most of the exercises have the
opposite effect; they make the mind lazy, sluggish and
slow. Furthermore, although those exercises worked at in
the wrestling school may lead to many victors' crowns,
they also bring excess flesh rather than cultivation of virtue. Anyway, many were so burdened in this way that they
had difficulty breathing. Such men would not be able to
become good generals for a war or administrators of royal
or civil matters. Anyone at all would sooner rely on pigs
than on these men.

But perhaps you might think I approve of running and
other such exercises that thin the body. This is not so, for   906K
I find fault with disproportion everywhere; I say it is necessary to practice every art with moderation, and if some-

---

²⁶ *post* πονοῦσιν: εἰς πολλοὺς στεφανίτας ἀγῶνας, W;
πολλοὺς στεφανίτας ἀγῶνας *om.* Ku, M

²⁷ δύναινθ᾽ *add.* W

μέτρου στερεῖται, τοῦτ᾽ οὐκ εἶναι καλόν. οὔκουν οὐδὲ δρόμους ἐπαινῶ τῷ τε καταλεπτύνειν τὴν ἕξιν καὶ τῷ μηδεμίαν ἄσκησιν ἀνδρείας ἔχειν. οὐ γὰρ δὴ τῶν ὠκέως φευγόντων τὸ νικᾶν, ἀλλὰ τῶν συστάδην κρατεῖν δυναμένων, οὐδὲ διὰ τοῦτο Λακεδαιμόνιοι πλεῖστον ἠδύναντο τῷ τάχιστα θεῖν, ἀλλὰ[28] τῷ μένειν θαρροῦντες. εἰ δὲ καὶ πρὸς ὑγίειαν ἐξετάζοις, ἐφ᾽ ὅσον ἀνίσως γυμνάζει τὰ μέρη τοῦ σώματος, ἐπὶ τοσοῦτον οὐδ᾽ ὑγιεινόν. ἀνάγκη γὰρ αὐτῷ τὰ μὲν ὑπερπονεῖν, τὰ δ᾽ ἀργεῖν παντελῶς. οὐδέτερον δ᾽ αὐτῶν ἀγαθόν, ἀλλ᾽ ἄμφω καὶ νόσων ὑποτρέφει σπέρματα καὶ δύναμιν ἄρρωστον ἐργάζεται.

4. Μάλιστ᾽ οὖν ἐπαινῶ γυμνάσιον, ὃ καὶ σώματος ὑγίειαν[29] ἱκανὸν ἐκπορίζειν καὶ μερῶν εὐαρμοστίαν καὶ ψυχῆς ἀρετήν,[30] ἃ πάντα τῷ διὰ τῆς σμικρᾶς σφαίρας ὑπάρχει. καὶ γὰρ ψυχὴν εἰς πάντα δυνατὸν ὠφελεῖν καὶ τοῦ σώματος τὰ μέρη δι᾽ ἴσου[31] πάντα γυμνάζειν· ὃ καὶ μάλιστ᾽ εἰς ὑγίειαν συμφέρει καὶ συμμετρίαν ἕξεως ἐργάζεται, μήτ᾽ ἄμετρον πολυσαρκίαν μήθ᾽ ὑπερβάλλουσαν ἰσχνότητα φέρον, ἀλλ᾽ εἴς τε τὰς ἰσχύος δεομένας πράξεις ἱκανὸν καὶ ὅσαι τάχους χρῄζουσιν ἐπιτήδειον. οὕτω μὲν οὖν ὅσον ἐν αὐτῷ[32] τὸ σφοδρότατον οὐδενὸς τῶν πάντων κατ᾽ οὐδὲν ἀπολείπεται.

907K

---

[28] post ἀλλὰ: τῷ μένειν θαρροῦντες. W, M; τῷ μένοντες ἀναιρεῖν. Ku

[29] post ὑγίειαν: ἱκανὸν ἐκπορίζειν W, M; ἐκπορίζει Ku

thing is deprived of moderation, this is not good. I do not, therefore, commend running because it thins the bodily state and because it affords no training of the manly spirit. Certainly victory does not await those who flee quickly; it is rather for those who are able to prevail in hand to hand combat. The Spartans were not great because they were able to run very quickly; it was because they had the courage to stand their ground. If you also examine closely the matter of health, someone is unhealthy to the extent that he exercises the parts of the body unequally. For inevitably some parts in him are worked excessively, while some are entirely idle. Neither of these things is good; both nurture the seeds of disease and create a weak capacity.

4. Therefore, I particularly commend exercise which provides sufficient health for the body, concord of the parts, and excellence of the soul—all things which exist with the small ball. For truly it is able to benefit the soul 907K in all respects and to exercise all the parts of the body equally. It is especially beneficial to health and creates a proper balance of the bodily state (*hexis*); it brings neither a disproportionate excess of flesh nor excessive thinness, but is sufficient for the actions needing strength and suitable for those that require speed. In this way, then, the great vigor in this [form of exercise] is such that it lacks nothing compared to the others.

---

30 *post* ἀρετήν: ἃ πάντα τῷ W, M; παρὰ τούτοις· τοῦτο δέ τὸ Ku   31 *post* ἴσου: πάντα γυμνάζειν· W; πάντα γυμνάζει· M; μάλιστα γυμνάζει πάντα, Ku

32 *post* αὐτῷ: τὸ σφοδρότατον οὐδενὸς W; σφοδρότητος οὐδενὶ Ku; σφοδρότατον, οὐδενός M

τὸ δὲ πραότατον ἴδωμεν αὖθις. ἔστι γὰρ ὅτε καὶ
τούτου δεόμεθα διά θ᾽ ἡλικίαν ἢ μηδέπω φέρειν ἰσχυ-
ροὺς πόνους ἢ μηκέτι δυνάμενοι[33] καὶ κάματον ἐπ-
ανεῖναι βουληθέντες ἢ ἐκ νόσων ἀνακομιζόμενοι. δοκεῖ δέ μοι κἂν τούτῳ πλέον ἔχειν ἑτέρου παντός· οὐδὲν
γὰρ οὕτω πρᾷον[34] ὡς αὐτὸ τοῦτ᾽, εἰ πρᾴως αὐτὸ μετα-
χειρίζοιο. δεῖ δὲ[35] μέσῳ μὲν οὖν τηνικαῦτα χρῆσθαι
μηδενὶ σύμμετρον ἀποστάντα, τὰ μὲν ἠρέμα προβαί-
νοντα, τὰ δὲ καὶ κατὰ χώραν μένοντα, μὴ πολλὰ
διαγωνισάμενα, ἐπὶ τῷδε δὲ τρίψεσι μαλακαῖς δι᾽
908K   ἐλαίου καὶ λουτροῖς θερμοῖς χρῆσθαι. τοῦτο μὲν οὖν
ἁπάντων ἐστὶ πραότατον, ὥστε καὶ[36] ἀναπαύσασθαι
δεομένῳ συμφορώτατον εἶναι καὶ ἀρρώστου δύναμιν
ἀνακαλέσασθαι[37] δυνατώτατον καὶ γέροντι καὶ παιδὶ
συμφορώτατον.

ὅσα δὲ τούτου μὲν ἰσχυρότερα, τοῦ δ᾽ ἄκρως σφο-
δροῦ πραότερα διὰ τῆς σμικρᾶς σφαίρας ἐνεργεῖται,
χρὴ καὶ ταῦτα γιγνώσκειν, ὅστις γ᾽ ὀρθῶς βούλεται
διὰ παντὸς αὐτὴν[38] μεταχειρίζεσθαι. καὶ γὰρ εἴ ποτε
δι᾽ ἀναγκαίων ἔργων, οἷά τινα πολλὰ πολλάκις ἡμᾶς
καταλαμβάνει, πονήσειας ἀμέτρως ἢ τοῖς ἄνω μέρεσι
καὶ τοῖς κάτω πᾶσιν[39] ἢ ποσὶ μόνοις ἢ χερσίν, ἔνεστί

---

33 δυνάμενοι W; δυναμένην Ku, M
34 post πρᾷον: ὡς αὐτὸ τοῦτ᾽ add. W
35 post δεῖ δὲ: μέσῳ μὲν οὖν τηνικαῦτα χρῆσθαι μηδενὶ
W; μέσως μένοντα τηνικαῦτα χρῆσθαι, μηδενὶ Ku; μέσως μὲν
ἔχοντα μηδὲ M   36 post ὥστε καί: ἀναπαύσασθαι δεομένῳ W, M; ἀναπαύεσθαι δεομένοις Ku

392

However, let us look at the great gentleness again, for sometimes we also require this due to being of an age when we are not yet, or no longer, able to bear heavy labors, or we wish to recover from heavy labor or recuperate from disease. It seems to me that even in this, it offers more than any other exercise, for nothing is as gentle in the way this is, if you do it gently. Therefore, under these circumstances, one needs to take a median position,[4] not departing at all from the right measure, sometimes advancing slowly and sometimes standing one's ground, not contending much, and using soft massage with olive oil and warm baths. This, then, is the most gentle [exercise] 908K of all, so that it is most suitable for someone who needs to rest and restore to health a weak capacity, and most able to be effective for an old person and a child.

There are those exercises, stronger than this but more gentle than the exceedingly strong, which are carried out with the small ball. Anyone who wishes to pursue this correctly in its entirety must know these. For surely, if sometimes due to essential tasks, many of which often fall to us, there is disproportionate work through the lower parts, or all the parts above, or the feet alone, or the hands, it is

[4] Taking "a median position" is understood here as referring to moderation in the practice of the exercise. It has, however, been suggested that it applies to the exercise itself, as described in section 2, paragraph 5 above—see Singer, *Galen: Selected Works,* 429.

---

37 ἀνακαλέσασθαι W, M; ἀναλαβέσθαι Ku

38 αὐτὴν W; αὐτὰ Ku, M   39 post πᾶσιν: ἢ ποσὶ μόνοις ἢ χερσὶν, W; ἢ χερσὶ μόναις, ἢ ποσὶν, M, Ku

σοι διὰ τοῦδε τοῦ γυμνασίου τὰ μὲν ἀναπαῦσαι, τὰ
πρότερον κεκμηκότα, τὰ δ' εἰς τὴν ἴσην ἐκείνοις κίνησιν καταστῆσαι, τὰ πρότερον ἀργὰ παντελῶς μεμε-
νηκότα. τὸ μὲν γὰρ ἐκ διαστήματος ἱκανοῦ βάλλειν
εὐτόνως, ἢ οὐδὲν τοῖς σκέλεσιν ἢ παντάπασιν ὀλίγα
χρώμενον, ἀναπαύει μὲν τὰ κάτω, τὰ δ' ἄνω κινεῖ
σφοδρότερον· τὸ δ' ἐπὶ πλέον διαθέοντα καὶ ὠκέως ἐκ
πολλῶν διαστημάτων ὀλιγάκις προσχρῆσθαι τῇ
βολῇ τὰ κάτω μᾶλλον διαπονεῖ. καὶ τὸ μὲν ἠπειγμέ-
909K νον ἐν αὐτῷ καὶ ταχὺ χωρὶς συντονίας ἰσχυρᾶς τὸ
πνεῦμα μᾶλλον γυμνάζει· τὸ δ' εὔτονον⁴⁰ ἐν ταῖς ἀντιλήψεσι καὶ βολαῖς καὶ λαβαῖς, οὐ μὴν ταχύ γε, τὸ
σῶμα μᾶλλον ἐντείνει τε καὶ ῥώννυσιν· εἰ δ' εὔτονόν
θ' ἅμα καὶ ἠπειγμένον εἴη, διαπονήσει τοῦτο μεγάλως
καὶ τὸ σῶμα καὶ τὸ πνεῦμα καὶ πάντων ἔσται γυμνασίων σφοδρότατον. ἐφ' ὅσον δὲ δεῖ καθ' ἑκάστην
χρείαν ἐπιτείνειν τε καὶ ἀνιέναι, γράψαι μὲν οὐχ οἷόν
τε, τὸ γὰρ ἐν ἑκάστῳ ποσὸν ἄρρητον, ἐπ' αὐτῶν δὲ
τῶν ἔργων εὑρεῖν τε καὶ διδάξαι δυνατόν, ἐν ᾧ δὴ καὶ
μάλιστα τὸ πᾶν κῦρος· οὐδὲ γὰρ ἡ ποιότης ἐστὶ χρήσιμος, εἰ τῷ ποσῷ διαφθείροιτο. τοῦτο μὲν δὴ τῷ παι-
δοτρίβῃ μεθείσθω⁴¹ τῷ μέλλοντι τὸ γυμνάσιον ὑφη-
γεῖσθαι.⁴²

5. Τὸ δ' ὑπόλοιπον τοῦ λόγου περαινέσθω. βούλομαι γὰρ ἐφ' οἷς εἶπον ἀγαθοῖς προσεῖναι τῷδε τῷ
γυμνασίῳ μηδ'⁴³ οἵων τε καὶ ὅσων ἐκτός ἐστι κιν-

⁴⁰ εὔτονον W, M; ἕτερον Ku

possible for you, through this same exercise, to rest those parts that have previously worked, while bringing those that have previously remained completely idle to an equal movement with the former. For if you throw [the ball] energetically from a considerable distance, this either does not use the legs at all, or very little, so resting the parts below, while moving those above quite vigorously. On the other hand, running to and fro still more and quickly, while availing oneself of throwing a few times from long distances works out the lower parts more. And when there is urgency in this and swiftness apart from strong exertions, it exercises the breath (*pneuma*) more, while vigor in the grips, throws and holds, which are not in fact rapid, strains and strengthens the body more. If there is vigor and at the same time also urgency, this will greatly work out both the body and the breath, and will be the most vigorous of all exercises. The extent to which straining and letting go is needed in each use is impossible to write down, for the amount cannot be expressed in each case, whereas it is possible to discover and teach in the case of the actions themselves, in which, in truth, there is particularly the whole force. For the quality is not useful if it is weakened in quantity. This should certainly be left to the physical trainer who is going to be in charge of the exercise.

909K

5. Let us put an end to what remains of the discussion. For I wish to add to the goods I stated for this exercise that, regardless of kinds and amounts, it is free of external

---

41 μεθείσθω W; μετέστω Ku, M
42 ὑφηγεῖσθαι W, Ku; ἀφηγεῖσθαι M
43 *post* μηδ': οἴων τε καὶ ὅσων W; ὡς Ku, M

δύνων παραλιπεῖν, οἷς τὰ πλεῖστα τῶν ἄλλων περιπίπτει. δρόμοι μὲν γὰρ ὠκεῖς πολλοὺς ἤδη διέφθει-
ραν, ἀγγεῖον ἐπίκαιρον ῥήξαντες. οὕτω δὲ καὶ φωναὶ
910K μεγάλαι θ' ἅμα καὶ σφοδραὶ καθ' ἕνα χρόνον ἀθρόως
ἐκφωνηθεῖσαι μεγίστων κακῶν οὐκ ὀλίγοις αἴτιαι⁴⁴
κατέστησαν. καὶ μέντοι καὶ ἱππασίαι σύντονοι τῶν τε
κατὰ νεφροὺς ἔρρηξάν τι καὶ τῶν κατὰ θώρακα πολλάκις ἔβλαψαν, ἔστι δ' ὅτε καὶ τοὺς σπερματικοὺς
πόρους, ἵνα τὰ τῶν ἵππων ἁμαρτήματα παραλείπω,
δι' ἅ γε⁴⁵ πολλάκις ἐκπεσόντες τῆς ἕδρας οἱ ἱππεῖς
παραχρῆμα διεφθάρησαν. οὕτω δὲ καὶ τὸ ἅλμα⁴⁶ καὶ
ὁ δίσκος καὶ τὰ διὰ τοῦ σκάπτειν⁴⁷ γυμνάσια⁴⁸ πολλοῖς μελῶν τι διέστρεψε γυμνασθεῖσιν. τοὺς δ' ἐκ τῆς
παλαίστρας τί δεῖ καὶ λέγειν, ὡς ἅπαντες λελώβηνται
τῶν Ὁμηρικῶν Λιτῶν οὐδὲν μεῖον; ὡς γὰρ ἐκείνας
φησὶν ὁ ποιητὴς "χωλάς τε ῥυσάς τε παραβλῶπάς τ'
ὀφθαλμώ," οὕτω τοὺς ἐκ τῆς παλαίστρας ἴδοις ἂν ἢ
χωλοὺς ἢ διεστραμμένους ἢ τεθλασμένους ἢ πάντως
γέ τι μέρος πεπηρωμένους. εἰ δὴ πρὸς οἷς εἶπον ἀγαθοῖς ἔτι καὶ τοῦθ' ὑπάρχει τοῖς διὰ τῆς σμικρᾶς σφαίρας γυμνασίοις, ὡς μηδὲ κινδύνῳ πελάζειν, ἁπάντων
ἂν εἴη πρὸς ὠφέλειαν ἄριστα παρεσκευασμένα.

---

⁴⁴ post αἴτιαι: κατέστησαν. καὶ μέντοι add. W; γεγένηνται
add. M   ⁴⁵ ἅ γε W; ὧν Ku; οὓς M
⁴⁶ post τὸ ἅλμα: πολλοὺς ἔβλαψε add. M
⁴⁷ σκάπτειν W; κάμπτειν M, Ku
⁴⁸ post γυμνάσια: πολλοῖς μελῶν τι διέστρεψε γυμνασθεῖσιν add. W

dangers with which the majority of the other exercises are beset. Thus, running swiftly has already killed many, when they rupture an important blood vessel. In the same way too, loud and at the same time strong sounds, when produced at one time and all at once, are established as causes 910K of very great harms in quite a few people. And indeed also, vigorous horse riding has caused rupture of those structures in relation to kidneys, and has often harmed structures in the chest, and sometimes also the spermatic ducts. I omit here the faults of the horses due to which the riders, when they fall from the back of the horse, immediately die. The same also applies to leaping, discus throwing, and the exercises due to digging that have twisted one of the limbs in many of those being exercised. What do I need to say about those from the wrestling school, who were maimed no less than those from the Homeric hymns! For as the poet says of them: "Lame, wrinkled and with blind eyes."[5] In this way you may see those from the wrestling school either lame, twisted, crushed or with some part altogether maimed. If, in addition to the goods I spoke of, there is also this in the exercise with the small ball—that it does not come near to danger—then, in terms of benefit, it would be the best preparation of all.

[5] Homer, *Iliad* 9.503.

INDEXES

The index is divided into four sections covering both volumes (marked I and II), as follows: (a) Personal and Place Names, (b) Books and Treatises, (c) Foods and Medications, and (d) General. Entries relating to the General Introduction (vol. I), and the specific introductions to the two short treatises (vol. II), are given first according to the actual book page. The entries relating to the translations are then given according to the Kühn page, divided into those for the *Hygiene,* marked H; *Thrasybulus,* marked T; and *On Exercise with a Small Ball,* marked SB.

a
PERSONAL AND PLACE NAMES

b
BOOKS AND TREATISES

INDEXES

c
FOODS AND MEDICATIONS